AF353002

Remote Sensing of Flow Velocity, Channel Bathymetry, and River Discharge

Remote Sensing of Flow Velocity, Channel Bathymetry, and River Discharge

Editors

Carl J. Legleiter
Tamlin Pavelsky
Michael Durand
George Allen
Angelica Tarpanelli
Renato Frasson
Inci Guneralp
Amy Woodget

MDPI • Basel • Beijing • Wuhan • Barcelona • Belgrade • Manchester • Tokyo • Cluj • Tianjin

Editors

Carl J. Legleiter
United States Geological Survey
USA

George Allen
Texas A&M University
USA

Inci Guneralp
Texas A&M University
USA

Amy Woodget
Loughborough University
UK

Tamlin Pavelsky
University of North Carolina at
Chapel Hill
USA

Angelica Tarpanelli
Research Institute for
Geo-Hydrological Protection
Italy

Michael Durand
The Ohio State University
USA

Renato Frasson
The Ohio State University
USA

Editorial Office
MDPI
St. Alban-Anlage 66
4052 Basel, Switzerland

This is a reprint of articles from the Special Issue published online in the open access journal *Remote Sensing* (ISSN 2072-4292) (available at: https://www.mdpi.com/journal/remotesensing/special_issues/river_discharge).

For citation purposes, cite each article independently as indicated on the article page online and as indicated below:

LastName, A.A.; LastName, B.B.; LastName, C.C. Article Title. *Journal Name* **Year**, *Article Number, Page Range.*

ISBN 978-3-03936-900-3 (Hbk)
ISBN 978-3-03936-901-0 (PDF)

Contents

About the Editors

Carl J. Legleiter earned a Ph.D. in Geography from the University of California Santa Barbara in 2008, completed a brief post-doc with the US Geological Survey (USGS), and became an Assistant Professor in the Department of Geography at the University of Wyoming in 2009. He returned to the USGS Geomorphology and Sediment Transport Laboratory in 2016 as Research Hydrologist. Legleiter's research focuses on the remote sensing of rivers, specifically spectrally based mapping of water depth, estimating surface flow velocities via particle image velocimetry of thermal and optical image sequences, and inferring concentrations of visible tracer dye to support dispersion studies in rivers. In addition to developing remote sensing methods, Legleiter also applies these techniques in numerous contexts including habitat assessment for various aquatic species, flow modeling to understand sediment transport and channel change, and predicting the movement of contaminants and biota in river systems.

Tamlin Pavelsky (Professor, University of North Carolina) earned a B.A. in geography from Middlebury College and an M.A. and Ph.D., also in geography, from UCLA. He is currently a professor of global hydrology at the University of North Carolina, where he leads a group that focuses on using satellite imaging methods to understand the quantity and quality of water in the world's rivers and lakes, among other topics. He has published more than 70 articles in peer-reviewed journals and was a 2012 recipient of the Presidential Early Career Award for Scientists and Engineers (PECASE), the highest award given by the U.S. government to early career researchers. Tamlin was born in Fairbanks, Alaska and grew up in cabins without electricity or running water.

Michael Durand (Ph.D. UCLA 2007) studies land surface hydrology processes, with a focus on using spaceborne remote-sensing measurements to answer open science questions in snow hydrology and in fluvial hydrology. He is a Professor in the School of Earth Sciences and the Byrd Polar and Climate Research Center.

George Allen (Assistant Professor) received his Ph.D. in Geological Sciences from the University of North Carolina at Chapel Hill in 2017. He completed a Caltech Postdoc at the Jet Propulsion Laboratory before joining the Department of Geography at Texas A&M in 2018. His research focuses on surface water resources with an emphasis on river systems at the global scale. He combines satellite remote sensing data, hydrologic models, and fieldwork to conduct his research. Dr. Allen uses satellites, models, and fieldwork to understand how Earth's rivers and streams are responding to climate and land surface change. As a member of the NASA/CNES Surface Water and Ocean Topography (SWOT) satellite Science Team, he studies the hydrological interaction between rivers and surface water reservoirs. Allen also is a member of the NSF Dry Rivers Research Coordination Network, in which he leads an effort to map ephemeral rivers and intermittent streams globally. He is involved in the Global Carbon Project, where he contributes to planetary evaluations of greenhouse gas emissions from inland waters. Allen is also involved in several collaborative efforts to develop global, freely available hydrography datasets for flood modeling and other applications. He has authored/co-authored articles in Nature, Nature Communications, and Science. As an advocate for STEM education, Allen teaches inclusive courses in Earth Surface Remote Sensing and Geographic Data Analysis. He also enjoys mentoring student research at the undergraduate and graduate levels and promoting diversity in STEM at Texas A&M and throughout the surrounding community. Allen's work is motivated by a desire to further the conservation, management and understanding of freshwater resources worldwide.

Angelica Tarpanelli, Ph.D., received her M.S. degree and the Ph.D. in Hydraulic Engineering at University of Perugia in 2006 and 2014, respectively. Since September 2012, she has been a Researcher at National Research Council, Research Institute for Geo-Hydrological Protection of (CNR-IRPI) in Perugia. She has participated (being responsible for and as a coordinator of) to several research projects in the frame of Italian and European programs and funded by International space agencies, in collaboration with Italian and international institutions. Her research interests include the use of remote sensing for hydrological and hydraulic applications, especially on river discharge estimation.

Renato Frasson earned a Ph.D. in Civil and Environmental Engineering from the University of Iowa in 2011 and completed a post-doc at the Civil, Environmental and Geodetic Engineering department at Ohio State University before joining the Byrd Polar and Climate Research Center at the Ohio State University as a Senior Research Associate in 2015, where he remained until 2020, when he joined the Jet Propulsion Laboratory at the California Institute of Technology as a Research Scientist. Frasson's research focuses on remote sensing of river discharge and geomorphology with the purpose of improving the current state of knowledge of global water fluxes. Frasson also aims to increase the use of remote sensing data for flood detection to support the design of the next generation of early warning systems.

Inci Guneralp earned a Ph.D. in Geography from the University of Illinois in 2007. She became an Assistant Professor in 2008 and an Associate Professor in 2015 in the Department of Geography, College of Geosciences, at the Texas A&M University. Güneralp's research focuses on the hydrodynamics and biomorphodynamics of river channels and floodplains in lowland environments and the impacts of human activities and climatic changes. Güneralp utilizes remote sensing data and techniques in her research, with a particular emphasis on multispectral and hyperspectral remote sensing and LiDAR, for river channel and water boundary extraction, flood inundation modeling, geomorphic change assessment and modeling, and riparian vegetation mapping.

Amy Woodget is an early-career researcher in geospatial science at Loughborough University, UK, with expertise in the use of unmanned aerial systems and their application to quantitative river habitat monitoring. Amy's research in this field is underpinned by wide-ranging experience in both academic and consultancy work, led by her passion for driving forward innovative approaches to high-resolution environmental monitoring which are reliable, sustainable and effective. Since 2016, Amy has been the chairperson of the Remote Sensing and Photogrammetry Society's Unmanned Aerial Vehicles Special Interest Group.

 *remote sensing*

Editorial

Editorial for the Special Issue "Remote Sensing of Flow Velocity, Channel Bathymetry, and River Discharge"

Carl J. Legleiter [1,*], Tamlin Pavelsky [2], Michael Durand [3], George H. Allen [4], Angelica Tarpanelli [5], Renato Frasson [6], Inci Guneralp [7] and Amy Woodget [8]

[1] U.S. Geological Survey, Integrated Modeling and Prediction Division, Golden, CO 80403, USA
[2] Department of Geological Sciences, University of North Carolina at Chapel Hill, Chapel Hill, NC 27599, USA; pavelsky@unc.edu
[3] School of Earth Sciences, The Ohio State University, Columbus, OH 43210, USA; durand.8@osu.edu
[4] Department of Geography, Texas A & M University, College Station, TX 77843, USA; geoallen@tamu.edu
[5] Research Institute for Geo-Hydrological Protection, National Research Council, 06128 Perugia, Italy; a.tarpanelli@irpi.cnr.it
[6] Byrd Polar and Climate Research Center, The Ohio State University, Columbus, OH 43210, USA; frasson.1@osu.edu
[7] College of Geosciences, Texas A & M University, College Station, TX 77843, USA; iguneralp@geos.tamu.edu
[8] Department of Geography, Loughborough University, Loughborough, Leicestershire B60 4AZ, UK; a.woodget@lboro.ac.uk
*   Correspondence: cjl@usgs.gov; Tel.: +1-303-271-3651

Received: 30 June 2020; Accepted: 3 July 2020; Published: 17 July 2020

Keywords: remote sensing; rivers; discharge; flow velocity; bathymetry; hydrology; geomorphology

1. Introduction

River discharge is a fundamental hydrologic quantity that summarizes how a watershed transforms the input of precipitation into output as channelized streamflow. Accurate discharge measurements are critical for a wide range of applications including water supply, navigation, recreation, management of in-stream habitat, and prediction and monitoring of floods and droughts. However, the traditional, in situ stream gage networks that provide such data are sparse and declining, even in developed nations, and absent in many parts of the world (e.g., [1]). Moreover, establishing and maintaining these gages is expensive, labor-intensive, and can place personnel at risk (e.g., [2]).

For all these reasons, remote sensing represents an appealing alternative means of obtaining streamflow information. Potential advantages of a remote sensing approach include greater efficiency, expanded coverage, increased measurement frequency, lower cost, and reduced risk to field hydrographers. In addition, remote sensing techniques provide exciting opportunities to examine not just isolated cross sections but long segments of rivers with continuous coverage and high spatial resolution. To realize these benefits, further research is needed to focus on the remote measurement of flow velocity, channel geometry, and, most critically, their product – river discharge.

This special issue was motivated by our desire to foster the development of novel methods for retrieving discharge and its components and thus stimulate progress toward an operational capacity for streamflow monitoring. Our goals as guest editors were to encourage studies on this topic and to compile high-quality, peer-reviewed articles in a special issue of *Remote Sensing* dedicated to this theme. We solicited manuscripts concerned with all aspects of the remote measurement of streamflow—including estimation of flow velocity, channel bathymetry (or water depth), and discharge—from various types of remotely sensed data (active or passive) acquired from a range of platforms (manned or unmanned aircraft, satellites, or ground-based imaging systems).

Papers describing past, present, or future missions devoted to various aspects of fluvial remote sensing were welcomed.

A total of 16 manuscripts were submitted to this special issue and subjected to a rigorous peer review process that involved a total of 41 anonymous, conscientious reviewers. Of these 16 manuscripts, 10 papers achieved the level of quality and innovation expected by *Remote Sensing* and ultimately were published in this special issue. A total of 53 authors contributed to these 10 articles and hailed from six different nations: Germany (one author), Italy (one), the United Kingdom (two), Austria (five), the Netherlands (five), and the United States of America (41).

2. Themes Represented in this Special Issue

The papers published in this special issue span a wide range of topics, from specific measurement techniques [3] and rigorous evaluation of a particular sensor [4] to studies illustrating the diversity of ways in which fluvial remote sensing can be applied [5,6]. We identified five emergent themes from the work presented in this special issue and allocated each of the published papers to one or more of these themes, as some articles addressed more than one of the following topics:

1. Measuring surface flow velocities via various non-contact methods [3,7,8]
2. Mapping water depth using both active and passive remote sensing approaches [4,8,9]
3. Deriving estimates of river discharge from various types of remotely sensed data [7,8,10]
4. Characterizing flow frequency and flooding using image-derived data products [11,12]
5. Applying remote sensing techniques to characterize flow-related spatial and temporal heterogeneity of key river attributes [5,6]

In addition to this special issue focused on discharge and its components, we also want to direct the reader's attention to another special issue on "Remote Sensing of Large Rivers" published in *Remote Sensing* in 2020. This editorial was inspired by and is modeled after the overview of "Remote Sensing of Large Rivers" presented by Alcantara and Park [13].

2.1. Surface Flow Velocities

Beginning with the velocity theme, Fulton et al. [7] described a near-field remote sensing approach to measuring surface flow velocities using Doppler radars. In this context, the term near-field refers to fixed, ground-based platforms such as bridges; such techniques thus are well-suited for deployment at established stream gages and can be readily incorporated into operational streamflow monitoring programs. Radar-based measurements of surface flow velocity were made at 10 U.S. Geological Survey (USGS) gaging stations and a probability concept used to estimate the cross-sectional mean velocity on the basis of a single surface velocity measurement at the cross-channel location where the maximum velocity occurs. Mean velocities computed using this method agreed closely ($R^2 = 0.993$) with observations from conventional, in situ instrumentation, with an average error of −1.1%.

Another paper in this special issue also focused on measuring surface flow velocities at a local scale, but from a mobile airborne platform rather than a permanent installation. Kinzel and Legleiter [8] used a small unmanned aircraft system (sUAS) equipped with a cooled, mid-wave infrared camera to acquire thermal image time series from which surface flow velocities were inferred via particle image velocimetry (PIV). This technique involved tracking the motion of turbulent structures expressed at the water surface as small differences in temperature and thus did not require seeding the flow with artificial tracers or assuming the presence of floating foam or debris, as in typical PIV applications based on optical images. Comparison of image-derived velocity estimates with field measurements yielded good accuracy ($R^2 = 0.82$) at one cross section and a moderate level of agreement ($R^2 = 0.64$) at another transect. A significant source of uncertainty was the velocity index used to convert remotely sensed surface velocities to depth-averaged velocities comparable to those measured in situ.

A third velocity-themed paper further increased the scale of observation by using image sequences acquired from a helicopter deployed above two large rivers in Alaska, USA. Rather than thermal data, Legleiter and Kinzel [3] showed that surface flow velocities could be inferred from standard red-green-blue (RGB) images in sediment-laden rivers, again without introducing tracers to facilitate PIV. In this case, plumes of suspended sediment upwelled from within the water column produced boils and vortices that were expressed as differences in water color trackable via PIV. A parameter optimization framework was used to demonstrate that, for a 200-m-wide channel imaged from approximately 600 m above the river, relatively large PIV interrogation areas of 9.6–48 m and modest frame rates of 0.5–2 Hz were sufficient to yield strong agreement ($R^2 > 0.9$) between remotely sensed velocities and field measurements. Similarly, examining the effect of image sequence duration indicated that a high level of accuracy could be maintained with dwell times as short as 16 s at 1 Hz or as little as 8 s at 2 Hz. This study demonstrated that data from inexpensive video cameras could enable reach-scale mapping of flow fields and introduced a modular workflow to support such analysis.

Also note that in addition to the three papers published in this special issue, non-contact measurement of flow velocities has garnered increased coverage in *Remote Sensing* recently. For example, Tauro et al. [14] chose this journal to introduce an optical flow-based technique for detecting, tracking, and filtering feature displacements. Similarly, several studies focusing on PIV algorithms [15], their application [16], and their refinement [17] have already appeared in *Remote Sensing* in 2020.

2.2. Bathymetry (Water Depth)

Our second theme, remote sensing of river bathymetry, was specifically addressed by three of the papers appearing in this special issue. Two of these studies employed an active form of remote sensing that has drawn considerable interest in recent years: topo-bathymetric lidar [4,8]. These systems feature water-penetrating green wavelength lasers and have been miniaturized to such a degree as to enable deployment from sUAS platforms. For example, Mandlburger et al. [4] described a novel lightweight laser scanner and assessed its ability to measure submerged elevations along a gravel-bed river in Austria. A pulse repetition rate of up to 200 kHz yielded point densities as high as 50 points/m^2 and full laser waveforms were captured for both online and post-flight processing. Laser pulses penetrated up to two times the Secchi depth, but a depth-dependent bias was reported and attributed to forward scattering of the laser beam by suspended materials. Overall, the system's high spatial resolution and depth measurement accuracy lead the authors to conclude that the new laser scanner is well-suited for a range of river-oriented applications.

The second paper to evaluate a novel bathymetric lidar deployed from a sUAS was that of Kinzel and Legleiter [8]. In this case, the lidar system considered not only the travel time of laser pulses but also their polarization state. This additional source of information allowed returns from the channel bed to be distinguished from pulses reflected from the water surface. This approach enabled more precise depth measurements in shallower water than conventional bathymetric lidars. Comparison of lidar-derived and field-surveyed depths was favorable for a relatively shallow cross section with a maximum depth of 0.7 m ($R^2 = 0.95$) but less encouraging for a transect with greater depths, up to a maximum of 1.2 m ($R^2 = 0.61$).

The third paper focused on remote sensing of water depth using a different, passive optical approach that has seen increasingly widespread application in river research: Structure-from-Motion (SfM) photogrammetry. Woodget et al. [9] used a sUAS to acquire multiple, overlapping images from a small river in the United Kingdom for two different time periods and showed that the level of topographic accuracy achieved in submerged areas was similar to that in exposed areas, even without separate calibration data and different SfM processing methods for within the wetted channel. Importantly, these findings imply that multiple techniques are not required to map both the subaqueous channel bed and dry bar surfaces, nor are extensive in-channel survey data. Moreover, the selection of a refraction correction method had little impact on results derived from near-nadir imagery. Instead, Woodget et al. [9] identified improved estimation of water surface elevations as the most direct means of increasing the accuracy of SfM-based bed elevation measurements. The paper also introduced a machine learning framework for producing continuous, high-resolution maps of geomorphic change that include spatially variable error estimates. This approach could facilitate efforts to characterize channel morphodynamics in shallow, clear-flowing streams.

2.3. River Discharge

The third theme is the primary, unifying topic of this special issue: remote sensing of river discharge. All of the papers in this special issue involved both remote sensing and river discharge (or one of its components) in some capacity, but three papers directly focused on this subject. Of these three, two were discussed previously in the context of inferring surface flow velocities but also went on to estimate river discharge. First, in addition to evaluating the utility of Doppler radars for velocity measurement as described above, Fulton et al. [7] applied the probability concept described above to infer cross-sectional mean velocities and compute river discharge. Because the data were collected at established gaging stations, information on channel geometry was available from prior field surveys used to define the rating curves that relate water level (i.e., stage) to discharge for each gage. For the 10 sites evaluated by Fulton et al. [7], observed discharges ranged from 0.17 to 4890 m^3/s and agreement between radar-derived and conventional in situ discharges was very strong ($R^2 = 0.999$), with an average error of −1.1%. The radars also were deployed in a continuous mode to provide time series with a 15-min resolution, implying that this approach could be used for real-time streamflow monitoring on an operational basis.

A second paper in this special issue also performed discharge calculations, but the analysis presented by Kinzel and Legleiter [8] was based entirely on remotely sensed data collected from two sUAS. In this case, discharges were computed by combining surface flow velocity estimates inferred via PIV of thermal image time series with information on cross-sectional area derived from the polarizing bathymetric lidar. Compared to direct field measurements made with an acoustic Doppler current profiler, the remotely sensed discharge estimates were 22% greater than the field observations at one cross section and within 1% at a second transect. Although acquiring useful thermal images required collecting data at dawn to maximize the air-water temperature contrast and the bathymetric lidar had limited penetration, important advantages of this approach include the ability to perform PIV under natural conditions without seeding the flow and to obtain information on channel geometry without field measurements of depth for calibration.

The third paper to explicitly consider remote sensing of river discharge (RSQ) did so in a comprehensive and thought-provoking manner. The review contributed by Gleason and Durand [10] points out that although widespread innovation has allowed RSQ to advance rapidly, this new subfield has become somewhat non-cohesive, leading to confusion amongst the broader hydrologic community regarding the role of RSQ and its potential to contribute to the discipline. Gleason and Durand [10] attempt to provide clarity by summarizing the literature and organizing work on RSQ first by application area and then by methodology. More specifically, a distinction was made between methods appropriate for gaged, semi-gaged, regionally gaged, and totally ungaged basins, but categorization by sensor was not considered useful. Instead, Gleason and Durand [10] emphasize the need to provide proper context for research on RSQ as a means of fostering hydrologic understanding. For example, clarifying what the term 'ungaged' means as it relates to RSQ and defining which techniques are appropriate for such basins would be helpful. The review concludes by lauding the diversity that has become a hallmark of RSQ and encouraging further 'methodological proliferation'.

2.4. Flow Frequency and Floods

The fourth theme addressed within this special issue involves using remotely sensed data to characterize the frequency with which different river discharges occur and to provide critical information on rare but important flow events. The first paper in this category sought to answer a key question regarding the ability of remote sensing to yield hydrologic insight: to what degree do archived satellite data effectively capture the overall population of river flow frequency? Allen et al. [11] used archives of Landsat data to determine when cloud-free images were available over USGS gaging stations on rivers large enough to be observed by Landsat. The flow frequency distributions derived from the cloud-free Landsat overpasses were then compared to those from the in situ stream gages. This analysis indicated that these two frequency distributions were not significantly different from one another except for hydrologic extremes, such as the maximum flow. Allen et al. [11] also reported that the degree to which a Landsat-based sample can be used to characterize the flow frequency distribution varies by location but concluded that the Landsat archive is, on average, representative of the temporal frequencies of discharges along large rivers.

The second paper within this theme focused on documenting the impacts of the extreme events that might not be captured by satellite archives. Forbes et al. [12] established the need for basic information on inundation extent and peak flood stage to support analysis of flood events by providing compelling statistics on the loss of life and property due to flooding. The study then demonstrated the potential of sUAS and close-range remote sensing techniques to identify high-water marks (i.e., indicators of peak stage) after a flood and to obtain the topographic input data required for hydraulic modeling. Remotely sensed data were compared to traditional, ground-based Global Positioning System (GPS) surveys of two small streams, one in a semiarid and the other in a temperate environment. Mean elevation errors were greater in the more humid setting (0.14 m), due to the presence of vegetation that obscured the ground surface, than in the drier climate (0.07 m), but these results were similar to the accuracy that can be achieved via GPS surveys. Forbes et al. [12] thus concluded that sUAS-based identification of high-water marks and measurement of channel cross sections can be an efficient and effective alternative to conventional field methods of characterizing flood impacts.

2.5. Broader Applications

The fifth and final theme covered in this special issue is concerned with the application of remote sensing techniques in the broader context of river systems. For example, by examining the effects of dam operations on spatial and temporal thermal heterogeneity, Mejia et al. [5] illustrated the significance of flow management, which must be informed by reliable discharge data that in turn could be obtained via remote sensing. More specifically, Mejia et al. [5] focused on cold-water refuges for salmonids in a large, regulated river in the northwestern USA and used thermal image data and generalized additive models to characterize the occurrence of cool-water areas. Importantly, a remote sensing-based approach allowed this analysis to be conducted across scales ranging from reaches to sub-catchments. The results of this study indicated that lateral contributions from tributaries were the primary control on thermal heterogeneity, which thus peaked at confluences, and that cool areas were associated with channel morphology and distance from the dam. These insights, along with remotely sensed information on river discharge and spatial patterns of flow velocity, could help to guide habitat management for salmonids.

A second application-oriented paper in this special issue further illustrates the many ways in which remote sensing can contribute to our understanding of river systems, all the way to their termini in estuaries. Leuven et al. [6] focused on intertidal areas where human activities and rising sea levels impact the distribution of depths and thus habitat conditions. Although numerical modeling can provide information on the spatial pattern and duration of inundation, as well as peak flow velocities and salinity, the requirements of these models in terms of both data and computing power can be prohibitive. As an alternative, Leuven et al. [6] presented a Python-based software tool that predicts hydrodynamics, bed elevations, and the distribution of channels and bars at minimal computational expense. These predictions are based on empirical relations derived from natural estuaries and the only inputs required to use the tool are an along-channel width profile and tidal amplitude, both of which can be derived from remotely sensed data. The approach is thus useful for rapid assessment of potential habitat when only images of the estuarine environment are available.

3. Concluding Remarks

We wish to extend our sincere gratitude to all 53 authors who elected to contribute their research to this special issue. Similarly, we want to explicitly thank the 41 reviewers for performing such an essential, but all too often, thankless task. The thoughtful, timely comments provided by the reviewers improved each of the papers published in this special issue, which came to fruition only because they were willing to volunteer their time and attention. Finally, we appreciate the efforts of Nelson Peng and the entire MDPI editorial team to support the guest editors in efficiently processing each manuscript.

Remote sensing of flow velocity, channel bathymetry, and river discharge is an important and timely topic in the fields of hydrology and geomorphology and although the 10 papers compiled in this special issue represent some meaningful progress, additional work in this area is needed. We hope that the studies published herein will help the river research and management communities to more effectively characterize and more thoroughly understand river systems through the informed use of remote sensing technologies.

Author Contributions: All authors have read and agreed to the published version of this manuscript.

Acknowledgments: The guest editors would like to thank the authors who contributed to this special issue and the reviewers who helped to improve the quality of the special issue by providing constructive recommendations to the authors.

Conflicts of Interest: The guest editors declare no conflict of interest.

References

1. Hannah, D.M.; Demuth, S.; van Lanen, H.A.; Looser, U.; Prudhomme, C.; Rees, G.; Stahl, K.; Tallaksen, L.M. Large-scale river flow archives: Importance, current status and future needs. *Hydrol. Process.* **2011**, *25*, 1191–1200. [CrossRef]
2. Conaway, J.; Eggleston, J.; Legleiter, C.; Jones, J.; Kinzel, P.; Fulton, J. Remote sensing of river flow in Alaska—New technology to improve safety and expand coverage of USGS streamgaging. *U.S. Geol. Surv. Fact Sheet 2* **2019**, *2019*, 4. [CrossRef]
3. Legleiter, C.J.; Kinzel, P.J. Inferring Surface Flow Velocities in Sediment-Laden Alaskan Rivers from Optical Image Sequences Acquired from a Helicopter. *Remote Sens.* **2020**, *12*, 1282. [CrossRef]
4. Mandlburger, G.; Pfennigbauer, M.; Schwarz, R.; Flöry, S.; Nussbaumer, L. Concept and Performance Evaluation of a Novel UAV-Borne Topo-Bathymetric LiDAR Sensor. *Remote Sens.* **2020**, *12*, 986. [CrossRef]
5. Mejia, F.H.; Torgersen, C.E.; Berntsen, E.K.; Maroney, J.R.; Connor, J.M.; Fullerton, A.H.; Ebersole, J.L.; Lorang, M.S. Longitudinal, Lateral, Vertical, and Temporal Thermal Heterogeneity in a Large Impounded River: Implications for Cold-Water Refuges. *Remote Sens.* **2020**, *12*, 1386. [CrossRef]
6. Leuven, J.; Verhoeve, S.; van Dijk, W.; Selaković, S.; Kleinhans, M. Empirical Assessment Tool for Bathymetry, Flow Velocity and Salinity in Estuaries Based on Tidal Amplitude and Remotely-Sensed Imagery. *Remote Sens.* **2018**, *10*, 1915. [CrossRef]
7. Fulton, J.W.; Mason, C.A.; Eggleston, J.R.; Nicotra, M.J.; Chiu, C.L.; Henneberg, M.F.; Best, H.R.; Cederberg, J.R.; Holnbeck, S.R.; Lotspeich, R.R.; et al. Near-Field Remote Sensing of Surface Velocity and River Discharge Using Radars and the Probability Concept at 10 U.S. Geological Survey Streamgages. *Remote Sens.* **2020**, *12*, 1296. [CrossRef]
8. Kinzel, P.; Legleiter, C. sUAS-Based Remote Sensing of River Discharge Using Thermal Particle Image Velocimetry and Bathymetric Lidar. *Remote Sens.* **2019**, *11*, 2317. [CrossRef]
9. Woodget, A.S.; Dietrich, J.T.; Wilson, R.T. Quantifying Below-Water Fluvial Geomorphic Change: The Implications of Refraction Correction, Water Surface Elevations, and Spatially Variable Error. *Remote Sens.* **2019**, *11*, 2415. [CrossRef]
10. Gleason, C.J.; Durand, M.T. Remote Sensing of River Discharge: A Review and a Framing for the Discipline. *Remote Sens.* **2020**, *12*, 1107. [CrossRef]
11. Allen, G.H.; Yang, X.; Gardner, J.; Holliman, J.; David, C.H.; Ross, M. Timing of Landsat Overpasses Effectively Captures Flow Conditions of Large Rivers. *Remote Sens.* **2020**, *12*, 1510. [CrossRef]

12. Forbes, B.T.; DeBenedetto, G.P.; Dickinson, J.E.; Bunch, C.E.; Fitzpatrick, F.A. Using Small Unmanned Aircraft Systems for Measuring Post-Flood High-Water Marks and Streambed Elevations. *Remote Sens.* **2020**, *12*, 1437. [CrossRef]

13. Alcântara, E.; Park, E. Editorial for the Special Issue "Remote Sensing of Large Rivers". *Remote Sens.* **2020**, *12*, 1244. [CrossRef]

14. Tauro, F.; Tosi, F.; Mattoccia, S.; Toth, E.; Piscopia, R.; Grimaldi, S. Optical Tracking Velocimetry (OTV): Leveraging Optical Flow and Trajectory-Based Filtering for Surface Streamflow Observations. *Remote Sens.* **2018**, *10*, 2010. [CrossRef]

15. Pearce, S.; Ljubičić, R.; Peña-Haro, S.; Perks, M.; Tauro, F.; Pizarro, A.; Dal Sasso, S.; Strelnikova, D.; Grimaldi, S.; Maddock, I.; et al. An Evaluation of Image Velocimetry Techniques under Low Flow Conditions and High Seeding Densities Using Unmanned Aerial Systems. *Remote Sens.* **2020**, *12*, 232. [CrossRef]

16. Strelnikova, D.; Paulus, G.; Käfer, S.; Anders, K.H.; Mayr, P.; Mader, H.; Scherling, U.; Schneeberger, R. Drone-Based Optical Measurements of Heterogeneous Surface Velocity Fields around Fish Passages at Hydropower Dams. *Remote Sens.* **2020**, *12*, 384. [CrossRef]

17. Dal Sasso, S.F.; Pizarro, A.; Manfreda, S. Metrics for the Quantification of Seeding Characteristics to Enhance Image Velocimetry Performance in Rivers. *Remote Sens.* **2020**, *12*, 1789. [CrossRef]

 *remote sensing*

Article

Near-Field Remote Sensing of Surface Velocity and River Discharge Using Radars and the Probability Concept at 10 U.S. Geological Survey Streamgages

John W. Fulton [1,*], Christopher A. Mason [1], John R. Eggleston [1], Matthew J. Nicotra [1], Chao-Lin Chiu [2], Mark F. Henneberg [1], Heather R. Best [1], Jay R. Cederberg [1], Stephen R. Holnbeck [1], R. Russell Lotspeich [1], Christopher D. Laveau [1], Tommaso Moramarco [3], Mark E. Jones [1], Jonathan J. Gourley [4] and Daniel Wasielewski [4]

[1] U.S. Geological Survey, U.S. Geological Survey, Denver Federal Center, Box 25046, MS 415, Denver, CO 80225, USA; camason@usgs.gov (C.A.M.); jegglest@usgs.gov (J.R.E.); mnicotra@usgs.gov (M.J.N.); mfhenneb@usgs.gov (M.F.H.); hbest@usgs.gov (H.R.B.); cederber@usgs.gov (J.R.C.); holnbeck@usgs.gov (S.R.H.); rlotspei@usgs.gov (R.R.L.); cdlaveau@usgs.gov (C.D.L.); mejones@usgs.gov (M.E.J.)

[2] Civil and Environmental Engineering, University of Pittsburgh, 724 Field Club Rd, Pittsburgh, PA 15238, USA; chaolin.chiu@gmail.com

[3] IRPI-Consiglio Nazionale delle Ricerche, Research Institute for Hydrogeological Protection, National Research Council, via della Madonna Alta 126, 06128 Perugia, Italy; tommaso.moramarco@irpi.cnr.it

[4] NOAA National Severe Storms Laboratory, NOAA/National Severe Storms Laboratory, National Weather Center, 120 David L. Boren Blvd., Norman, OK 73072 USA; jj.gourley@noaa.gov (J.J.G.); daniel.wasielewski@noaa.gov (D.W.)

* Correspondence: jwfulton@usgs.gov; Tel.: +1-303-236-6890

Received: 15 February 2020; Accepted: 17 April 2020; Published: 20 April 2020

Abstract: Near-field remote sensing of surface velocity and river discharge (discharge) were measured using coherent, continuous wave Doppler and pulsed radars. Traditional streamgaging requires sensors be deployed in the water column; however, near-field remote sensing has the potential to transform streamgaging operations through non-contact methods in the U.S. Geological Survey (USGS) and other agencies around the world. To differentiate from satellite or high-altitude platforms, near-field remote sensing is conducted from fixed platforms such as bridges and cable stays. Radar gages were collocated with 10 USGS streamgages in river reaches of varying hydrologic and hydraulic characteristics, where basin size ranged from 381 to 66,200 square kilometers. Radar-derived mean-channel (mean) velocity and discharge were computed using the probability concept and were compared to conventional instantaneous measurements and time series. To test the efficacy of near-field methods, radars were deployed for extended periods of time to capture a range of hydraulic conditions and environmental factors. During the operational phase, continuous time series of surface velocity, radar-derived discharge, and stage-discharge were recorded, computed, and transmitted contemporaneously and continuously in real time every 5 to 15 min. Minimum and maximum surface velocities ranged from 0.30 to 3.84 m per second (m/s); minimum and maximum radar-derived discharges ranged from 0.17 to 4890 cubic meters per second (m^3/s); and minimum and maximum stage-discharge ranged from 0.12 to 4950 m^3/s. Comparisons between radar and stage-discharge time series were evaluated using goodness-of-fit statistics, which provided a measure of the utility of the probability concept to compute discharge from a singular surface velocity and cross-sectional area relative to conventional methods. Mean velocity and discharge data indicate that velocity radars are highly correlated with conventional methods and are a viable near-field remote sensing technology that can be operationalized to deliver real-time surface velocity, mean velocity, and discharge.

Keywords: surface velocity; river discharge; Doppler radar; pulsed radar; probability concept

---

1. Introduction

The U.S. Geological Survey (USGS) operates approximately 10,330 streamgages in the United States (U.S.), Puerto Rico, and Virgin Islands. The streamgage network constitutes one of the largest in the world [1]. Approximately 80,000 river discharge (discharge or streamflow) measurements are made annually to create or maintain stage-discharge ratings, which relate a stage (water-level elevation) to a discharge. These data are used by various end users for water-resources planning, water-supply/flood/drought forecasting, and early warning networks [1]. Traditional measurements of velocity and discharge are recorded using sensors that are deployed in the water column. At most of these sites, water levels are measured using stilling wells or pressure transducers and converted to discharges that are empirically derived from a rating curve. Because of temporal changes in hydraulics and channel morphology, periodic measurements are needed to maintain the stage-discharge ratings. Typically, data are collected every 6 weeks and during select hydraulic events using mechanical current meters [2,3] or hydroacoustic instrumentation [4–11], including acoustic Doppler velocimeters (ADVs) and acoustic Doppler current profilers (ADCPs). These measurements record velocity and cross-sectional area, which are used to compute discharge.

The near-field remote sensing of streams and rivers has the potential to transform streamgaging operations in the USGS and other agencies around the world. Traditional remote sensing studies focus on satellite-based platforms—which rely largely on imagery, radar altimeters, and physically based fluvial models—to measure top widths and surface-water elevations and compute reach-based river discharge. This research focuses on near-field remote sensing, which offers an alternative worldview, to measure surface velocities and compute at-a-section river discharge using an efficient algorithm based on the probability concept (PC). The probability concept solves the assumption of a monotonic velocity distribution, which serves as the basis for other remote sensing studies such as particle image velocimetry (PIV) and particle tracking velocimetry (PTV). This research demonstrates that regardless of the method selected (physically based, field based, or probability based) computing river discharge can be achieved using multiple methods. It is the end user's choice to decide which methods are most appropriate for their application.

Particle image and tracking velocimetry has emerged as a method for measuring surface velocity and computing discharge from the ground [12–14], from small, unmanned aircraft systems (sUAS) [15–17], and from manned aircraft systems [18]. Satellite-based sensors [19–47] can be used to derive parameters such as water elevation, slope, and top width. These parameters are used to compute discharge in regions where streamgage networks are lacking. Pulsed Doppler and coherent Doppler velocity radars have been deployed from river banks [48], helicopters [49], sUAS [50], and can be coupled with pulsed (stage) radars from portable and fixed platforms such as bridges and cable stays [51–53]. Although this collection of research has advanced the field of remote sensing of rivers, their principle objective was not to operationalize the process and deliver real-time, continuous time series of velocity and discharge. Rather, this research demonstrates that the real-time delivery of mean-channel (mean) velocity and discharge is possible by combining radars with an efficient discharge algorithm, the probability concept.

The advantages of velocity and stage radars combined with the probability concept include: (1) improved safety, (2) reduced operational costs, (3) improved data delivery, and (4) increased operational efficiency. Safety is improved because hydrographers spend less time in the water and are able to work from stable platforms such as bridges, cable ways, and cable stays. Operational costs are less than stage-discharge streamgages, because fewer site visits are required. Data delivery is improved with radars, because discharge is computed immediately after radar siting and installation. Some stage-discharge ratings require months or years of data collection to establish the requisite number of stage and discharge values to develop a reliable rating. Operational efficiency is increased, because site visits are driven by event, rather than frequency. It should be noted that event-based and a nominal number of visits are needed: (a) to confirm stage-area ratings, (b) to confirm the location (stationing) of the maximum velocity in a cross section, (c) to confirm ϕ, which represents the ratio between the mean velocity and maximum velocity, and (d) to maintain operational functionality, including trouble shooting data transmission mishaps and power checks. The disadvantages of radars include: (1) initial capital costs ($5K to $10K U.S. dollars), (2) restrictive siting requirements, including the need for infrastructure such as a bridge or light cableway, (3) measurement uncertainty associated with poor-surface scattering, (4) wind-influenced (drift) surface velocities, and (5) an inability to reliably detect low surface-water (surface) velocities (less than 0.15 m/s).

This study examines the results of 10 fixed-mount velocity and stage radars collocated with USGS continuous streamgages and addresses the following questions: (1) Can radars accurately measure surface velocities? (2) Can velocity and stage radars be used to accurately compute the mean velocity and discharge when compared to conventional methods that serve as validation (truth)? (3) Can near-field remote sensing deliver real-time mean velocity, stage, and discharge operationally and in the absence of historical data? (4) What environmental and hydraulic factors influence radar-derived surface velocities?

2. Materials and Methods

2.1. Site Selection

The radar gages used in this study are located in the U.S. (Figure 1), and their characteristics are summarized in Tables 1–3. The sites were selected using the following criteria: (1) radars were collocated with existing USGS continuous streamgages, (2) radar locations, cross-sectional areas, velocities, and heights above the water surface (air gap) support radar operations, and (3) radars were operated for a sufficient duration of time to capture low and high streamflow conditions. The basins, where the 10 radar gages are located, range in size from 381 to 66,200 kilometers (Table 1) and are characterized by: (1) a variety of land-use types, including agricultural, desert, forest, mixed, and high-gradient mountain environments, (2) operating periods ranging from 2 weeks to 19 months, (3) river top widths (Table 2) ranging from 7 to 380 meters (m), and (4) river hydraulic depths ranging from 0.09 to 7 m. Each of these characteristics has a profound effect on velocity and discharge.

Table 1. U.S. Geological Survey (USGS) streamgages with radar gages used in this study in the United States. Note: Streamgage identification numbers will be omitted from subsequent tables. (DA = basin drainage area; km^2 = square kilometers; mtn = mountain).

USGS Streamgage	USGS Streamgage Identification Number	Setting	DA (km^2)	Radar Deployment Period		
				Start	End	Duration (Months)
Blackfoot River near Bonner, Montana	12340000	High-gradient mtn stream	5930	May 2013	Aug 2013	4
Cherry Creek at Denver, Colorado	06713500	Urban	1060	Aug 2017	Apr 2019	19
Clear Creek near Lawson, Colorado	06716500	High-gradient mtn stream	381	Apr 2019	Sep 2019	6
Gunnison River near Grand Junction, Colorado	09152500	Mixed	20,500	Aug 2018	Aug 2019	12
NF Shenandoah River near Strasburg, Virginia	01634000	Mixed	1990	Feb 2015	Jan 2016	11
Red River of the North at Grand Forks, North Dakota	05082500	Agricultural	66,100	Apr 2013	May 2013	< 1
Rio Grande at Embed, New Mexico	08279500	Desert	19,300	Apr 2014	Sep 2015	17
Susquehanna River at Bloomsburg, Pennsylvania	01538700	Mixed	27,400	Apr 2011	May 2011	< 1
Tanana River at Nenana, Alaska	15515500	Forest	66,200	May 2018	Oct 2019	18
Yellowstone River near Livingston, Montanan	06192500	High-gradient mtn stream	9200	May 2013	Aug 2013	4

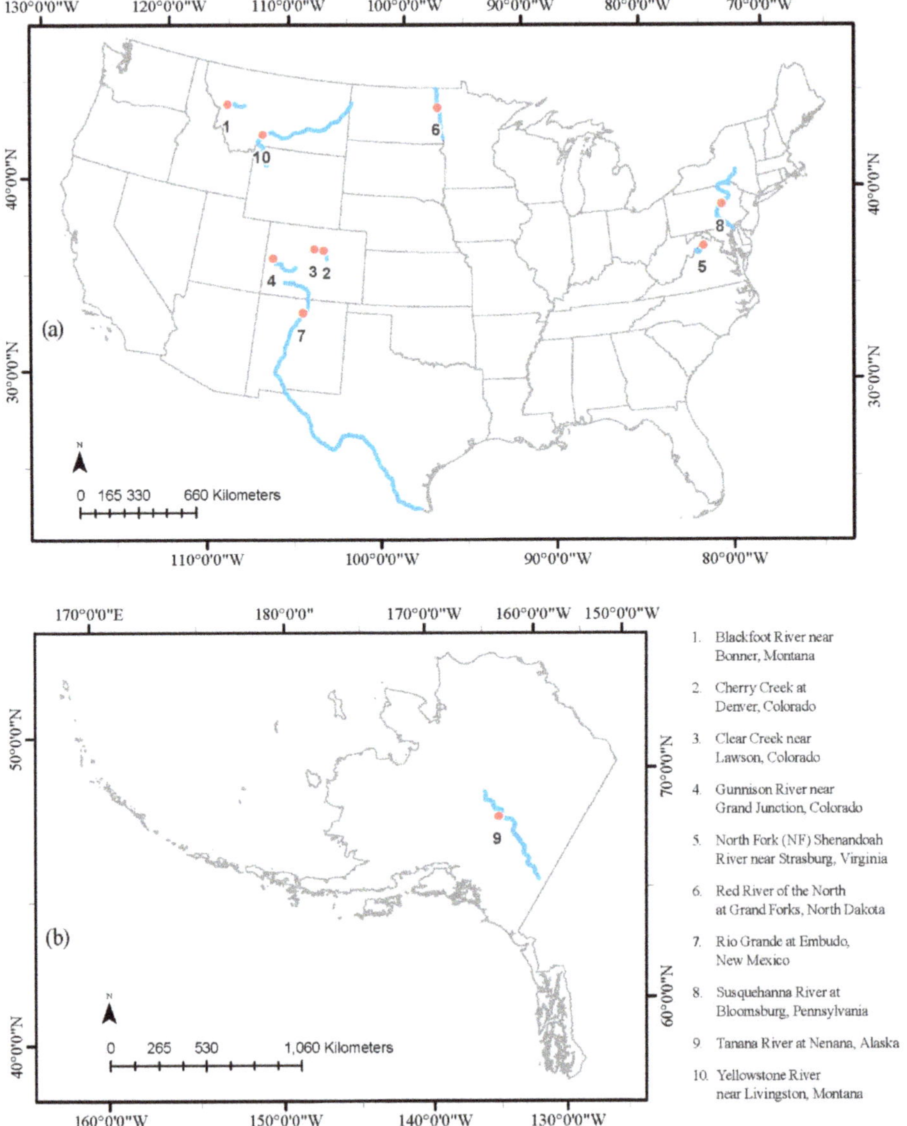

Figure 1. Map illustrating the 10 velocity and stage radars deployed by the U.S. Geological Survey (USGS) in the (**a**) conterminous United States and (**b**) Alaska. Each was collocated with an existing USGS streamgage. The blue line represents the river reach, and the red dots represent the radar gage and USGS streamgage location.

Table 2. Minimum and maximum river top widths and hydraulic depths measured during radar gage deployments in the United States based on periodic measurements with conventional methods. (m = meters; min = minimum; max = maximum).

USGS Streamgage	Number of Visits	Top Width (m)		Hydraulic Depth (m)	
		min	max	min	max
Blackfoot River near Bonner, Montana	2	44	58	0.52	2.1
Cherry Creek at Denver, Colorado	6	7.3	13	0.19	0.82
Clear Creek near Lawson, Colorado	8	7.0	19	0.34	0.88
Gunnison River near Grand Junction, Colorado	1	49	85	0.09	2.9
NF Shenandoah River near Strasburg, Virginia	1	56	67	0.40	1.9
Red River of the North at Grand Forks, North Dakota	2	87	160	5.5	7.0
Rio Grande at Embed, New Mexico	2	20	37	0.61	0.85
Susquehanna River at Bloomsburg, Pennsylvania	6	320	340	2.7	5.2
Tanana River at Nenana, Alaska	7	38	380	1.3	3.7
Yellowstone River near Livingston, Montana	1	40	104	0.98	1.9

Table 3. Minimum, median, and maximum stage-discharge measured during the operational phase of the radar gage deployments. Streamflow exceedance represents the magnitude of the peak streamflow measured while the radar was deployed. Streamflow exceedances were calculated as the percentage of the daily streamflow recorded at a continuous USGS streamgage during the radar deployment relative to the maximum discharge recorded for the historical period of record. (m^3/s = cubic meters per second; min = minimum; med = median; max = maximum).

USGS Streamgage	Stage-Discharge (m^3/s)			Streamflow Exceedance
	min	med	max	% exceeded
Blackfoot River near Bonner, Montana	17.8	52.6	168	5.10
Cherry Creek at Denver, Colorado	0.12	0.79	41.6	0.00
Clear Creek near Lawson, Colorado	1.19	6.41	32.1	0.48
Gunnison River near Grand Junction, Colorado	13.2	32.4	486	0.67
NF Shenandoah River near Strasburg, Virginia	2.10	11.2	190	0.46
Red River of the North at Grand Forks, North Dakota	308	875	1,240	0.40
Rio Grande at Embudo, New Mexico	7.65	17.1	117	3.00
Susquehanna River at Bloomsburg, Pennsylvania	1,250	1,970	4,950	0.00
Tanana River at Nenana, Alaska	452	1,410	2,850	0.48
Yellowstone River near Livingston, Montana	57.1	78.1	113	5.40

To demonstrate the extensibility of the radar methods, streamflow exceedances were calculated as the percentage of the daily streamflow recorded at a continuous USGS streamgage during the radar deployment relative to the maximum discharge recorded for the historical period of record (Table 3). Streamflow exceedances range from approximately 0% to 5.4%. These metrics indicate the sites and streamflows represent relevant variations in discharge for various hydrologic and hydraulic conditions.

2.2. Doppler Radar Technology

Radars measure velocity and stage remotely [51,52]. When sited properly, near-field remote sensing compares favorably to conventional methods described in Section 2.1. Velocity radars leverage the Doppler effect to translate radar frequencies to a surface velocity [50]. For example, velocity radars used for this research transmit a prescribed, 24-gigahertz (GHz) electromagnetic wave, which encounters small-scale surface waves (scatterers). These small-scale scatterers ride advectively on top of larger surface waveforms. If they are incident to the transmitted radar signal, the movement of these scatterers modify the frequency of the transmitted signal [51], and the frequency of the backscattered signal is shifted. The difference in the transmitted and returned frequencies is the Doppler shift; however, to detect the complete spectrum, the transmitted signal must be coherent where there are no phase discontinuities [51]. A spectral analysis is performed on the reflected signal to assess the quality of the spectrum and compute a value for the surface velocity.

To collect useful surface-velocity measurements, the radar signal is transmitted at an angle to the water surface. This angle is measured internally and is automatically corrected to compute the velocity. Only waves traveling toward or away from the radar serve as effective scatterers. The surface velocity represents the mean value over the footprint of the radar on the water surface and is calculated using Equation (1):

$$v_h = (\lambda f_c)/(2 \sin \theta) \tag{1}$$

where v_h = the wave velocity in the look (streamwise) or along-track direction of the radar antenna; λ = wavelength of the transmitted radar signal; f_c = shifted frequency; and θ = incidence angle of the radar [51].

When the transmitted signal originating from the radar encounters small-scale waves at incidence angles that are not too large or small, the signal is scattered back to the antennae. Scatterer size and quality is a function of the frequency of the transmitted radar and environmental factors such as wind drift, turbulence, and rain. The length of these small-scale surface waves can be computed using Equation (2):

$$\lambda_b = \lambda/(2 \sin \theta) \tag{2}$$

where λ_b = wavelength of the water wave; GHz, λ = wavelength of the transmitted radar signal; GHz, θ = incidence angle, $\lambda = c/v$, c = speed of light = 3×10^8 m per second in meter per second (m/s), v = frequency in hertz (Hz), given a K-band radar transmission frequency = 24 GHz, which has a wavelength of approximately 1.25 centimeters (cm) and an incidence angle of 56 degrees from horizontal. The small-scale surface waves that serve as scatterers are computed using Equation (3):

$$\lambda_b = 3 \times 10^8 \text{ m/s} \times 100 \text{ cm/m/24E9 Hz}/(2 \times \sin 56 \text{ degrees}) = 0.75 \text{ cm} \tag{3}$$

2.3. Validation Measurements

Velocity, stage (stage-area ratings) and discharge were measured using conventional methods to assist with: (1) siting the radars and (2) serving as truth to determine the accuracy of the radar-derived discharge at the time of installation and during operation. Vertical velocity profiles were recorded to compute the PC parameters, mean velocity, and discharge described in subsequent sections. Stage-discharge, time-series data were recorded simultaneously with radar data.

Conventional methods were used to assess the accuracy of the radar data and include: (1) instantaneous measurements—in-water, event-based conventional mechanical current meter and hydroacoustic measurements of velocity, stage, and discharge and (2) continuous measurements—empirical stage-discharge measurements of river stage and discharge. The USGS requires the standard error for stage to be ± 3 mm accuracy or better, which is needed to calculate discharge [54]. Of the two conventional methods, instantaneous measurements deliver the lowest expected error in velocity, stage, and discharge. Standard errors for individual discharge measurements vary from 2% for ideal conditions to 19% and greater for conditions dominated by flood flows, wind, ice, boundary effects, streamflow obstructions, or improper equipment and procedures [55–58]. However, most measurements exhibit standard errors ranging from 3 to 6% [54]. Continuous measurements of stage-discharge offer validation; however, they possess an inherent measure of uncertainty depending on the stability of the stage-discharge rating [57] where uncertainties (95% uncertainty intervals) can range from ±50 to 100% for low streamflow, ±10 to 20% for medium to high streamflow, and ±40% for out-of-bank flow [58].

2.4. Radar Siting

Velocity, cross-sectional area, and discharge were measured and analyzed to assist with data processing [3,4,59–61] and siting [62,63]. The same principles used to site conventional streamgages [4] were adopted for siting radar gages, which: (1) consist of straight channels with parallel streamlines, (2) include streambeds free of large rocks, weeds, or obstructions that would create turbulence or slack water, (3) comprise cross sections that are parabolic, trapezoidal, or rectangular, (4) avoid variable streamflow conditions downstream from piers or channel obstructions, (5) avoid highly turbulent conditions, (6) target sections where stream velocities exceed 0.15 m/s and depths greater than 0.15 m,

(7) avoid tributaries or contributing drainage, and (8) avoid wind-dominated reaches, eddies, secondary flows, and macroturbulence.

2.4.1. Velocity

Velocity data were collected at cross sections of interest and in the vicinity of the radar footprint. The radar's ability to record a surface velocity is influenced by: (1) the quality of scatterers or waveforms on the water surface, (2) the air gap or the distance between the bridge deck and the water surface, and (3) the potential bias imposed by wind drift, eddies, secondary flows, and macroturbulence. Current meters, ADVs, ADCPs, and surface velocity radars can be used to characterize the velocity distribution and assist in the computation of the PC parameters discussed in Section 2.6.2. A template for siting and collecting channel and hydraulic data associated with radar deployments was developed to assist with data collection [62].

The SonTek FlowTracker®(FT) is a handheld ADV that is (1) used widely for wading measurements, (2) capable of measuring velocity with an accuracy of ±0.25 cm/s, and (3) capable of sampling a 25-cubic-centimeter (cm^3) water volume, which permits velocity sampling close to the air–water interface. When wading was possible, the FT was used to collect velocity and depth data from 25 to 30 verticals that comprise the measured section. The vertical that contains the maximum velocity is referred to as the y-axis and was revisited during each site visit. A vertical velocity profile was acquired at the y-axis location beginning near the channel bottom and extending to the water surface. Depending on the water depth (D_i), a minimum of 6 point velocities were measured (channel bottom, $0.2\,D_i$, $0.3\,D_i$, ... , $0.8\,D_i$, and near the water surface, while minimizing air entrainment). This procedure was repeated to river left (LEW, to the left facing downstream) and river right (REW, to the right facing downstream) of the y-axis to confirm its location.

When wading was not possible, velocities were collected at the y-axis using an ADCP and a Stationary Bed Test (SBT) by boat, light cableway, or tethered from a bridge or a river bank. When possible, surface velocities were collected concurrently with SBTs using a handheld or portable velocity radar. As with the FT, SBTs were conducted to river left and river right of the y-axis. Discharge was computed using QRev software [10] or an equivalent. Alternatively, when coupled with a Global Positioning System (GPS) receiver, the latitude and longitude of the y-axis can be established using tools such as the Velocity Mapping Toolbox [64,65]. The SBT is processed in WinRiver II software [63] and is used to assess the velocity distribution, which is ingested by R-scripts to compute the PC parameters described in Section 2.6.2. The WinRiver II workflow for processing SBTs included the following steps: (1) select "Moving Bed Test" of interest, (2) select "Playback" and "Reprocess Selected Transect," (3) select "Configure" and "Averaging Data" and identify the number of ensembles needed to average the SBT to reduce noise in the data, (4) select "View" and "Tabular," and "Earth Velocity Magnitude and Direction," and (5) copy the depth and velocity data to a .txt or .csv file.

2.4.2. y-axis

The y-axis (Figure 2) is the location in a stream cross section that contains the maximum velocity and maximum surface velocity [65–76]. More importantly, once established, all velocity and depth data, which are needed to translate a surface velocity to a mean velocity [71], are obtained at the y-axis location. It is important to note that the y-axis location rarely coincides with the channel thalweg in open or engineered channels and is determined during the siting phase [62]. Assuming site selection is well posed, research indicates the location of the y-axis is stable and does not vary with stage, velocity, discharge, changes in channel geometry, bed form and material, slope, or alignment [73]. During this research, periodic field verification was conducted to verify the location of the y-axis [62].

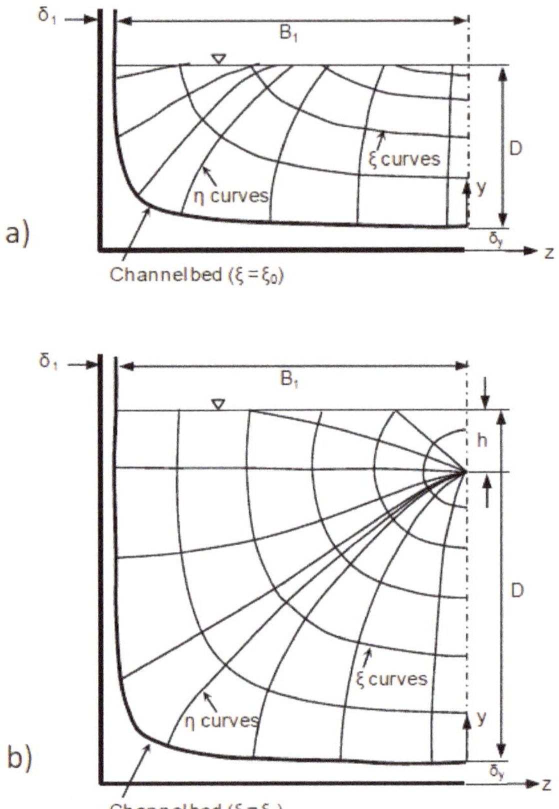

Figure 2. Location of the y-axis and maximum velocity and maximum surface velocity in a channel cross section defined by the *x-h* orthogonal coordinate system [66], when (**a**) $h \geq 0$ and (**b**) $h < 0$. (y = incremental distance from the channel bottom to water surface; D = total distance from the channel bottom to the water surface at the y-axis; y-axis = the vertical in a cross section that contains the maximum velocity; h = distance from the water surface to the maximum velocity, u_{max}; h = orthogonal trajectory of x; x = dimensionless coordinate related to velocity and h; ξ_0 is the value of ξ in the point (or points) where the velocity $u = 0$; B_i = top-half width of channel).

2.4.3. Cross-sectional Area

Cross-sectional area surveys were conducted during the siting phase to establish a stage-area rating, which is maintained throughout the radar gage operation. Included in the surveys were the wetted perimeter and above-water points of interest in the floodplain. The stage-area rating is used to compute area, based on the stage recorded by a pulsed radar, bubbler, or pressure transducer. Horizontal and vertical surveys were acquired relative to the collocated USGS streamgage datum using a total station survey, GPS receiver, or an equivalent. AreaComp2 software [77] was used to generate a synthetic stage-area rating to account for areas above the water surface that were measured during the day of siting.

2.4.4. Wind Drift

Wind-corrected discharge can be computed for sites where the surface velocity is affected by wind. The correction (1) relies on measurements from a wind sensor located at a nominal height (10 m) above the water surface and adjacent to the radar, (2) assumes a vertical wind profile based on the Prandtl von Karman (PvK) velocity distribution equation, and (3) assumes a surface-roughness height for water equal to 0.01 cm [78]. Wind-drift corrections are based on the measured wind speed and direction recorded by the sensor every 15 min and are temporal with velocity and stage data. A theoretical wind speed can be computed near the water surface using the PvK distribution and a roughness height of 0.01 cm. Based on the orientation of the river reach (e.g., azimuth equal to 130 degrees), the resultant wind vector can be subtracted from the surface velocity if the winds are concordant (in the direction) of streamflow. If the winds are counter (in the opposite direction) to streamflow, wind velocity is added to the surface velocity.

2.5. Radar Deployments

Radars were installed on bridges at the y-axis location coincident with the maximum surface velocity established during the siting phase. Generally, the maximum velocity will occur at the same vertical as the maximum surface velocity [53,71]. The location of the y-axis was determined initially using a hand-held velocity radar and confirmed using mechanical current meters or hydroacoustics. A minimum of three surface velocities were collected to confirm the y-axis location and magnitude of the maximum surface velocity by sampling river left and river right of the y-axis. Velocity radars can point upstream or downstream from a bridge or walkway. For this research, eight of the 10 deployments were oriented facing upstream to avoid complex streamflow patterns including eddies, secondary flows, and macroturbulence. The remaining two sites were not influenced by bridge piers or streamflow complexities and were deployed facing downstream. Radar deployments can occur from bridges, light cableways, or cable stays (Figure 3). Fixed-mount radars (RQ-30 and RG-30 models) were manufactured and provided by Sommer Messtechnik, and the hand-held radar was manufactured by Applied Concepts, Inc.

(a) (b)

Figure 3. Photographs of typical radar deployments from (**a**) bridge along Clear Creek near Lawson, Colorado, and (**b**) cable stays in Waldo Canyon, Colorado, USA.

Reaches dominated by wind fetch were avoided in an effort to minimize uncertainties in surface velocities influenced by wind drift. Radar locations were geo-referenced using a GPS receiver or measured from known bridge stationing, were oriented parallel to steamflow lines to avoid the need for a cosine correction based on the angle of attack, and were tilted downward (from horizontal) at a nominal 56-degree incidence angle.

2.6. Discharge Algorithms

2.6.1. Conventional Methods

Conventional methods (stage-discharge ratings; current meters/ADVs/mid-section method; and ADCPs) are widely accepted as industry standards [2–11,63,79] and were used as validation of the velocity and stage radar results. If surface velocities are measured at 25 to 30 verticals from LEW to REW, the mean vertical velocity ($u_{vertical}$) or depth-averaged velocity at a vertical in a cross section can be computed using Equation (4):

$$u_{vertical} = u_D \times \text{coefficient (typically ranging from 0.84 to 0.90)} \tag{4}$$

where $u_{vertical}$ = mean vertical velocity at a given station in a cross section and u_D = surface velocity. This equation assumes the vertical velocity profile can be characterized by a logarithmic or 1/6th or 1/7th power law [11,79]. The coefficient referenced in Equation (4) is needed to convert a surface velocity to a mean vertical velocity [3,4]; however, these coefficients are generally difficult to determine reliably and may vary with stage, depth, and stationing in the cross section. Experience has shown that the coefficient generally ranges from about 0.84 to about 0.90, depending on the shape of the vertical velocity curve [3] and the proximity of the vertical to channel walls where secondary currents may develop causing the maximum velocity to dip below the water surface [80–86]. In these instances, the velocity distribution is non-standard, and an alternative velocity distribution is needed to characterize the velocity profile and compute the mean vertical velocity, mean velocity, or discharge.

2.6.2. Probability Concept

The probability concept (radar-derived mean velocity and discharge) in many ways is similar to conventional, physically based methods, which rely on integrating depth and velocity data to compute the mean velocity and discharge. For example, velocity and depth are recorded at 25 to 30 verticals in a cross section using mechanical current meters and ADVs [3]; whereas, ADCPs record velocity and depth at tens to hundreds of verticals in a cross section. The probability concept relies on these same instruments to measure velocity and corresponding depth at a single vertical; however, only that vertical, which contains the most information content ($u_{min} = 0$ and $u = u_{max}$) in a cross section, is used.

Surface velocities derived from radars are translated to a mean velocity using a velocity distribution equation based on the probability concept [66–76], which was pioneered by C.-L. Chiu and is based on Shannon's Information Entropy [87]. Equation (5) through Equation (10) offer a summary of the Chiu velocity distribution equation, and the workflow needed to compute a mean velocity. The velocity distribution at the y-axis in probability space is represented by Equation (5):

$$u = \frac{u_{max}}{M} \ln\left[1 + \left(e^M - 1\right) F(u)\right] \tag{5}$$

where u = velocity as a function of depth at the y-axis; u_{max} = maximum velocity at the y-axis; M = parameter of the probability distribution used to describe the velocity distribution; and $F(u) = \int_0^u f(u)du$, which represents the cumulative distribution function, or the probability of a randomly sampled point velocity that is less than or equal to u. At those cross sections where u_{max} occurs below the water surface, the velocity distribution at the y-axis can be characterized by Equation (6):

$$u = \frac{u_{max}}{M} \ln\left[1 + \left(e^M - 1\right) \frac{y}{D-h} \exp\left(1 - \frac{y}{D-h}\right)\right] \tag{6}$$

where D = total distance from the channel bottom to the water surface at the y-axis, y = incremental distance from the channel bottom to water surface, h = distance from the water surface to the maximum velocity. The orthogonal coordinate system (Figure 2) is used to translate the velocity distribution from probability space to physical space and is used to describe the variables h, D, and y in Equation (6) through Equation (8). In those instances when u_{max} occurs at the water surface, the velocity distribution at the y-axis can be characterized by Equation (7):

$$u = \frac{u_{max}}{M} \ln\left[1 + \left(e^M - 1\right) \frac{y}{D} \exp\left(1 - \frac{y}{D}\right)\right] \tag{7}$$

The probability distribution $f(u)$ is resilient and is invariant with time and water level at a channel cross section and subsequently M and h/D are constant at a channel cross section [66–73]. In practice, Equation (6) is used only when there is a clear and observed velocity distribution where u_{max} occurs below the water surface. In those instances, radar-derived surface velocities are used to estimate u_{max} and assumes u is equal to u_D, which is the velocity at which y equals D [71] and is represented by Equation (8):

$$u_{max} = uD \times M \times \left\{ \ln\left[1 + \left(e^M - 1\right) \frac{1}{1 - \frac{h}{D}} \exp\left(1 - \frac{1}{1 - \frac{h}{D}}\right)\right] \right\}^{-1} \tag{8}$$

The parameter ϕ, which is a function of M, can be computed using two methods. First, point velocities are measured along the y-axis beginning near the channel bed and extending to the water surface. An R-script was developed to compute u_{max}, $M(\phi)$, and h/D by incorporating the depth and velocity data recorded by conventional instruments. The "nls" algorithm [88,89] is a nonlinear (weighted) least-squares estimator of the parameters u_{max}, $M(\phi)$, and h/D. By default, Equations (6) and (7) are solved using a Gauss–Newton nonlinear least-squares method. A typical velocity profile located at the y-axis is illustrated in Figure 4 for the Tanana River at Nenana, Alaska. Second, historical pairs of mean and maximum velocities are plotted, where ϕ represents the slope of the pairs and is the preferred method for computing ϕ [74,75] and is represented by Equation (9):

$$\phi = \frac{u_{mean}}{u_{max}} = \frac{e^M}{\left(e^M - 1\right)} - \frac{1}{M} \tag{9}$$

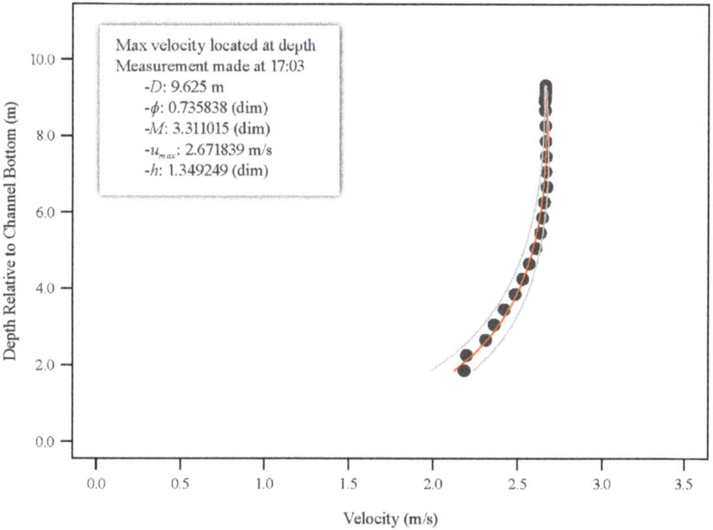

Figure 4. Typical velocity distribution profile measured using an acoustic Doppler current profiler (ADCP) in July 2018 for the Tanana River at Nenana, Alaska, USA. The dots represent the mean velocities derived from the binned ADCP velocities recorded during the stationary bed test at the y-axis, the red line represents the nonlinear curve fit based on Equation (6), and the gray lines represent the 95% confidence interval.

Discharge is computed using ϕ, u_{max} and the cross-sectional area derived from a stage-area rating using Equation (10):

$$Q = \phi \times u_{max} \times A \tag{10}$$

where, Q = discharge; $\phi = u_{mean}/u_{max}$ = function of M; u_{max} = maximum velocity; A = cross-sectional area. The parameter ϕ generally ranges from 0.58 to 0.82 [69].

3. Results

During the siting and validation phase, channel hydraulics and cross-sectional area were measured to: (1) parameterize the PC algorithm, (2) validate the resiliency of ϕ and location of the y-axis, and (3) establish a stage-area rating. During the operational phase, a workflow was established to: (1) record, compute, and transmit a radar-derived discharge in real-time and (2) compare the findings against the stage-discharge, which served as validation [90].

3.1. Siting and Validation Phase

3.1.1. Velocity

Instantaneous surface, maximum, and mean velocities were measured using velocity radars and conventional measurements and represent a variety of hydraulic conditions from small, high-gradient mountain and urban streams to wide, low gradient large rivers (Table 4). Radar-derived surface velocities ranged from 0.45 to 2.96 m/s for the NF Shenandoah River and the Yellowstone River; maximum velocities ranged from 0.09 to 2.96 m/s for the Red River of the North and the Yellowstone River; and mean velocities ranged from 0.05 to 2.11 m/s for the Red River of the North and the Yellowstone River. The lowest maximum velocity (0.09 m/s) was measured at the Red River of the North and was expected given the low hydraulic gradient associated with the site. The highest maximum velocity (2.96 m/s) was reported at the Yellowstone River, which is a high-gradient mountain stream. Percent differences in the radar and conventional mean velocities ranged from −0.1% to −11% for the Red River of the North and the Tanana River, respectively. The composite average percent error and absolute percent error of the mean velocity for the 10 sites was biased low −1.1% and high 3.6%, respectively. A velocity radar was not available during the siting phase at Clear Creek. Subsequently, a surface velocity was not recorded concurrently with the hydroacoustic data. Similarly, the Red River of the North was ice affected, and a surface velocity was not possible during the siting visit.

Subsequent validation visits were conducted at each of the 10 USGS streamgages to collect event-based instantaneous measurements. Figure 5 summarizes the results of 36 site visits (10 siting and 26 validation) and demonstrates general agreement between radar and conventional mean velocities. The average and absolute percent error was computed for the Blackfoot River (−1.4% and 2.5%), Cherry Creek (1.1% and 5.6%), Clear Creek (2.1% and 6.9%), Gunnison River (4.9% and 4.9%), NF Shenandoah River (−8.0% and 8.0%), Red River of the North (−8.6% and 8.6%), Rio Grande (4.6% and 4.6%), Susquehanna River (−0.4% and 3.8%), Tanana River (−2.5% and 3.9%), and Yellowstone River (−0.7% and 0.7%). The composite average and absolute percent error of all 36 site visits was −0.3% and 5.1%. The data show strong agreement (correlation coefficient, R^2, equal to 0.993) between instantaneous radar-derived and conventional mean velocities (Figure 5).

Table 4. Probability-concept-based and conventional-based metrics and velocities measured during the siting phase at 10 U.S. Geological Survey streamgages in the United States. (PC = probability concept; Conv = Conventional; M = parameter characterizing the velocity distribution and is dimensionless, dim; ϕ = a function of M equal to the ratio of the mean velocity to the maximum velocity; u_D = surface velocity in meters per second, m/s; u_{max} = maximum velocity; y-axis = vertical depth in a cross section that contains the maximum velocity in meters, m; u_{mean} = mean velocity = $\phi \times u_{max}$ or stage-discharge/area;% error = percent error = (PC discharge − Conventional Discharge)/Conventional Discharge × 100; na = data not available).

USGS Streamgage	Date Collected	PC Metrics					PC	Conv	% error in u_{mean}
		M (dim)	ϕ (dim)	u_D (m/s)	u_{max} (m/s)	Water Depth at y-axis (m)	u_{mean} (m/s)	u_{mean} (m/s)	
Blackfoot River near Bonner, Montana	05-20-2013	2.10	0.664	2.70	2.70	2.57	1.79	1.78	1.1
Cherry Creek at Denver, Colorado	08-25-2017	2.32	0.678	0.75	0.75	0.26	0.51	0.50	1.2
Clear Creek near Lawson, Colorado	04-19-2019	0.883	0.573	na	1.09	0.43	0.62	0.61	2.3
Gunnison River near Grand Junction, Colorado	03-27-2019	0.266	0.522	1.12	1.02	1.55	0.53	0.51	4.9
NF Shenandoah River near Strasburg, Virginia	12-04-2014	1.03	0.584	0.45	0.49	0.77	0.29	0.31	−8.0
Red River of the North at Grand Forks, North Dakota	02-05-2004	0.60	0.550	na	0.09	6.34	0.05	0.05	−0.1
Rio Grande at Embudo, New Mexico	03-21-2014	1.49	0.620	1.45	1.45	0.70	0.90	0.87	3.2
Susquehanna River at Bloomsburg, Pennsylvania	06-27-2002	4.35	0.783	0.73	0.73	1.83	0.57	0.59	−4.0
Tanana River at Nenana, Alaska	05-07-2015	2.98	0.718	1.41	1.41	2.84	1.01	1.14	−11
Yellowstone River near Livingston, Montana	05-22-2013	2.92	0.715	2.96	2.96	2.86	2.11	2.13	−0.7
						Average percent error			−1.1
						Absolute average percent error			3.6

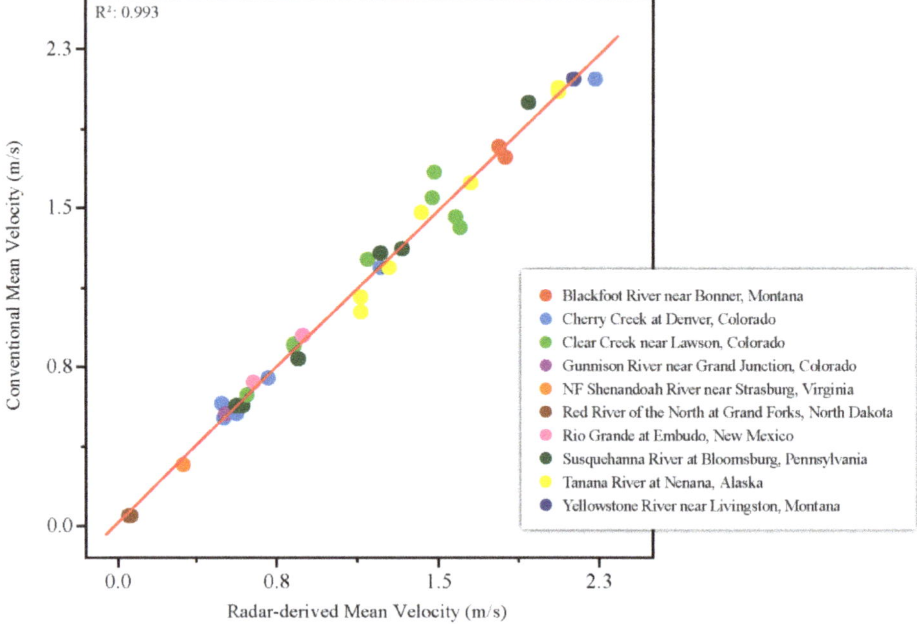

Figure 5. Line of agreement of conventional mean velocity (mechanical current meters or hydroacoustics) versus radar-derived mean velocity for 10 velocity and stage radars deployed by the U.S. Geological Survey (USGS) and collocated with USGS streamgages in the United States.

3.1.2. Values of ϕ and Location of the y-axis

The values of ϕ were computed for all 10 gages and were derived using vertical velocity profiles obtained during the siting phase. Values ranged from 0.522 to 0.783 (Table 4) for the Gunnison River and the Susquehanna River, respectively, and are generally consistent with the findings of other researchers [62–76,86], who report ϕ values ranging from 0.58 to 0.82. In addition, ϕ was derived using historical pairs of the mean and maximum velocities for Cherry Creek (0.6896), Clear Creek (0.5752), Susquehanna River (0.8114), and Tanana River (0.7639). These values were obtained from site visits and the USGS National Water Information System: Web Interface NWISWeb [91], which is a database available to USGS cooperators and the general public. The ϕ values derived during the siting phase and historical pairs for Cherry Creek (0.678 and 0.6896), Clear Creek (0.573 and 0.5752), Susquehanna River (0.783 and 0.8114), and Tanana River (0.718 and 0.7639) resulted in percent differences equal to −1.7%, −0.3%, −3.5%, and −6.0% and exhibited strong agreement based on R^2 values equal to 0.984, 0.957, 0.999, and 0.992, respectively. Figure 6a–d illustrates the variability of mean and maximum velocities for Cherry Creek (0.51 to 2.1 m/s and 0.74 to 3.2 m/s), Clear Creek (0.62 to 1.7 m/s and 1.1 to 2.8 m/s), Susquehanna River (0.56 to 2.0 m/s and 0.72 to 2.5 m/s), and Tanana River (1.0 to 2.1 m/s and 1.4 to 2.7 m/s) and the stability of ϕ with time and velocity.

The location of the y-axis at each radar gage was verified during each site visit using the methods described in Section 2.4.1. At no time or during no specific streamflow condition did the y-axis differ from the siting phase. In addition, previous research [66–76] indicates the location of the y-axis in stable reaches is resilient and does not vary with stage, velocity, discharge, changes in channel geometry, bed form and material, slope, or alignment.

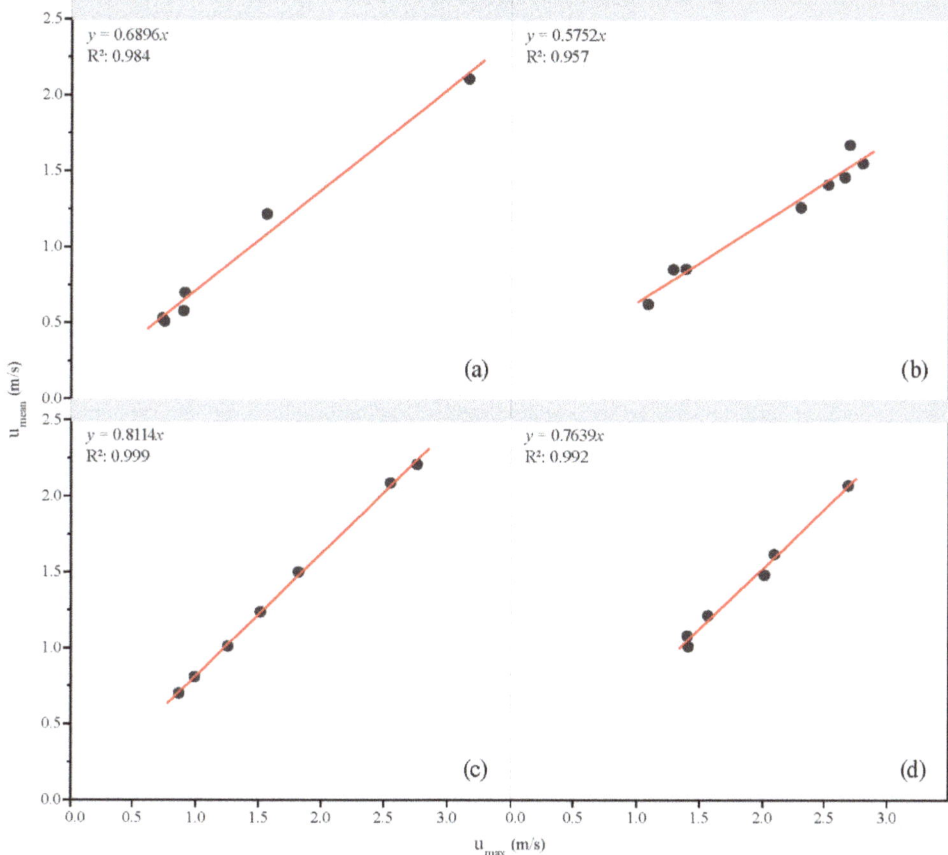

Figure 6. Historical pairs of mean (u_{mean}) and maximum (u_{max}) velocities obtained during site visits at (**a**) Cherry Creek at Denver, Colorado, (**b**) Clear Creek at Lawson, Colorado, (**c**) Susquehanna River at Bloomsburg, Pennsylvania, and (**d**) Tanana River at Nenana, Alaska, where velocity radars were deployed by the U.S. Geological Survey (USGS) and collocated with USGS streamgages in the United States. (R^2 = correlation coefficient, dim; m/s, meters per second; y = mean velocity = u_{mean}, in m/s; x = maximum velocity = u_{max}, m/s; 0.6896, 0.5752, 0.8114, 0.7639 = slope = ϕ = u_{mean}/u_{max}).

3.1.3. Discharge

Radar-derived discharges (Table 5) ranged from 0.85 to 615 cubic meters per second (m^3/s) at Cherry Creek and the Tanana River. Similarly, stage-discharges (Table 5) ranged from 0.84 to 690 m^3/s at Cherry Creek and the Tanana River. Because radar-derived discharge is calculated as a linear function of velocity (Equation (10)), discharge biases were identical to the mean velocity findings, where the composite average percent error was biased low −1.1%, and the absolute percent error was biased high 3.6% (Table 5). Figure 7 summarizes the results of 36 site visits (10 siting and 26 validation), where the data show strong agreement between the radar-derived and conventional methods, as demonstrated by an R^2 = 0.999.

Table 5. Probability concept (PC)-based and stage-discharge measured during the siting phase of the velocity-radar installation at 10 USGS streamgages in the United States. (m^3/s = cubic meters per second; ADCP = acoustic Doppler current profiler; FT = FlowTracker velocimeter;% error = percent error = (PC discharge – Conventional Discharge)/Conventional Discharge × 100).

USGS Streamgage	Date Collected	Method	Discharge (m^3/s)		% Error in Discharge
			PC	Conventional	
Blackfoot River near Bonner, Montana	05-20-2013	ADCP	144	142	1.1
Cherry Creek at Denver, Colorado	08-25-2017	FT	0.85	0.84	1.2
Clear Creek near Lawson, Colorado	04-19-2019	FT	1.43	1.39	2.3
Gunnison River near Grand Junction, Colorado	03-27-2019	ADCP	30.5	29.1	4.9
NF Shenandoah River near Strasburg, Virginia	12-04-2014	ADCP	11.7	12.7	−8.0
Red River of the North at Grand Forks, North Dakota	02-05-2004	ADCP	12.6	12.6	−0.1
Rio Grande at Embudo, New Mexico	03-21-2014	ADCP	19.3	18.7	3.2
Susquehanna River at Bloomsburg, Pennsylvania	06-27-2002	AA	292	306	−4.0
Tanana River at Nenana, Alaska	05-07-2015	ADCP	615	690	−11
Yellowstone River near Livingston, Montana	05-22-2013	ADCP	208	209	−0.7
				Average percent error	−1.1
				Absolute average percent error	3.6

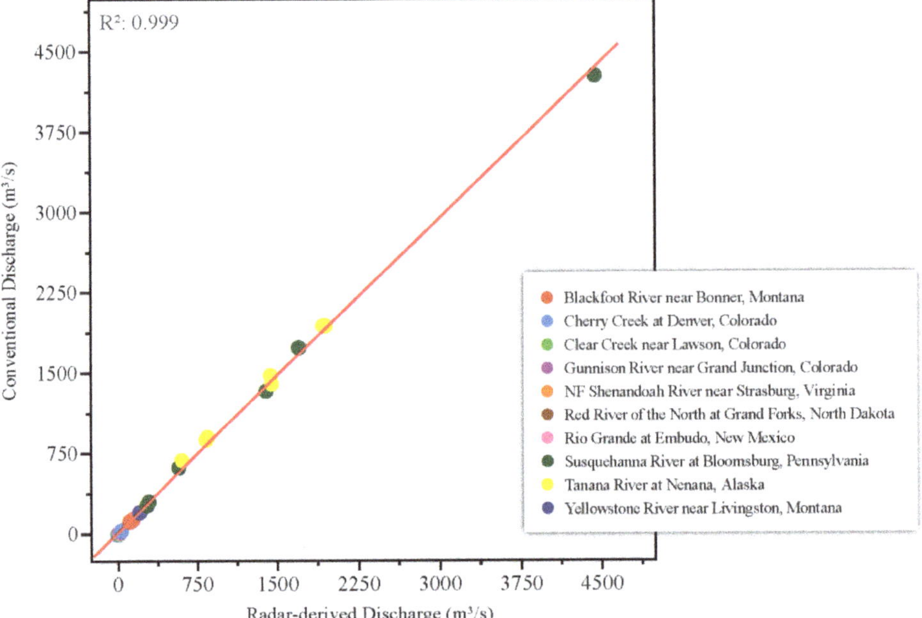

Figure 7. Conventionally measured discharge versus radar-derived discharge for 10 velocity and stage radars deployed by the U.S. Geological Survey (USGS) and collocated with USGS streamgages in the United States. These data were collected during siting and validation visits (m^3/s = cubic meters per second).

3.2. Operational Phase

3.2.1. Velocity and Discharge

Radar-derived surface velocities (Table 6) and discharges (Table 7) ranged from 0.30 to 3.84 m/s at Cherry Creek and the Yellowstone River, respectively, and 0.17 to 4890 m^3/s at Cherry Creek and the Susquehanna River, respectively. Similar to the radar gages, stage-discharges (Table 7) ranged from 0.12 to 4950 m^3/s at Cherry Creek and the Susquehanna River. The lowest radar-derived surface velocity (0.30 m/s) and discharge (0.17 m^3/s) was reported at Cherry Creek, which is an engineered urban channel with upstream streamflow regulation. The greatest radar-derived surface velocity

(3.84 m/s) was reported at the Yellowstone River, a high-gradient mountain stream. The greatest radar-derived discharge (4890 m^3/s) was recorded at the Susquehanna River, a large river system. The lowest stage-discharge (0.12 m^3/s) was reported at Cherry Creek, and the highest stage-discharge (4950 m^3/s) was reported for the Susquehanna River. These results support the contention that the velocity and stage radars captured a broad range of streamflow conditions.

Table 6. Radar-derived surface velocities recorded during the operational phase at 10 U.S. Geological Survey streamgages in the United States. (m/s = meters per second; min = minimum; med=median; max = maximum).

USGS Streamgage	Radar-Derived Surface Velocity (m/s)		
	min	med	Max
Blackfoot River near Bonner, Montana	0.66	2.03	3.01
Cherry Creek at Denver, Colorado	0.30	0.62	3.33
Clear Creek near Lawson, Colorado	0.84	1.85	3.78
Gunnison River near Grand Junction, Colorado	0.99	1.16	3.37
NF Shenandoah River near Strasburg, Virginia	0.41	0.66	1.76
Red River of the North at Grand Forks, North Dakota	0.55	1.19	1.78
Rio Grande at Embudo, New Mexico	0.51	0.92	2.88
Susquehanna River at Bloomsburg, Pennsylvania	0.64	1.62	2.53
Tanana River at Nenana, Alaska	0.70	1.81	3.02
Yellowstone River near Livingston, Montana	1.39	2.49	3.84

Table 7. Radar-derived discharge and stage-discharge recorded during the operational phase at 10 U.S. Geological Survey streamgages in the United States. (m^3/s = cubic meters per second; min = minimum; med=median; max = maximum).

USGS Streamgage	Radar-Derived Discharge (m^3/s)			Stage-Discharge (m^3/s)		
	min	med	max	min	med	max
Blackfoot River near Bonner, Montana	14.3	63.5	177	17.8	52.6	168
Cherry Creek at Denver, Colorado	0.17	0.73	38.6	0.12	0.79	41.6
Clear Creek near Lawson, Colorado	1.40	6.98	38.2	1.19	6.41	32.1
Gunnison River near Grand Junction, Colorado	21.0	31.1	450	13.2	32.4	486
NF Shenandoah River near Strasburg, Virginia	2.87	9.69	167	2.10	11.2	190
Red River of the North at Grand Forks, North Dakota	213	916	1470	308	875	1250
Rio Grande at Embudo, New Mexico	5.77	16.8	138	7.65	17.1	117
Susquehanna River at Bloomsburg, Pennsylvania	603	1780	4890	1250	1970	4950
Tanana River at Nenana, Alaska	381	1280	2640	452	1410	2850
Yellowstone River near Livingston, Montana	53.3	71.9	102	57.1	78.1	113

The radar gages were operated continuously for various time periods ranging from 2 weeks to 19 months. Discharge time series and companion scatter plots for Cherry Creek, Rio Grande, Tanana River, and Gunnison River demonstrate the utility of applying the PC method to small, moderate, and large river systems by comparing radar-derived discharges to stage-discharges (Figure 8a–h). It is important to note that hindcasting was not conducted. Subsequently, continuous radar discharges share the same time stamp as stage-discharges and were transmitted contemporaneously to the USGS NWIS database [91]. The minimum, median, maximum, and the number of radar-derived discharges are summarized in Table 8. The general timing and magnitude of peak and low-streamflow recorded by the radars and stage-discharge ratings were similar for all 10 sites based on the goodness-of-fit (GOF) metrics; however, only Cherry Creek, Rio Grande, Tanana River, and Gunnison River are illustrated graphically, because they span the hydraulic extremes tested. Several radar-peak discharges for Cherry Creek and the Rio Grande are biased low when compared to the stage-discharge ratings, and the radar-derived discharge for the Tanana River is biased low. To assess the value of radars to measure velocity and stage and the probability concept to compute discharge, GOF metrics were compiled for each gage, including the mean absolute error (MAE), percent bias (PBIAS), log-transformed Nash–Sutcliffe efficiency (logNSE), Nash–Sutcliffe efficiency (NSE), and volumetric efficiency (VE). These metrics represent accepted standards for comparing truth to modeled data [92–94]. For this application, the radar-derived discharge represents the model, and the stage-discharge represents truth.

Discharge was transmitted every 5 to 15 min for each radar gage resulting in data populations ranging from 448 to 151,161 samples. The entire record of each radar and conventional streamgage was analyzed for GOF, except for the Yellowstone River, where a split in streamflow occurred at discharges greater than approximately 113 m^3/s. Although the conventional Yellowstone River streamgage measures the entire streamflow regime, the radar gage measures streamflow less than 113 m^3/s. Subsequently, the record size was reduced from 7655 to 3099 samples. In general, model simulations can be considered satisfactory if the NSE exceeds 0.50 [93–95], PBIAS are within ±25% for streamflow [92–94], and VE approaches 1.0 [92–94]. The MAE and VE are most effective when used as an objective function for both low flows and high flows. The MAE ranged from 0.13 m^3/s for Cherry Creek to 286 m^3/s for the Susquehanna River. PBIAS ranged from 13.5% for the Blackfoot River to −13.9% for the NF Shenandoah River. The values for NSE varied from 0.59 for the Tanana River to 0.98 for the Gunnison River. The logNSE, which accounts for data bias, varied from 0.60 for the Tanana River to 0.99 for the Gunnison River. The VE ranged from 0.81 for Clear Creek to 0.94 for the Rio Grande. The scatter plots provide a measure of agreement and identify outliers between the radar-derived and stage-discharges. The data show strong agreement for Cherry Creek (R^2 = 0.947), Rio Grande (R^2 = 0.975), and the Gunnison River (R^2 = 0.998) and moderate agreement for the Tanana River (R^2 = 0.696).

Table 8. Goodness-of-fit statistics between operationally measured stage-discharge (Table 7) and radar-derived discharge based on the probability concept at 10 USGS streamgages in the United States. (n = number of samples; MAE = mean absolute error; PBIAS = percent bias, dimensionless; NSE = Nash-Sutcliffe Efficiency, dimensionless; logNSE = log-transformed Nash-Sutcliffe Efficiency, dimensionless; VE = Volumetric Efficiency, dimensionless; 1= number of samples were reduced from the complete discharge time series prior to the goodness-of-fit analysis to account for a split in streamflow caused by an island).

USGS streamgage	n	MAE (m^3/s)	PBIAS	NSE	(log) NSE	VE
Blackfoot River near Bonner, Montana	7499	9.12	13.5	0.95	0.88	0.86
Cherry Creek at Denver, Colorado	151,161	0.13	−9.2	0.91	0.90	0.86
Clear Creek near Lawson, Colorado	13,222	1.91	12	0.85	0.96	0.81
Gunnison River near Grand Junction, Colorado	34,628	7.85	−8.6	0.98	0.99	0.90
NF Shenandoah River near Strasburg, Virginia	29,563	2.53	−13.9	0.95	0.97	0.85
Red River of the North at Grand Forks, North Dakota	1558	122	−3.0	0.71	0.70	0.86
Rio Grande at Embudo, New Mexico	41,219	1.13	−2.9	0.97	0.97	0.94
Susquehanna River at Bloomsburg, Pennsylvania	448	286	−10.6	0.79	0.61	0.87
Tanana River at Nenana, Alaska	28,047	247	−10.6	0.59	0.60	0.82
Yellowstone River near Livingston, Montana	3099 [1]	6.17	−7.8	0.79	0.81	0.92

3.2.2. Wind Drift

Wind can bias surface velocity measurements depending on site characteristics (e.g., incised river valley), hydraulics (e.g., low hydraulic gradient), and surface velocities (less than 0.15 m/s). A wind sensor was deployed at the Red River of the North and recorded wind magnitudes ranging from near zero to 15.2 m/s and variable wind direction [90]. The Red River of the North flows generally south to north at an azimuth of approximately 130 degrees. Winds originating from the north (220 degrees < wind direction < 40 degrees) would impede surface velocities and produce a low bias in discharge. Conversely, winds originating from the south (220 degrees > wind direction > 40 degrees) would promote greater than expected surface velocities and produce a high bias in discharge. Figure 9 illustrates the wind-corrected radar-derived discharge (blue line) relative to (1) the stage-discharge rating (red line), (2) instantaneous confirmation measurements (black dots), which were made during the radar deployment, (3) radar-derived discharge (orange line), and (4) wind speed (purple line). Using the wind-correction algorithm described in Section 2.4.4., the wind-corrected radar-derived discharge is an improvement when compared to the uncorrected radar-derived discharge.

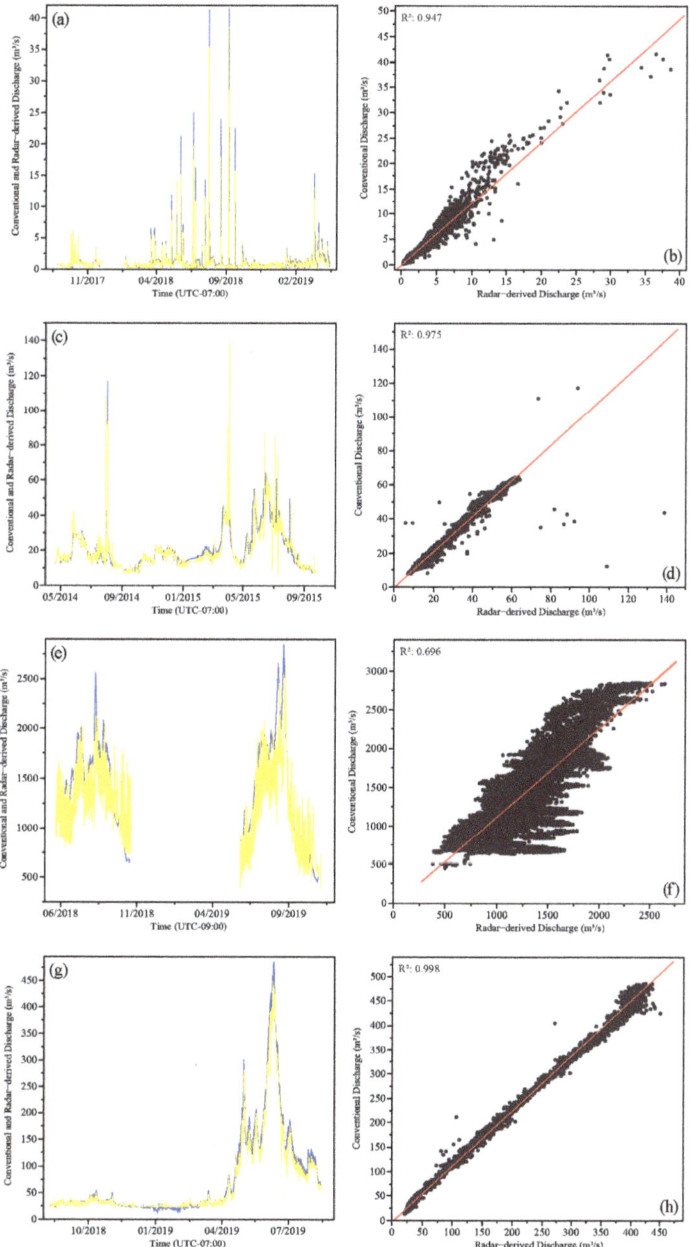

Figure 8. Continuous time series (**a,c,e,g**) and scatter plots (**b,d,f,h**) of radar-derived discharge and stage-discharge (conventional discharge) recorded concurrently during the operational period for (**a,b**) Cherry Creek at Denver, Colorado, (**c,d**) Rio Grande at Embudo, New Mexico, (**e,f**) Tanana River at Nenana, Alaska, and (**g,h**) Gunnison River near Grand Junction, Colorado. The velocity radars were deployed by the U.S. Geological Survey and collocated with existing USGS streamgages in the United States. (yellow line = radar-derived discharge; blue line = stage-discharge; red line = line of agreement; black dots = radar-derived and conventional discharges).

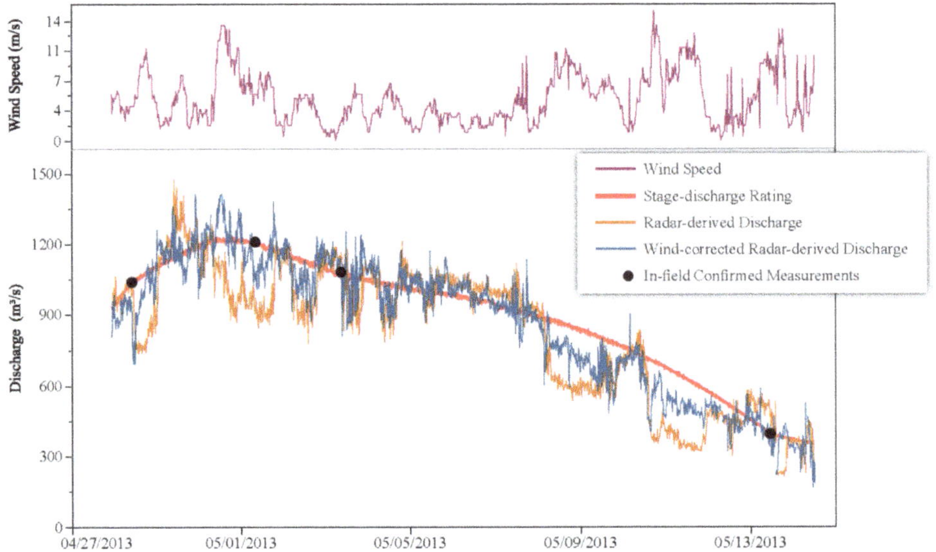

Figure 9. Time series of radar-derived discharge, without wind-drift correction (orange line), with wind-drift correction (blue line) relative to stage-discharge (red line) and confirmation measurements (black dots) and wind speed (purple line) recorded concurrently during the operational period for the Red River of the North at Grand Forks, North Dakota.

4. Discussion

4.1. Velocity and Discharge Computations

The composite average and absolute percent error in the instantaneous radar-derived mean velocities and discharges as compared to conventional methods for all 36 site visits was −0.3% and 5.1%, respectively. The data show strong agreement between the mean velocities (R^2 = 0.993) and discharges (R^2 = 0.999). Subsequently, when radars are properly sited, and the PC algorithm is properly parameterized, the near-field remote sensing of mean velocity and discharge can be computed and telemetered in real-time. Parameter calibration and hindcasting is not necessary. Both the timing and magnitude of peak and low-streamflow compared favorably to stage-discharge time series. An ensemble approach, which relied on a suite of GOF metrics including MAE, PBIAS, logNSE, NSE, and VE, was used to assess the potential for radars to measure discharge within an acceptable range of uncertainty. The extensibility of near-field remote sensing is based on the contention that: (1) 10 radar gages operated for sufficiently long periods of time and (2) hydraulic extremes were captured by radar gages deployed in a variety of hydrologic basins. On the basis of data presented, Doppler velocity and pulsed stage radars coupled with the PC algorithm can be operationalized to deliver real-time velocity and discharge within acceptable uncertainties common to all streamgaging methods.

The range of radar-derived surface, maximum, and mean velocities were expected based on the hydraulic settings, which were gaged. Low velocities were common in rivers that have low hydraulic gradients such as the Red River of the North. High velocities were reported in engineered urban and high-gradient mountain streams such as Cherry Creek, Clear Creek, Blackfoot River, and Yellowstone River. Because radar-derived discharge is a function of ϕ, maximum velocity, and the cross-sectional area, errors in any of these parameters will introduce uncertainties in the computed discharge, and it is important to acknowledge those biases. Although the results indicate that ϕ is generally resilient and invariant to changes with time or hydraulic conditions, temporal variations in ϕ were nominal and percent differences between the siting and operational phases were minimal for Cherry Creek

(0.678 and 0.6896, −1.7%), Clear Creek (0.573 and 0.5752, −0.3%), Susquehanna River (0.783 and 0.8114, −3.5%), and Tanana River (0.718 and 0.7639, −6.0%), respectively. For the Tanana River, the results indicate that the mean velocity and discharge can be computed in the absence of historical data to within an average percent error of about −6.0%. This assumes that the stage-area rating is stable, which must be periodically confirmed. Any deviation in the rating will affect the reported discharge. There is an operational advantage with radars relative to conventional methods, because a radar gage can be stood-up immediately after siting is complete, rather than waiting for a stage-discharge rating to be established. This has profound implications for ungaged basins, which lack historical stage and discharge measurements. Rather than wait months or years to develop a stage-discharge rating, real-time discharge can be delivered straight away using near-field remote sensing radars and the PC algorithm. The level of effort associated with radar gage operation and maintenance can be reduced to event-based visits only. Site visits would be justified prior to, during, and after extreme events to validate ϕ, the location of the y-axis (does its location vary with discharge), and the stage-area rating. This approach is consistent with conventional USGS methods, where periodic site visits are used to maintain stage-discharge ratings.

Based on the GOF metrics [92–94], the remote sensing of the mean velocity, stage, and discharge via radars is a reality. The PBIAS for each site was less than the ± 25% recommended for streamflow prediction; the logNSE and NSE exceeded the recommended value of 0.50; and the VE approaches a value of 1.0, indicating the predictive potential of radars to measure surface velocity and stage and to compute discharge using the probability concept. Cherry Creek ($R^2 = 0.947$), Rio Grande ($R^2 = 0.975$), and the Gunnison River ($R^2 = 0.998$) demonstrate that radar gages are capable of measuring streamflow in real-time. Outliers exist at Cherry Creek and the Gunnison River at greater discharges and may be related to unsteady streamflow conditions. In general, the Tanana River exhibited the lowest GOF metrics, and the time-series comparisons and scatter plots ($R^2 = 0.696$) support that contention. The Tanana River is dominated by large-scale, secondary currents that may bias surface velocities and subsequently discharge. This is reflected by the spikey nature of the time-series data illustrated in Figure 8e,f. Additional research is needed for large river systems, which may be influenced by environmental factors such as wind drift and hydraulic factors such as eddies, secondary flows, and macroturbulence. The Gunnison River and the Rio Grande are moderate-sized river systems and exhibited some of the highest logNSE, NSE, and VE values. It is important to note that hindcasting or model calibration was not conducted for any of the 10 radar gages. Rather, continuous, radar-derived velocity, stage, and discharge were transmitted contemporaneously to the USGS NWIS database [91] and shared the same time stamp as stage-discharge.

4.2. Uncertainty

4.2.1. Variability in ϕ

It was demonstrated that ϕ is relatively constant for a channel cross section regardless of discharge or other hydraulic factors (Figure 6a–d). When operationalizing a radar gage, it was assumed that the PC model (u_{max} occurs at the water surface versus u_{max} occurs at depth) selected during the siting phase was used to forecast streamflow during the operational phase. Additional research is needed to quantify the uncertainty in ϕ in the event that u_{max} occurs at the surface for specific discharges, and if u_{max} occurs at depth for other streamflow events at the same cross-section. Based on Tables 1, 2 and 4, no correlation exists between the magnitude of ϕ and drainage area; rather, ϕ is a function of the velocity distribution measured at a cross-section of interest.

4.2.2. Variability in y-axis

As discussed in Section 3.1.2., the location of the y-axis at each radar gage was verified during each site visit using the methods described in Section 2.4.1. At no time or during no specific streamflow condition did the y-axis differ from the siting phase. For example, the location of the y-axis at the

Susquehanna River was inferred from historical streamgaging records based on conventional 0.2D and 0.8D velocity measurements, which were used to compute the theoretical maximum velocity. The location of the y-axis at Cherry Creek, Clear Creek, and the Tanana River was confirmed in the field using mechanical current meters or hydroacoustics. Previous investigations indicate the location of the y-axis is stable [66–76,95]; however, additional research is needed to corroborate the uncertainty of the shifts in the y-axis as a function of discharge. As such, research is being conducted on the Gunnison River, where video images are being collected and processed for PIV. The camera and velocity radar footprint intersect, and the resulting velocity data will provide data to validate the y-axis variability.

4.2.3. Wind drift, Sample Duration and Frequency, and Minimum-Surface Velocities

Wind drift is caused by shear at the air–water interface and can serve as a source of error [51,62,96–99]. The coherent, continuous-wave radars that were deployed are capable of measuring surface scatterers; however, the energy transmitted by the radar is attenuated rapidly after it encounters the water surface. The effective depth at which the velocity is measured is a function of the scatterer wavelength or approximately 0.044 λ, where λ = the wavelength of the resonant water wave [96]. For λ = 0.75 cm, this depth is about 0.03 cm. The wind drift layer at this depth has not completely decayed, and the surface velocity requires correction or qualification to accurately estimate discharge. The results from the Red River of the North demonstrate the need to (1) implement theoretical correction such as the one described in Section 3.2.2, (2) qualify the data as excellent, good, fair, or poor depending on the magnitude of wind versus the surface velocity, or (3) develop an empirical relation to correct the wind-affected surface velocities as a function of other metrics such as a change in water level. Regardless, wind can be a prominent bias in radar-derived velocities and discharges.

The sample duration and frequency has an effect on the radar return and can be optimized in the field based on the quality of the radar spectrum. Typically, when deploying velocity radars and depending on variations in velocity with time, spot dwells should range from 30 s to 2 min to achieve stable measurements [59,62]. For example, the PBIAS, NSE, and logNSE reported for the Tanana River may be a function of the radar sample duration and frequency. The Tanana River is dominated by large-scale secondary currents that exhibit themselves as eddies on the water surface. If the eddies are sufficiently large, and the sample duration is small, the radar will sample the eddy-dominated velocities rather than the scatterers that ride on the surface of large-scale waves characteristic of ambient velocities. This effect may bias the velocities depending on the magnitude and direction of the eddy-dominated currents and deviate from truth. Similarly, the noise (data spikes) data observed in Figures 8a,c,e,g and 9 may be a composite of each of these factors.

Various filters are being investigated to reduce the noise observed in some radar gages, which may be physically correct or may be associated with wind drift, sample duration, and temporal and spatial variability in surface velocities. Filters could include high- and low-pass filters, moving average, LOESS, Savitsky-Golay, and Kalman filters. The Savitsky-Golay filter has been incorporated in the USGS NWIS database [91] for Cherry Creek, Clear Creek, and the Tanana River; however, the filtered results are exploratory and were not included in this research.

Velocity radars operate best when surface velocities are greater than 0.15 m/s; particularly when surface-water scatterers (small waveforms) are ill-defined and wind drift is dominant.

5. Conclusions

To evaluate the efficacy of radars and algorithms used to measure and compute real-time discharge, 10 radar gages representing a wide range of hydraulic conditions were selected. The radars were deployed in near-field environments from fixed platforms on bridge decks. A range of hydrologic and hydraulic characteristics were considered in the study design and include basin variability (flashy-urban basins, high-gradient mountain streams, mixed use, and large rivers), drainage area (381 to 66,200 km^2), top width (7 to 380 m), surface velocities (0.3 to 3.84 m/s), and discharge (0.12 to 4950 m^3/s). Each of these characteristics has a profound effect on velocity and discharge magnitudes. To demonstrate the extensibility of near-field remote sensing, extremes in surface velocity and discharge were sought when siting a radar gage. All radar gages were collocated with an existing USGS streamgage, which served as truth, and were deployed for a sufficient period of time so that streamflow extremes could be captured and analyzed. Streamflow exceedances were computed and ranged from approximately 0% to 5.4%. These metrics indicate that the radar deployments are representative of the streamflow magnitudes and temporal variations, which were monitored at the collocated USGS streamgages.

The results show that Doppler velocity and stage-radar gages can produce continuous time series of mean velocity, stage, and discharges that: (1) compare favorably to stage-discharge streamgages and (2) can be computed in the absence of historical data. This assumes that the stage-area ratings developed for each site are stable or can be maintained. These methods offer an operational advantage relative to conventional methods, because unlike stage-discharge streamgages, they do not require months or years of rating development and maintenance. Rather, real-time data can be collected, computed, and transmitted immediately after the radars are deployed. This has profound implications for data analysis, delivery, and costs in settings such as ungaged basins, burn scars, flood-alert networks, and infrastructure design projects. Real-time discharge can be delivered using the PC algorithm and near-field remote sensing via radars. The level of effort associated with radar gage operation and maintenance can be reduced to event-based visits only, rather than recurring visits that are common with conventional streamgage maintenance. For example, a site visit would be conducted to validate ϕ, the location of the y-axis (does its location vary with discharge), and the stage-area rating during and after an extreme streamflow event.

Caution must be followed when deploying radars at sites where: (1) surface velocities are less than 0.15 m/s, (2) surface scatterers (small waveforms) are lacking, (3) wind is dominant, and (4) cross sections of interest are not compliant with traditional streamgaging methods. Although there is an operational advantage to near-field remote sensing, sources of uncertainty include gravity driven waves, wind drift, and sample duration, all of which can bias surface velocities and subsequently discharge.

During the siting phase, the composite average and absolute percent error in the instantaneous radar-derived mean velocities and discharges as compared to conventional methods for all 36 site visits was −0.3% and 5.1%, respectively. The data show strong agreement between the mean velocities (R^2 = 0.993) and discharges (R^2 = 0.999). During the operational phase, it is important to note the velocity, stage, and discharge time series represent two data streams—*near-field remote sensing data and conventional data*—that are independent and transmitted concurrently under the same time stamp. Hindcasting of the remote-sensing data was not conducted; rather, these two data streams represent independent, temporally equivalent, real-time products. The timing and magnitude of peak and low-streamflows measured by the velocity radar and reported in the USGS NWIS database [91] were comparable. The GOF metrics were compiled to demonstrate the comparability of the radar- and stage-discharge time series and included the number of samples, MAE, PBIAS, logNSE, NSE, and VE. Site operations for the 10 radar gages ranged from 2 weeks to 19 months, and the number of sample points per radar varied from 448 to 151,161. For the GOF parameters, MAE ranged from 0.13 to 286 m^3/s; PBIAS ranged from 13.5% to -13.9%; NSE varied from 0.59 to 0.98; logNSE varied from 0.60 to 0.99; and VE ranged from 0.81 to 0.94. In general, this research answers the following questions: (1) radars can accurately measure surface velocities, (2) velocity and stage radars can be used to accurately compute mean velocities and discharges when compared to conventional methods that serve as truth, (3) near-field remote sensing can deliver real-time mean velocity, stage, and discharge operationally and in the absence of historical data, and (4) environmental and hydraulic factors must be considered when processing radar-derived surface velocities. The results support the contention that the probability concept coupled with velocity and stage radars are a viable technology that can be operationalized to deliver real-time velocity, stage, and discharge.

Author Contributions: Conceptualization, J.W.F.; methodology, J.W.F.,C.-L.C., T.M., software, C.A.M., J.W.F.; validation, TheTeam; formal analysis, J.W.F.; investigation, TheTeam; resources, J.J.G, D.W.; writing—original draft preparation, J.W.F., J.R.E., C.A.M.; writing—review and editing, J.W.F., J.R.E., C.A.M.; visualization, C.A.M.;

supervision, J.W.F.; project administration, J.W.F.; funding acquisition, TheTeam. All authors have read and agreed to the published version of the manuscript.

Funding: This research was supported by the U.S. Geological Survey's Water Mission Area Hydrologic Remote Sensing Branch and the U.S. Geological Survey Colorado Water Science Center.

Acknowledgments: The authors would like to recognize Jeff Conaway (USGS), Michelle Kang (USGS), Mike Kohn (USGS) and Jason Lambrecht (USGS) for their support in the field. Peter Ward (Hydrological Services America) provided the Doppler velocity and stage radars. Without these contributions, this work would not have been possible. The field measurements and image datasets used in this study are available in a USGS data release [90]. Any use of trade, firm, or product names is for descriptive purposes only and does not imply endorsement by the U.S. Government.

Conflicts of Interest: The authors declare no conflict of interest, and the funders had no role in the design of the study; in the collection, analyses, or interpretation of data; in the writing of the manuscript, or in the decision to publish the results.

References

1. Eberts, S.; Woodside, M.; Landers, M.; Wagner, C. *Monitoring the Pulse of our Nation's Rivers and Streams—The U.S. Geological Survey Streamgaging Network*; U.S. Geological Survey Fact Sheet 2018–3081; 2018. Available online: https://pubs.er.usgs.gov/publication/fs20183081 (accessed on 19 April 2020).

2. Buchanan, T.J.; Somers, W.P. *Discharge Measurements at Gaging Stations*; U.S. Geological Survey Techniques of Water-Resources Investigations, Book 3, Chap A8; 1969; p. 65. Available online: http://pubs.usgs.gov/twri/twri3a8/ (accessed on 30 December 2019).

3. Rantz, S. *Measurement and Computation of Streamflow: Volume 1. Measurement of Stage and Discharge*; U.S. Geological Survey Water Supply Paper 2175; United States Government Printing Office: Washington, DC, USA, 1982; p. 284.

4. Turnipseed, D.P.; Sauer, V.B. *Discharge Measurements at Gaging Stations*; U.S. Geological Survey Techniques and Methods Book 3, Chap. A8; 2010; p. 87. Available online: http://pubs.usgs.gov/tm/tm3-a8/ (accessed on 30 December 2019).

5. Christensen, J.L.; Herrick, L.E. *Mississippi River Test: Volume 1. Final Rep.*; DCP4400/300; the U.S. Geological Survey, AMETEK/Straza Division: El Cajon, CA, USA, 1982.

6. Simpson, M.R. Evaluation of a vessel-mounted acoustic Doppler current profiler for use in rivers and estuaries. In Proceedings of the 3rd Working Conference on Current Measurement, IEEE, Washington, DC, USA, 22–24 January 1986; pp. 106–121.

7. Gordon, R.L. Acoustic measurement of river discharge. *J. Hydraul. Eng.* **1989**, *115*, 925–936. [CrossRef]

8. Simpson, M.R.; Oltmann, R.N. *Discharge-Measurement System Using an Acoustic Doppler Current Profiler with Applications to Large Rivers and Estuaries*; U.S. Geological Survey Water Supply Paper 2395; U.S. Geological Survey: Reston, VA, USA, 1993; p. 32.

9. Oberg, K.A.; Morlock, S.E.; Caldwell, W.S. *Quality Assurance Plan for Discharge Measurements Using Acoustic Doppler Current Profilers*; U.S. Geological Survey Scientific Investigations Report 2005–5135; U.S. Geological Survey: Reston, VA, USA, 2005; p. 35.

10. Mueller, D.S. *QRev—Software for Computation and Quality Assurance of Acoustic Doppler Current Profiler Moving-Boat Streamflow Measurements—User's Manual for Version 2.8*; U.S. Geological Survey Open-File Report 2016–1052; USGS: Reston, VA, USA, 2016; p. 50. Available online: https://pubs.er.usgs.gov/publication/ofr20161052 (accessed on 30 December 2019).

11. Mueller, D.S. Extrap: Software to assist the selection of extrapolation methods for moving-boat ADCP streamflow measurements. *Comput. Geosci.* **2013**, *54*, 211–218. [CrossRef]

12. Lin, D.; Grundmann, J.; Eltner, A. Evaluating Image Tracking Approaches for Surface Velocimetry With Thermal Tracers. *Water Resour. Res.* **2019**, *55*, 3122–3136. [CrossRef]

13. Hauet, A.; Morlot, T.; Daubagnan, L. *Velocity Profile and Depth-Averaged to Surface Velocity in Natural Streams: A Review over a Large Sample of Rivers*. E3S Web of Conferences 40 06015 River Flow, EDP Sciences, France. 2018. Available online: https://www.e3s-conferences.org/articles/e3sconf/abs/2018/15/e3sconf_riverflow2018_06015/e3sconf_riverflow2018_06015.html (accessed on 20 January 2020).

14. Fujita, I.; Komura, S. Application of video image analysis for measurements of river-surface flows. *Proc. Hydraul. Eng. JSCE* **1994**, *38*, 733–738. [CrossRef]

15. Tauro, F.; Pagano, C.; Phamduy, P.; Grimaldi, S.; Porfiri, M. Large-scale particle image velocimetry from an unmanned aerial vehicle. *IEEE/ASME Trans. Mechatron.* **2015**, *20*, 3269–3275. [CrossRef]

16. Tauro, F.; Porfiri, M.; Grimaldi, S. Surface flow measurements from drones. *J. Hydrol.* **2016**, *540*, 240–245. [CrossRef]

17. Kinzel, P.J.; Legleiter, C.J. sUAS-based remote sensing of river discharge using thermal particle image velocimetry and bathymetric lidar. *Remote Sens.* **2019**, *11*, 2317. [CrossRef]

18. Dugan, J.P.; Anderson, S.P.; Piotrowski, C.C.; Zuckerman, S.B. Airborne infrared remote sensing of riverine currents. *IEEE Trans. Geosci. Remote Sens.* **2014**, *52*, 3895–3907. [CrossRef]

19. Smith, L.C.; Isacks, B.L.; Bloom, A.L.; Murray, A.B. Estimation of discharge from three braided rivers using synthetic aperture radar satellite imagery. *Water Resour. Res.* **1996**, *32*, 2021–2034. [CrossRef]

20. Leon, J.G.; Calmant, S.; Seyler, F.; Bonnet, M.P.; Cauhope, M.; Frappart, F. Rating curves and estimation of average water depth at the upper Negro River based on satellite altimeter data and modeled discharges. *J. Hydrol.* **2006**, *328*, 481–496. [CrossRef]

21. Getirana, A.C.V.; Bonnet, M.P.; Calmant, S.; Roux, H.; Filho, O.C.R.; Mansur, W.J. Hydrological monitoring of poorly gauged basins based on rainfall–runoff modeling and spatial altimetry. *J. Hydrol.* **2009**, *379*, 205–219. [CrossRef]

22. Birkinshaw, S.J.; O'Donnell, G.M.; Moore, P.; Kilsby, C.G.; Fowler, H.J.; Berry, P.A.M. Using satellite altimetry data to augment flow estimation techniques on the Mekong River. *Hydrol. Process.* **2010**, *24*, 3811–3825. [CrossRef]

23. Getirana, A.C.V.; Peters-Lidard, C. Estimating water discharge from large radar altimetry datasets. *Hydrol. Earth Syst. Sci.* **2013**, *17*, 923–933. [CrossRef]

24. Paiva, R.C.D.; Buarque, D.C.; Colischonn, W.; Bonnet, M.-P.; Frappart, F.; Calmant, S.; Mendes, C.A.B. Large-scale hydrological and hydrodynamics modelling of the Amazon River basin. *Water Resour. Res.* **2013**, *49*, 1226–1243. [CrossRef]

25. Tarpanelli, A.; Barbetta, S.; Brocca, L.; Moramarco, T. River discharge estimation by using altimetry data and simplified flood routing modeling. *Remote Sens.* **2013**, *5*, 4145–4162. [CrossRef]

26. Birkinshaw, S.J.; Moore, P.; Kilsby, C.G.; O'Donnell, G.M.; Hardy, A.J.; Berry, P.A.M. Daily discharge estimation at ungauged river sites using remote sensing. *Hydrol. Process.* **2014**, *28*, 1043–1054. [CrossRef]

27. Pavelsky, T.M. Using width-based rating curves from spatially discontinuous satellite imagery to monitor river discharge. *Hydrol. Process.* **2014**, *28*, 3035–3040. [CrossRef]

28. Paris, A.; Dias de Paiva, R.; Santos da Silva, J.; Medeiros Moreira, D.; Calmant, S.; Garambois, P.-A.; Collischonn, W.; Bonnet, M.-P.; Seyler, F. Stage-discharge rating curves based on satellite altimetry and modeled discharge in the Amazon basin. *Water Resour. Res.* **2016**, *52*, 3787–3814. [CrossRef]

29. Bjerklie, D.M.; Birkett, C.M.; Jones, J.W.; Carabajal, C.; Rover, J.A.; Fulton, J.W.; Garambois, P.-A. Satellite remote sensing estimation of river discharge: Application to the Yukon River Alaska. *J. Hydrol.* **2018**, *561*, 1000–1018. [CrossRef]

30. Bogning, S.; Frappart, F.; Blarel, F.; Niño, F.; Mahé, G.; Bricquet, J.-P.; Seyler, F.; Onguéné, R.; Etamé, J.; Paiz, M.-C.; et al. Monitoring water levels and discharges using radar altimetry in an ungauged river basin: The case of the Ogooué. *Remote Sens.* **2018**, *10*, 350. [CrossRef]

31. Moramarco, T.; Barbetta, S.; Bjerklie, D.M.; Fulton, J.W.; Tarpanelli, A. River bathymetry estimate and discharge assessment from remote sensing. *Water Resour. Res.* **2019**, *55*, 6692–6711. [CrossRef]

32. Birkett, C.M. Contribution of the TOPEX NASA radar altimeter to the global monitoring of large rivers and wetlands. *Water Resour. Res.* **1998**, *34*, 1223–1239. [CrossRef]

33. Birkett, C.M.; Mertes, L.A.K.; Dunne, T.; Costa, M.H.; Jasinski, M.J. Surface water dynamics in the Amazon Basin: Application of satellite radar altimetry. *J. Geophys. Res.* **2002**, *107*, 8059. [CrossRef]

34. Kouraev, A.V.; Zakharova, E.A.; Samain, O.; Mognard, N.M.; Cazenave, A. Ob' river discharge from TOPEX/Poseidon satellite altimetry (1992–2002). *Remote Sens. Environ.* **2004**, *93*, 238–245. [CrossRef]

35. Papa, F.; Bala, S.K.; Kumar Pandey, R.; Durand, F.; Rahman, A.; Rossow, W.B. Ganga-Brahmaputra River discharge. *Geophys. Res.* **2012**, *117*, C11021. [CrossRef]

36. Tulbure, M.G.; Broich, M. Spatiotemporal dynamic of surface water bodies using Landsat time-series data from 1999 to 2011. *ISPRS J. Photogramm. Remote Sens.* **2013**, *79*, 44–52. [CrossRef]

37. Gleason, C.J.; Smith, L.C. Toward global mapping of river discharge using satellite images and at-many-stations hydraulic geometry. *Proc. Natl. Acad. Sci. USA* **2014**, *111*, 4788–4791. [CrossRef]

38. Kim, J.-W.; Lu, Z.; Jones, J.W.; Shum, C.K.; Lee, H.; Jia, Y. Monitoring Everglades freshwater marsh water level using L-band synthetic aperture radar backscatter. *Remote Sens. Environ.* **2014**, *150*, 66–81. [CrossRef]

39. Jones, J.W. Efficient wetland surface water detection and monitoring via Landsat: Comparison with in situ data from the Everglades Depth Estimation Network. *Remote Sens.* **2015**, *7*, 12503–12538. [CrossRef]

40. Carroll, M.; Wooten, M.; DiMiceli, C.; Sohlberg, R.; Kelly, M. Quantifying surface water dynamics at 30 meter spatial resolution in the North American high northern latitudes 1991–2011. *Remote Sens.* **2016**, *8*, 622. [CrossRef]

41. Brakenridge, G.R.; Nghiem, S.V.; Anderson, E.; Mic, R. Orbital microwave measurement of river discharge and ice status. *Water Resour. Res.* **2007**, *43*, W04405. [CrossRef]

42. Durand, M.; Neal, J.; Rodríguez, E.; Andreadis, K.M.; Smith, L.C.; Yoon, Y. Estimating reach-averaged discharge for the River Severn from measurements of river water surface elevation and slope. *J. Hydrol.* **2014**, *511*, 92–104. [CrossRef]

43. Biancamaria, S.; Lettenmaier, D.P.; Pavelsky, T. The SWOT Mission and its capabilities for land hydrology. *Surv. Geophys.* **2015**, *37*, 307–337. [CrossRef]

44. Garambois, P.-A.; Monnier, J. Inference of effective river properties from remotely sensed observations of water surface. *Adv. Water Resour.* **2015**, *79*, 103–120. [CrossRef]

45. Bonnema, M.G.; Sikder, S.; Hossain, F.; Durand, M.; Gleason, C.J.; Bjerklie, D.M. Benchmarking wide swath altimetry-based river discharge estimation algorithms for the Ganges river system. *Water Resour. Res.* **2016**, *52*, 2439–2461. [CrossRef]

46. Durand, M.; Gleason, C.J.; Garambois, P.A.; Bjerklie, D.M.; Smith, L.C.; Roux, H.; Rodriguez, E.; Bates, P.D.; Pavelsky, T.M.; Monnier, J.; et al. An intercomparison of remote sensing river discharge estimation algorithms from measurements of river height, width, and slope. *Water Resour. Res.* **2016**, *52*, 4527–4549. [CrossRef]

47. National Aeronautics and Space Administration (NASA). SWOT Surface Water and Ocean Topography. Available online: https://swot.jpl.nasa.gov/home.htm (accessed on 17 July 2018).

48. Costa, J.E.; Spicer, K.R.; Cheng, R.T.; Haeni, F.P.; Melcher, N.B.; Thurman, E.M.; Plant, W.J.; Keller, W.C. Measuring stream discharge by non-contact methods: A Proof-of-Concept Experiment. *Geophys. Res. Lett.* **2000**, *27*, 553–556. [CrossRef]

49. Melcher, N.B.; Costa, J.E.; Haeni, F.P.; Cheng, R.T.; Thurman, E.M.; Buursink, M.; Spicer, K.R.; Hayes, E.; Plant, W.J.; Keller, W.C. River discharge measurements by using helicopter-mounted radar. *Geophys. Res. Lett.* **2002**, *29*, 41–44. [CrossRef]

50. Fulton, J.W.; Anderson, I.E.; Chiu, C.-L.; Sommer, W.; Adams, J.D.; Moramarco, T.; Bjerklie, D.M.; Fulford, J.M.; Sloan, J.L. sUAS-based remote sensing of velocity and discharge using Doppler radar. (manuscript in preparation).

51. Plant, W.J.; Keller, W.C.; Hayes, K. Measurement of river surface currents with coherent microwave systems. *IEEE Trans. Geosci. Remote Sens.* **2005**, *43*, 1242–1257. [CrossRef]

52. Costa, J.E.; Cheng, R.T.; Haeni, F.P.; Melcher, N.; Spicer, K.R.; Hayes, E.; Plant, W.; Hayes, K.; Teague, C.; Barrick, D. Use of radars to monitor stream discharge by noncontact methods. *Water Resour. Res.* **2006**, *42*, 14. [CrossRef]

53. Fulton, J.W.; Ostrowski, J. Measuring real-time streamflow using emerging technologies: Radar, hydroacoustics, and the probability concept. *J. Hydrol.* **2008**, *357*, 1–10. [CrossRef]

54. Boning, C.W. *Policy Statement on Stage Accuracy*; U.S. Geological Survey, Office of Surface Water Technical Memorandum 93.07; 1992. Available online: https://water.usgs.gov/admin/memo/SW/sw93.07.html (accessed on 14 February 2020).

55. Sauer, V.B.; Meyer, R.W. *Determination of Error in Individual Discharge Measurements*; U.S. Geological Survey Open-File Report 92–144; U.S. Geological Survey: Reston, VA, USA, 1992; p. 21. [CrossRef]

56. Kiang, J.E.; Gazoorian, C.; McMillan, H.; Coxon, G.; Le Coz, J.; Westerberg, I.K.; Belleville, A.; Sevrez, D.; Sikorska, A.E.; Petersen-Øverleir, A.; et al. A Comparison of Methods for Streamflow Uncertainty Estimation. *Water Resour. Res.* **2018**, *54*, 7149–7176. [CrossRef]

57. Cohn, T.A.; Kiang, J.E.; Mason, R.R. Estimating Discharge Measurement Uncertainty Using the Interpolated Variance Estimator. *J. Hydraul. Eng.* **2013**, *139*, 502–510. [CrossRef]

58. McMillan, H.; Krueger, T.; Freer, J. Benchmarking observational uncertainties for hydrology: Rainfall, river discharge and water quality. *Hydrol. Process.* **2012**, *26*, 4078–4111. [CrossRef]

59. Welber, M.; Le Coz, J.; Laronne, J.B.; Zolezzi, G.; Zamler, D.; Dramais, G.; Hauet, A.; Salvaro, M. Field assessment of noncontact stream gauging using portable surface velocity radars (SVR). *Water Resour. Res.* **2016**, *2*, 1108–1126. [CrossRef]

60. Sauer, V.B. *Standards for the Analysis and Processing of Surface-Water Data and Information Using Electronic Methods*; U.S. Geological Survey Water-Resources Investigations Report 01–4044; U.S. Geological Survey: Reston, VA, USA, 2002; p. 91.

61. Levesque, V.A.; Oberg, K.A. *Computing Discharge Using the Index Velocity Method*; U.S. Geological Survey Techniques and Methods 3–A23; 2012; p. 148. Available online: https://pubs.usgs.gov/tm/3a23/ (accessed on 20 January 2019).

62. Fulton, J.W. Guidelines for Siting and Operating Surface-Water Velocity Radars. Available online: https://my.usgs.gov/confluence/display/SurfBoard/Guidelines+for+Siting+and+Operating+Surface-water+Velocity+Radars (accessed on 25 June 2018).

63. Moramarco, T.; Barbetta, S.; Tarpanelli, A. From surface flow velocity measurements to discharge assessment by the entropy theory. *Water.* **2017**, *9*, 120. [CrossRef]

64. Parsons, D.R.; Jackson, P.R.; Czuba, J.A.; Engel, F.L.; Rhoads, B.L.; Oberg, K.A.; Best, J.L.; Mueller, D.S.; Johnson, K.K.; Riley, J.D. Velocity Mapping Toolbox (VMT): A processing and visualization suite for moving-vessel ADCP measurements. *Earth Surf. Process. Landf.* **2013**, *38*, 1244–1260. [CrossRef]

65. U.S. Geological Survey (USGS), Office of Surface Water Hydroacoustics Page. Available online: https://hydroacoustics.usgs.gov/movingboat/WinRiverII.shtml (accessed on 30 January 2020).

66. Chiu, C.-L.; Chiou, J.-D. Structure of 3-D Flow in rectangular open channels. *J. Hydraul. Eng.* **1986**, *112*, 1050–1068. [CrossRef]

67. Chiu, C.-L. Entropy and probability concepts in hydraulics. *J. Hydraul. Eng.* **1987**, *113*, 583–600. [CrossRef]

68. Chiu, C.-L. Velocity distribution in open channel flow. *J. Hydraul. Eng.* **1989**, *115*, 576–594. [CrossRef]

69. Chiu, C.-L. *Probability and Entropy Concepts in Fluid Flow Modeling and Measurement*; National Taiwan University: Taipei, Taiwan, 15 March 1995.

70. Chiu, C.-L.; Tung, N.C.; Hsu, S.M.; Fulton, J.W. *Comparison and Assessment of Methods of Measuring Discharge in Rivers and Streams*; Research Report No. CEEWR-4; Dept. of Civil & Environmental Engineering, University of Pittsburgh: Pittsburgh, PA, USA, 2001.

71. Chiu, C.-L.; Tung, N.C. Velocity and regularities in open-channel flow. *J. Hydraul. Eng.* **2002**, *128*, 390–398. [CrossRef]

72. Chiu, C.-L.; Hsu, S.M.; Tung, N.C. Efficient methods of discharge measurements in rivers and streams based on the probability concept. *Hydrol. Process. Wiley Intersci.* **2005**, *19*, 3935–3946. [CrossRef]

73. Chiu, C.-L.; Hsu, S.-M. Probabilistic approach to modeling of velocity distributions in fluid flows. *J. Hydrol.* **2006**, *316*, 28–42. [CrossRef]

74. Moramarco, T.; Dingman, S.L. On the theoretical velocity distribution and flow resistance in natural channels. *J. Hydrol.* **2017**, *555*, 777–785. [CrossRef]

75. Moramarco, T.; Saltalippi, C.; Singh, V.P. Estimation of mean velocity in natural channel based on Chiu's velocity distribution equation. *J. Hydrol. Eng.* **2004**, *9*, 42–50. [CrossRef]

76. Fulton, J.W. Comparison of Conventional and Probability-Based Modeling of Open-Channel Flow in The Allegheny River, Pennsylvania, USA. Unpublished Thesis, Department of Civil and Environmental Engineering, University of Pittsburgh, Pittsburgh, PA, USA, April 1999.

77. Lant, J.; Mueller, D.S. Stage Area Rating Application – AreaComp2, USGS. Office of Surface Water. 2012. Available online: http://hydroacoustics.usgs.gov/software/AreaComp2_Users_Guide.pdf (accessed on 21 January 2020).

78. Chow, V.T.; Maidment, D.R.; Mays, L.W. *Applied Hydrology, International Edition 1988*; McGraw-Hill Inc.: Singapore, 1988; pp. 42–47.

79. González, J.A.; Melching, C.S.; Oberg, K.A. Analysis of open-channel velocity measurements collected with an acoustic Doppler current profiler. Reprint from RIVERTECH 96. In Proceedings of the 1st International Conference On New/Emerging Concepts for Rivers Organized by the International Water Resources Association, Chicago, IL, USA, 22–26 September 1996.

80. Murphy, E.C. *Accuracy of Stream Measurements*; Water-Supply and Irrigation Paper No. 95, Series M, General Hydrographic Investigations 10; United States Government Printing Office: Washington, DC, USA, 1904; pp. 58–163.

81. Guo, J.; Julien, P.Y. Application of the modified log-wake law in open-Channels. *J. Appl. Fluid Mech.* **2008**, *1*, 17–23.
82. Jarrett, R.D. Wading measurements of vertical velocity profiles. *Geomorphology* **1991**, *4*, 243–247. [CrossRef]
83. Kundu, S.; Ghoshal, K. Velocity Distribution in Open Channels: Combination of Log-law and Parabolic-law. *World Acad. Sci. Eng. Technol. Int. J. Math. Comput. Phys. Electr. Comput. Eng.* **2012**, *6*, 1234–1241.
84. Wiberg, P.L.; Smith, J.D. Velocity distribution and bed roughness in high-gradient streams. *Water Resour. Res.* **1991**, *27*, 825–838. [CrossRef]
85. Yang, S.-Q.; Xu, W.-L.; Yu, G.-L. Velocity distribution in gradually accelerating free surface flow. *Adv. Water Resour.* **2006**, *29*, 1969–1980. [CrossRef]
86. Blodgett, J.C. *Rock Riprap Design for Protection of Stream Channels Near Highway Structures, Volume 1: Hydraulic Characteristics of Open Channels*; U.S. Geological Survey Water-Resources Investigations Report 86–4127; U.S. Geological Survey: Reston, VA, USA, 1986; p. 60. [CrossRef]
87. Shannon, C.E. A mathematical theory of communication. *Bell Syst. Tech. J.* **1948**, *27*, 623–656. [CrossRef]
88. R Core Team. *R: A Language and Environment for Statistical Computing*; R Foundation for Statistical Computing: Vienna, Austria, 2019; Available online: https://www.R-project.org/ (accessed on 9 February 2020).
89. Zambrano-Bigiarini, M. HydroGOF: Goodness-of-Fit Functions for Comparison of Simulated and Observed Hydrological Time Series. R Package Version 0.3–10. 2017. Available online: http://hzambran.github.io/hydroGOF/ (accessed on 9 February 2020).
90. Fulton, J.W.; Mcdermott, W.R.; Mason, C.A. Radar-based field measurements of surface velocity and discharge from 10 USGS streamgages for various locations in the United States, 2002–2019. U.S. Geological Survey data release, (manuscript in review).
91. U.S. Geological Survey. *USGS Water Data for the Nation*; U.S. Geological Survey National Water Information System Database; 2016. Available online: https://waterdata.usgs.gov/nwis (accessed on 15 November 2016).
92. Criss, R.E.; Winston, W.E. Do Nash values have value? Discussion and alternate proposals. *Hydrol. Process.* **2008**, *22*, 2723–2725. [CrossRef]
93. Moriasi, D.N.; Arnold, J.G.; Van Liew, M.W.; Bingner, R.L.; Harmel, R.D.; Veith, T.L. Model evaluation guidelines for systematic quantification of accuracy in watershed simulations. *Am. Soc. Agric. Biol. Eng.* **2007**, *50*, 885–900. Available online: http://citeseerx.ist.psu.edu/viewdoc/download?doi=10.1.1.532.2506&rep=rep1&type=pdf (accessed on 14 February 2020).
94. Muleta, M.K. Model performance sensitivity to objective function during automated Calibrations. *J. Hydrol. Eng.* **2012**, *17*, 756–767. [CrossRef]
95. Bechle, A.J.; Wu, C.H. An entropy-based surface velocity method for estuarine discharge measurement. *Water Resour. Res.* **2014**, *50*, 6106–6128. [CrossRef]
96. Plant, W.J.; Keller, W.C. Evidence of Bragg scattering in microwave Doppler spectra of sea return. *J. Geophys. Res.* **1990**, *95*, 16299–16310. [CrossRef]
97. Plant, W.J. *Bragg Scattering of Electromagnetic Waves from The air/sea Interface, in Surface Waves and Fluxes: Current Theory and Remote Sensing*; Geernaert, G.L., Plant, W.J., Eds.; Kluwer: Dordrecht, The Netherlands, 1990; Volume 2, p. 336.
98. Wright, J.W. A new model for sea clutter. *IEEE Trans. Antennas Propag.* **1968**, *AP-16*, 217–223. [CrossRef]
99. Plant, W.J.; Wright, J.W. Phase speeds of upwind and downwind traveling short gravity waves. *J. Geophys. Res.* **1980**, *85*, 3304–3310. [CrossRef]

 *remote sensing*

Article

sUAS-Based Remote Sensing of River Discharge Using Thermal Particle Image Velocimetry and Bathymetric Lidar

Paul J. Kinzel [1,*] and Carl J. Legleiter [1,2]

[1]   U.S. Geological Survey, Integrated Modeling and Prediction Division, Golden, CO 80403, USA; cjl@usgs.gov
[2]   Department of Geography, University of Wyoming, Laramie, WY 82071, USA
*   Correspondence: pjkinzel@usgs.gov; Tel.: +1-303-278-7941

Received: 19 August 2019; Accepted: 1 October 2019; Published: 5 October 2019

Abstract: This paper describes a non-contact methodology for computing river discharge based on data collected from small Unmanned Aerial Systems (sUAS). The approach is complete in that both surface velocity and channel geometry are measured directly under field conditions. The technique does not require introducing artificial tracer particles for computing surface velocity, nor does it rely upon the presence of naturally occurring floating material. Moreover, no prior knowledge of river bathymetry is necessary. Due to the weight of the sensors and limited payload capacities of the commercially available sUAS used in the study, two sUAS were required. The first sUAS included mid-wave thermal infrared and visible cameras. For the field evaluation described herein, a thermal image time series was acquired and a particle image velocimetry (PIV) algorithm used to track the motion of structures expressed at the water surface as small differences in temperature. The ability to detect these thermal features was significant because the water surface lacked floating material (e.g., foam, debris) that could have been detected with a visible camera and used to perform conventional Large-Scale Particle Image Velocimetry (LSPIV). The second sUAS was devoted to measuring bathymetry with a novel scanning polarizing lidar. We collected field measurements along two channel transects to assess the accuracy of the remotely sensed velocities, depths, and discharges. Thermal PIV provided velocities that agreed closely (R^2 = 0.82 and 0.64) with in situ velocity measurements from an acoustic Doppler current profiler (ADCP). Depths inferred from the lidar closely matched those surveyed by wading in the shallower of the two cross sections (R^2 = 0.95), but the agreement was not as strong for the transect with greater depths (R^2 = 0.61). Incremental discharges computed with the remotely sensed velocities and depths were greater than corresponding ADCP measurements by 22% at the first cross section and <1% at the second.

Keywords: small unmanned aerial system (sUAS); river flow; thermal infrared imagery; particle image velocimetry; lidar bathymetry

---

1. Introduction

The U.S. Geological Survey (USGS) operates one of the largest streamflow information networks in the world, collecting water-level data at over 10,000 stations. In 2018, over 8200 of these stations continuously monitored streamflow (i.e., discharge) year-round and disseminated the data online [1]. USGS streamflow information is queried frequently by various agencies and the general public and is used for a broad range of purposes including flood hazard warning, water resource management, and recreation. Streamflow measurement stations (gaging stations) rely upon periodic measurements made by hydrographers, with an estimated 80,000 on-site streamflow measurements collected each year to develop rating curves that relate water level to discharge [1]. Developing remotely sensed approaches to on-site streamflow measurement could reduce or eliminate risk to hydrographers during extreme events, augment and economize current streamflow information networks, and facilitate expansion of networks into ungaged basins, including remote watersheds that are difficult to access.

Fundamentally, a streamflow computation requires measurement of two variables: the cross-sectional area of the channel and the velocity of the water flowing through that cross section. Acoustic Doppler current profilers (ADCPs) measure both of these variables. ADCP technology has evolved over the last few decades (e.g., [2,3]) and is now a widely used operational tool for hydrographers. Discharge and other relevant hydraulic characteristics can be viewed in real time as the ADCP traverses the channel. This real-time capability provides the hydrographer with immediate feedback in the field and allows the measurement to be repeated, if necessary, to ensure quality control and adhere to established standards of practice [4–6].

Although ADCPs are unlikely to be supplanted by an alternative technology in the near future, hydrographers already have experienced the transition from mechanical current meters to acoustic instrumentation; the next step in the evolution of hydrometry will be driven by remote sensing. A thorough review of the extensive literature on and rapid technological advances in remote sensing of streamflow is beyond the scope of this paper. To summarize in brief, research on remote sensing of discharge and other river characteristics is being actively pursued by many academic, governmental, and commercial institutions using a wide variety of platforms and sensors including ground-based approaches [7,8], manned [9,10] and unmanned aerial vehicles [11–13], and satellites [14,15].

Imagery acquired from fixed installations or airborne platforms can be used to remotely measure surface flow velocities in rivers. If floating objects (i.e., features) or textural patterns that are advected by the flow are present and the scaling (i.e., number of image pixels per meter of distance on the ground) of the imaging system and the time interval between images are known, the displacement of features or patterns between successive images can be tracked and their velocity computed. This technique, referred to as Large Scale Particle Image Velocimetry (LSPIV), has been used for decades to estimate water-surface velocity using sequences of visible imagery [7,16]. In addition to PIV, Particle Tracking Velocimetry (PTV) is a similar correlation-based technique for identifying and tracking particles in motion. An alternative approach relies on image-based methods (e.g., optical flow) that are applicable even in the absence of readily visible features and/or in the presence of non-stationary patterns [17,18].

Thermal image time series also can be used for estimating surface flow velocities. Water has a high heat capacity relative to the air or ground surface. Over the course of a day, solar radiation warms the water, and this energy is stored within the water to keep its bulk temperature relatively steady after sunset. The air and the ground adjacent to the channel, in contrast, cool much more rapidly overnight. Water also is a good emitter of infrared radiation, with an emissivity close to 1. Thermal cameras measure energy emitted from the thin (100 μm) surface layer of the water column [19], and surface features produced by flow turbulence can be detected if the camera is sufficiently sensitive to subtle differences in temperature [9,20–22].

Although approaches for image-based surface velocity measurement have matured through the use of new sensors and/or algorithms, acquiring bathymetric information remotely has proven to be more challenging. Sensors for mapping river bathymetry can be classified as passive or active. Commonly used passive optical techniques can be broadly characterized as semi-empirical [23,24] or through-water stereo photogrammetric [25,26]. A recent comparison of these approaches is presented in [27]. Because the amount of incoming solar radiation and the proportion of this energy reflected from the river bottom determine the signal returned to the imaging system, water clarity is a limiting factor in passive optical mapping of river bathymetry [24]. Similarly, although some alternative approaches to calibrating image-derived quantities to water depth have been proposed [28,29], spectrally based passive optical techniques typically require simultaneous field measurements to establish an empirical relationship between depth and reflectance; through-water stereo photogrammetric techniques do not require such calibration [25,26].

In the last 20 years, an active form of remote sensing known as airborne laser bathymetry that was initially developed in coastal environments [30] has seen increased application in river studies [31–34]. However, water clarity and bottom reflectivity can limit the success of any passive or active optical method applied in rivers, including bathymetric lidar surveys [33,35]. Another active remote sensing alternative is ground penetrating radar (GPR), which has been used to measure bathymetry from terrestrial [36,37] and airborne platforms [38]. While GPR is sensitive to the specific conductance of the water (values over 1000 microSiemens/cm absorb the signal [37]), the technology has been applied in rivers with suspended sediment values up to 10,000 mg/L [36]. Thus, GPR technology shows promise for non-contact bathymetric mapping in freshwater rivers with high concentrations of suspended sediment.

The objectives of this paper are to demonstrate the application of sUAS-based thermal velocimetry and polarizing lidar for measuring surface flow velocity and cross-sectional area, respectively. When combined, these two components provide an integral, fully non-contact method of measuring river discharge. The following sections detail how the two instruments were deployed, describe the field data collected for accuracy assessment, and compare remotely sensed estimates of river discharge to in situ flow measurements.

2. Materials and Methods

2.1. Study Area

This study was conducted in Grand County, Colorado, USA, along the Blue River upstream of its confluence with the Colorado River (Figure 1). The USGS operates a real-time streamflow measurement station on the Blue River approximately 18 km upstream of the study site (USGS 09057500 Blue River below Green Mountain Reservoir, CO). The field evaluation was conducted on 18 October 2018, when the upstream gage indicated a relatively low but steady discharge of approximately 6 m^3/s. The study area consisted of two transects (XS1 and XS2) selected to span the range of depths, velocities, and substrate present within this short section of the Blue River. The two cross sections were located 230 m apart in a straight reach of the river that lacked a regular sequence of alternate bars, although a large shallow submerged bar was located on the right side of the channel upstream of XS2. The wetted channel widths at the time of the survey were 26 m at XS1 and 40 m at XS2. The mean ± standard deviation of the depths measured in the field at transects 1 and 2, were 0.50 ± 0.20 m and 0.67 ± 0.42 m, respectively.

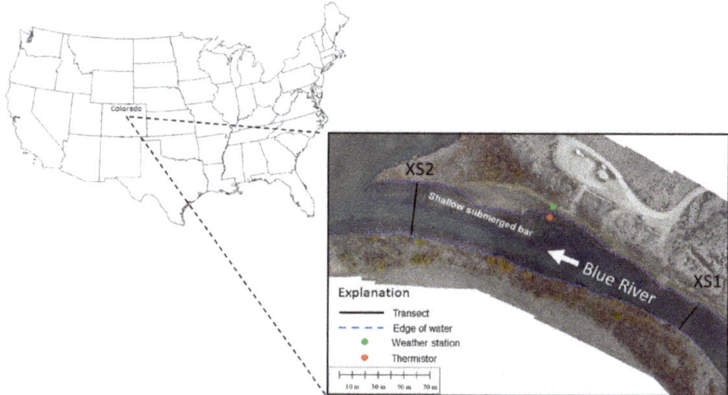

Figure 1. Location of study area and the two cross sections on the Blue River where discharge calculations were made. The edge of water, a shallow submerged bar, and the positions of the weather station and thermistor are also shown.

2.2. Field Data

To establish survey control for the thermal imagery, we placed 12 aluminum disks, 0.41 m in diameter, along the banks of the Blue River on 17 October 2018, the day prior to the sUAS flights. We used aluminum disks because the low emissivity of this material made these targets easily identifiable in both the thermal and visible images (Figure 2). At each of the two transects, three disks were placed along each bank. Real Time Kinematic Global Navigation Satellite System (RTK-GNSS) surveying equipment was used to determine the locations of the disks and to measure water-surface and bed elevations while wading between transect end points. Depths were obtained by subtracting each bed elevation from the average of the water-surface elevations measured on each side of the channel. Additional RTK-GNSS water-surface elevation points were collected along the right bank of the Blue River on 18 October 2018. These elevations were confirmed to be essentially the same as those collected during the previous day's survey.

Also on 17 October 2018, we used a Wet Labs ECO Triplet™ to measure several optical properties of the water column. This instrument recorded turbidity and concentrations of chlorophyll and colored dissolved organic matter (CDOM). The mean values of these three parameters were 1.94 NTU, 0.41 μg/L, and 6.87 μg/L, respectively.

ADCP profiles were collected along two transects on 18 October 2018. A SonTek S5 RiverSurveyor™ ADCP with RTK capability was mounted on a hydroboard; the ADCP's RTK base station was placed nearby on the shore. The hydroboard was tethered to an inflatable kayak that traversed the river four times at each transect.

A portable weather station installed at the site approximately 100 m upstream of transect 2 on 18 October 2018 recorded wind speed and air temperature. The wind speed sensor was positioned approximately 1.5 m above the ground surface. In addition, water temperature was logged within the channel near the weather station using a submerged thermistor (Figure 1) . The air temperature at the field site was –5.3 °C, wind speed was near zero, and the water temperature was 6.9 °C at 07:00.

All of these field measurements are included in a USGS ScienceBase data release [39].

2.3. Remote Sensing Measurements

2.3.1. Thermal Image Acquisition and Particle Image Velocimetry

The thermal camera was deployed from a DJI Matrice™ 600 Pro sUAS operated by an independent, third-party contractor. Flights began at civil twilight, approximately 06:50 local time on 18 October 2018. Prior to the sUAS flights, the transect locations were provided to the flight contractor for input to the flight management software as waypoints. The sUAS carrying the thermal camera was programmed to navigate to each waypoint, hover approximately 100 m above the river, and collect imagery for approximately one minute. Image sequences were saved to a microSD card on the sUAS at a rate of 0.5 Hz. Specifications of the thermal camera used in the study are listed in Table 1. Imagery was transmitted to the ground in real-time, and the operator was able to toggle between views from the thermal camera and a visible camera also mounted on the sUAS (Figure 2).

Table 1. Specifications of small Unmanned Aerial System (sUAS) thermal camera.

Model	ICI Mirage 640
Lens	25 mm lens (26° field of view)
Detector	Cooled Indium Antimonide
Pixel dimensions on focal array	15 μm
Number of pixels	640 × 512
Wavelength range	3.4 μm–5.1 μm
Noise-equivalent temperature difference	<0.012 °C at 30 °C
Bits per pixel	14
Camera dimensions	111 × 96 × 131 mm
Camera weight	765 g
Power	12 V

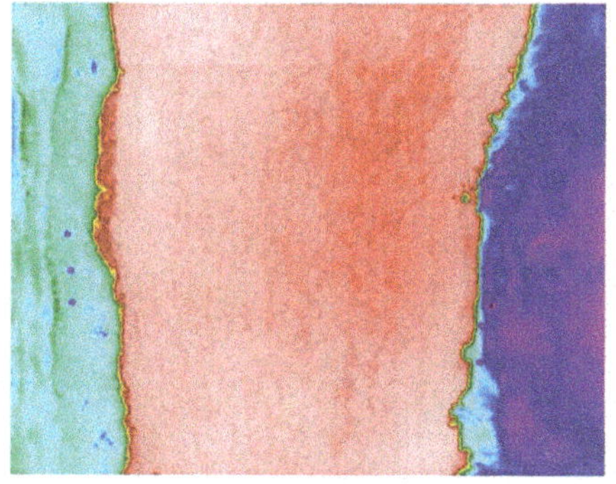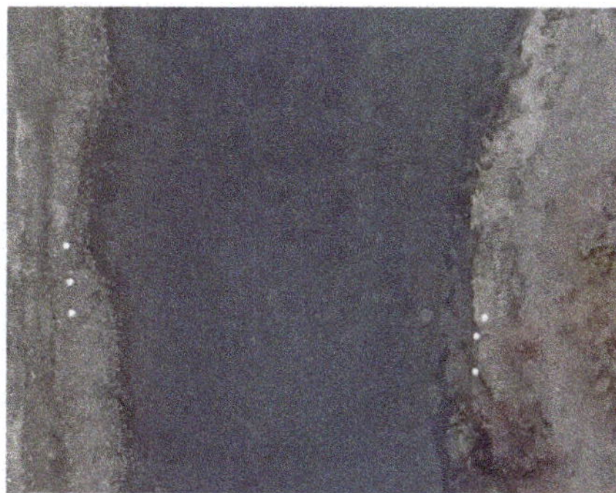

Thermal infrared Image   Visible Image (cropped to the same extent)

Figure 2. Imagery collected by sUAS camera platform over cross section 1. The aluminum disks used as ground control points appear as dark purple circular features in the thermal image and as white circles in the visible image.

We used the MATLAB-based software PIVlab [40–42] to compute the velocity of surface features imaged by the thermal camera using a work flow implemented entirely within MATLAB. A total of 26 frames were used for PIV analysis at each transect. The raw thermal images were first converted to grayscale, a Wiener spatial smoothing filter applied, and contrast-limited adaptive histogram equalization (CLAHE) used to enhance the imagery. Next, the raw image sequence was input to a video stabilization algorithm available within the MATLAB Computer Vision Toolbox. This procedure involved identifying point correspondences between successive pairs of images, applying an affine transformation based on these point correspondences to all images in the sequence, and warping the frames to produce a stabilized set of images. The spatial extent of each image (i.e., footprint) was plotted, and an area encompassing all footprints in the sequence was defined interactively. The image sequence was processed by calling the main PIVlab routine with the fast Fourier transform (FFT) window deformation option. The two key parameters of this cross-correlation algorithm are the interrogation window and the step size. In this study, we used a 128-pixel window and a 32-pixel step size. We did not apply any post-processing to the PIVlab output because the velocity vectors were of realistic magnitude and smoothly varying, without any obviously spurious vectors or noise that would have required filtering. Finally, all the velocity vectors produced by PIVlab were scaled (the resulting pixel size was 0.063 m) and transformed to real-world spatial coordinates (UTM zone 13) using an affine transformation based on the aluminum disks placed in the field to serve as ground control points (Figure 3).

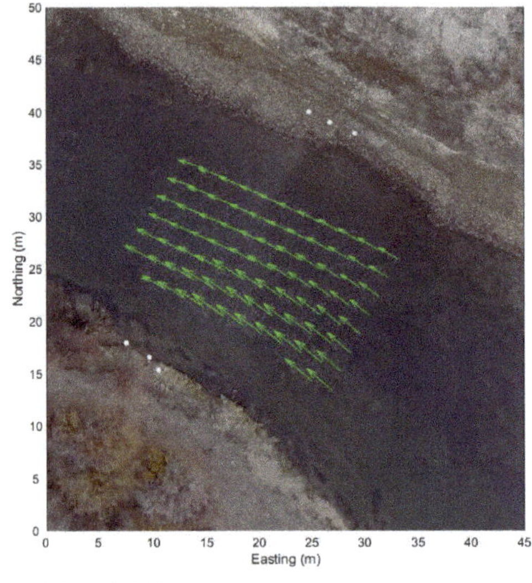

Figure 3. PIV-derived velocity vectors for cross section 1.

2.3.2. Bathymetric Lidar Acquisition and Processing

Bathymetric lidar systems record the travel time of laser pulses reflected from both the water surface and the bed, yielding both terrestrial and submerged elevations and thus water depth. Existing bathymetric lidars, however, might fail to provide reliable depths and bed elevations in shallow water (<0.15 m) due to the difficulty of distinguishing between laser returns from the water surface and bed when the two surfaces are in close proximity. To overcome this signal separation issue, researchers at the University of Colorado and Atmospheric & Space Technology Research Associates Lidar Technologies (ASTRALiTe) developed an innovative bathymetric lidar, based on an alternative signal processing technique known as INtrapulse Phase Modification Induced by Scattering, or INPHAMIS.

In contrast to typical lidar systems that consider only time-of-flight, the INPHAMIS technique also considers the polarization of the returned laser pulses. The technique is described in detail by Mitchell et al. [43] and Mitchell and Thayer [44] and is only briefly summarized herein. When traveling from the air, across the air-water interface, and through water column to the bottom, the 532-nm laser pulse is scattered in planes both parallel and perpendicular to the transmitted signal. Dual photomultiplier tubes are used to evaluate the polarization orientation of the return signals, isolate the differently polarized returns, and use their timing to compute the range of each returned pulse.

At approximately 11:00 local time on 18 October 2018 the ASTRALiTE bathymetric lidar was deployed above the Blue River, also from a DJI Matrice^{TM} 600 Pro sUAS operated by the same independent, third-party contractor. Specifications of the lidar are listed in Table 2. The lidar used an on-board inertial navigation system (INS) and a local GNSS base station to determine the position of the sUAS in post-processing. In order to ensure sufficient laser power to penetrate through the water column, the sUAS was operated at a relatively low altitude, approximately 4 m above the water surface. Due to potential hazards such as riparian vegetation along the channel margins, this flying height created difficulty in acquiring a transect that spanned the entire channel width while using an autonomous flight plan. For this reason, the sUAS was flown manually. The lidar system uses a whisk broom scanning technique (±15° off-nadir) to acquire a swath of data with a width of approximately one half the flying height above the water surface. In this study, a nominal height of 4 m above the water surface yielded a 2-m swath of data for each transect.

Table 2. Specifications of small Unmanned Aerial System (sUAS) bathymetric lidar.

Model	ASTRALiTe edge™
Laser Wavelength	532 nm
Laser Power	30 mW
Swath Width	1/2 flight altitude
Depth Penetration	$\sim$ 1.2 Secchi depth
Scan Technique	Whisk broom
Positioning	SBG Ellipse2-D™ INS
Weight	5 kg
Dimensions	0.18 m $\times$ 0.2 m $\times$ 0.23 m
Flight Duration	10–12 min on DJI Matrice™ 600 Pro

ASTRALiTe delivered a lidar point cloud that had been corrected for refraction and the change in the speed of the laser pulses in air versus water. The point cloud was smoothed by calculating the average elevation in each cell of a 10 cm by 10 cm grid. A single file including the easting and northing coordinates, average elevation, and a code to indicate whether a return was from the water surface or the bed was provided by ASTRALiTe for each lidar transect. The depth along the transect was obtained by using the code information to identify water surface and bed returns, finding coincident spatial locations for the two types of returns, and subtracting the bed elevation from the water-surface elevation. The lidar data were further processed to eliminate points where the depth was less than 0.15 m because water surface and bottom returns could not be distinguished unambiguously in these extremely shallow areas.

2.3.3. Comparison of Remotely Sensed Hydraulic Quantities with Field Measurements

We assessed the accuracy of remote sensing-derived estimates of velocity, depth, and discharge using field measurements of these hydraulic quantities. Making direct comparisons, however, was complicated by a number of different factors. For example, the thermal PIV output and the ADCP measurements did not have identical spatial and temporal coverage and resolution. The PIV algorithm produced a regular grid of surface velocity estimates, with a spacing dictated by the step size that provided detailed coverage of the entire channel area, with a spacing of approximately 2 m between velocity vectors. In addition, the PIV-derived velocity vectors were time averaged over the approximately 60 s duration of the image sequence. In contrast, ADCP measurements were collected along four transects that yielded an irregular spatial sampling of the flow field. Each ADCP measurement was an instantaneous profile and the data could not be time-averaged because the instrument was in motion. An additional complication in comparing the PIV output to the ADCP data was that the ADCP did not directly measure the velocity at the water surface. The ADCP itself generated a flow disturbance at the surface and a blanking distance beneath the surface was created by the time delay between when the transducer generated the acoustic signal and when the signal can be received. To reconcile and enable comparison between these two inherently different sampling methodologies and measurement types, the four ADCP traverses for each cross section were combined and spatially averaged to obtain a field-based data set more analogous to the PIV grid.

To facilitate spatial averaging, both types of data were transformed to a channel-centered frame of reference based on a centerline digitized on the thermal image. For the PIV output, the averaging window was 2 m wide in the cross-stream direction and 5 m in streamwise length (i.e., two rows of the PIV output grid). In addition to the average of the velocity magnitudes within this region, the minimum and maximum magnitudes were retained and used to define error bars. For the ADCP measurements, we used the same 2-m window width but used a greater streamwise extent to encompass all of the ADCP measurements at a given cross-stream position.

To enable comparison of the surface velocities inferred by PIV and the ADCP data, we used a velocity index approach to convert surface velocities to depth-averaged velocities. In open channel flows, the depth-averaged velocity is typically 0.85 of the velocity at the water surface, assuming a logarithmic velocity profile [16,37,45]. We applied this velocity index to the PIV magnitudes before comparing them to the depth-averaged velocities reported by the ADCP.

To compare the lidar-derived depths to the wading and ADCP measurements of depth, we projected both the remotely sensed and field data onto a common cross section. This transect was defined by fitting a line to the easting and northing coordinates of the lidar point cloud and discretizing the line with a 0.01-m increment in the cross-stream direction. The lidar-based depths that were within 0.5 m either up or downstream of the cross section were associated with the nearest cross-stream vertex along the line. This process served to project the three-dimensional lidar data onto a cross-sectional plane. We used a lowess (locally weighted scatterplot smoothing, [46,47]) fit to create a smooth line through the resulting two-dimensional point cloud of lidar depths. The lowess procedure was implemented with a span parameter of 0.1, specified as percentage of the total number of data points. The depths measured by wading the channel and by the ADCP were also projected onto the same cross section and a lowess fit applied to the ADCP point cloud. Whereas the wading measurements were collected in relatively close proximity to the lidar point cloud, many of the ADCP measurements were farther away from the cross section because the ADCP drifted in the current as it was towed behind the kayak (Figure 4). For this reason, we used wading depths to assess the accuracy of lidar-based depth estimates.

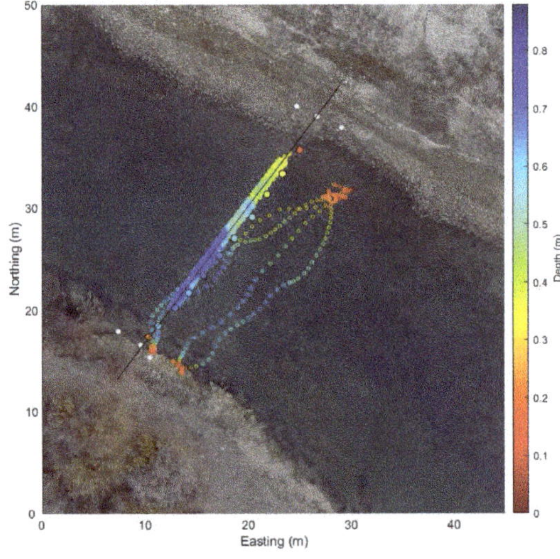

Figure 4. Depths measured by wading (large filled symbols with black circle outlines), ADCP (mid-size open symbols with colored circle outlines), and lidar (small filled symbols) collected at cross section 1.

A discharge calculation involves integrating the product of velocity and cross-sectional area across the channel, typically by multiplying the depth-averaged velocity measured at each one of a series of verticals by the depth at that vertical and the cross-stream distance between adjacent verticals. In this study we calculated discharge from the remotely sensed data as follows. First, we extracted the thermal PIV velocity magnitudes closest to the lidar cross section and multiplied by 0.85 to obtain a depth-averaged velocity. The locations of these velocities also served to define the mid-points of each vertical. To determine the width of each vertical, half the distance between adjacent midpoints was added to and subtracted from the cross-stream coordinate of the midpoint. The cross-sectional area associated with each vertical was then obtained by multiplying this width by the mean depth computed from the smoothed lowess fit to the lidar point cloud. The product of this cross-sectional area with the corresponding depth-averaged velocity magnitude yielded an incremental discharge that was summed laterally across the channel to produce the total river discharge.

To ensure consistency in comparing discharges calculated from the remotely sensed data to discharges computed from in situ ADCP measurements, we performed an independent discharge calculation rather than using the total discharge reported by the ADCP software. Instead, we applied the same incremental procedure outlined in the preceding paragraph to the depth-averaged velocities and depths measured directly by the ADCP. Another reason for performing our own ADCP-based discharge calculation is that the lidar coverage did not extend all the way across the channel for cross section 1 due to the presence of vegetation that posed a hazard to the low-flying sUAS. For this reason, the discharge calculation for the ADCP considered only that portion of the channel for which lidar data were available. For cross section 2, the sUAS was able to traverse the entire channel width.

3. Results

Although this study focused on measuring discharge via remote sensing, a discharge calculation requires information on both flow velocity and channel cross-sectional area. Therefore, in the following subsections we assess the accuracy of velocities and depths inferred from remotely sensed data before comparing their integrated product to field-based discharge measurements. Herein, we emphasize results from cross section 1, but an analogous set of figures for cross section 2 is provided in the Supplementary Materials.

3.1. Comparison of Thermal PIV with ADCP Measurements of Velocity

Figure 5a compares the magnitudes of time-averaged thermal PIV surface velocities and depth-averaged ADCP velocities along cross section 1. The error bars represent the total range of velocities within the streamwise extent of the spatial averaging window (at a given cross-stream position) for both the PIV output grid and the four passes made with the ADCP. Lower velocities were observed along the right side (facing downstream) of the channel (negative cross-stream coordinates relative to the channel centerline) and increased to a maximum PIV-based surface velocity of 0.84 m/s. The range of PIV-based surface velocities (i.e., error bars) for a given cross-stream coordinate varied from 0.01 to 0.1 m/s, whereas the ADCP velocities were less precise, with error bars from 0.16 to 0.32 m/s. Adjusting the PIV output with a velocity index of 0.85 brought the depth-averaged velocities from the PIV images within the range of velocities measured directly by the ADCP. Figure 5b quantifies the agreement between the PIV- and ADCP-based depth-averaged velocity magnitudes via linear regression. This analysis resulted in an R^2 of 0.82 and a slope approximately equal to 1. The positive intercept implied that, on average, the PIV overestimated the velocity by 0.05 m/s relative to the ADCP.

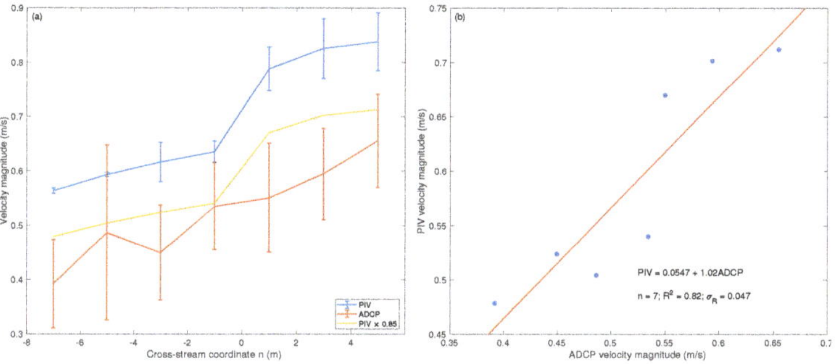

Figure 5. (**a**) Comparison of PIV-derived velocities and ADCP measurements along cross section 1. (**b**) Accuracy assessment of PIV-derived depth-averaged velocities via linear regression.

3.2. Comparison of Lidar-Derived Depths with Field Measurements

The correspondence between the wading, ADCP, and lidar-derived depths is shown in Figure 6a. The ADCP and lidar provided spatially distributed point clouds that were summarized and smoothed by fitting lowess lines to these data sets. The RTK GNSS-based wading measurements were discrete and fewer in number, and the individual points are simply connected by straight lines. All three methodologies captured the asymmetrical shape of cross section 1, with the depth increasing to a maximum of 0.7 m for the wading survey, 0.7 m for the ADCP, and 0.77 m for the lidar. For a given cross-stream location, the lidar-based depths typically varied by 0.10 to 0.12 m within the 1-m swath of data recorded by the ASTRALiTe edge^{TM}. Because wading provided the simplest and most precise ($\pm$0.02 m) field-based depth measurements, the RTK GNSS survey was used to evaluate the accuracy of the lidar depths. The results of this analysis are summarized in Figure 6b, which indicates a strong agreement ($R^2 = 0.95$) and a regression line with a slope near one. The intercept term of 0.06 m, however, implied that the lidar overestimated the depth relative to the wading surveys.

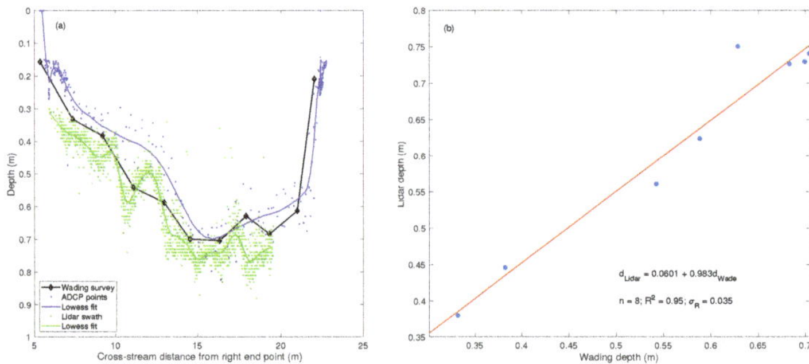

Figure 6. (a) Depth comparison between lidar, wading, and ADCP measurements along cross section 1. (b) Accuracy assessment of lidar-derived depths via linear regression.

3.3. Comparison of Remote Sensing-Based Discharge with Field Measurements

We combined the remotely sensed depths and velocities to calculate the discharge in the Blue River at the time these sUAS-based data sets were acquired. Figure 7 illustrates how both the PIV-derived velocities and lidar-based depths increased from right to left across the channel. Also shown as dashed lines are the left and right edges used to define the width of each vertical used in the incremental discharge calculation. This computation is summarized in Figure 8 for cross section 1. For each of the seven verticals, the cross-sectional area, depth-averaged velocity magnitude, and corresponding incremental discharge are represented by the bar symbols in each panel. Because both depth and velocity increase with distance from the right end point, most of the flow volume within the channel occurred closer to the left bank. Summing the discharge increments laterally across the channel resulted in a total discharge of 3.71 m^3/s. Applying the same procedure to the subset of the ADCP data that coincided with the lidar coverage resulted in a total discharge of 3.05 m^3/s. The remote sensing approach thus overestimated the discharge by 22% relative to the ADCP-based calculation for cross section 1. A similar analysis for cross section 2 (see Supplementary Materials Figures S3 and S4) produced much closer agreement, with remote sensing yielding a discharge estimate only 0.16% greater than the ADCP-based value of 6.32 m^3/s. The discharge for cross section 2 was greater than that for cross section 1 because the lidar was only able to traverse the entire channel width at the second cross section.

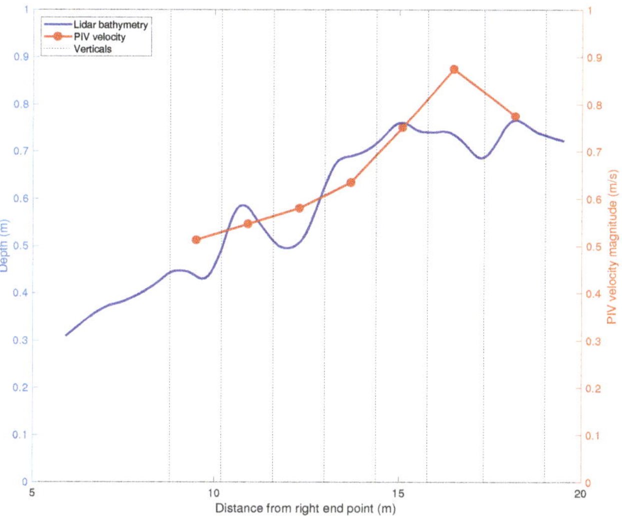

Figure 7. PIV-derived velocities, lidar depths, and left and right edges for each vertical used in the incremental discharge calculation for cross section 1 based entirely on remotely sensed data.

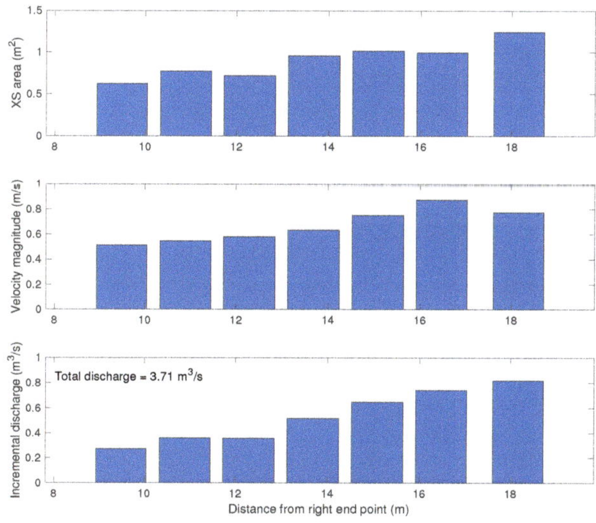

Figure 8. Illustration of how the cross sectional area associated with each vertical is multiplied by the corresponding velocity magnitude to yield a series of incremental discharges that are then summed laterally to obtain the total discharge. This example is based on remotely sensed data from cross section 1 on the Blue River.

4. Discussion

4.1. Potential Sources of Uncertainty in Remotely Sensing of Streamflow

Although various remote sensing technologies could potentially facilitate more efficient, safer measurement of streamflow, the uncertainties inherent to these methods must be acknowledged. Ultimately, our objective was to obtain information on river discharge entirely from remotely sensed data, but this process involved measuring the two fundamental hydraulic quantities that comprise discharge: velocity and depth. In the following subsections, we examine the uncertainties associated with each of these variables in turn and then discuss the implications of these sources of error for discharge calculations.

4.1.1. Uncertainties Associated with Thermal PIV-Based Velocity Estimates

Thermal PIV- and ADCP-based velocities observed along Blue River cross section 1 were compared in Figure 5. For both types of data, the width of the error bars represents the range of velocities present within the spatial-averaging window, providing an indication of measurement precision. The ADCP yielded wider error bars than the PIV because the ADCP data consisted of instantaneous velocity profiles collected as the instrument traversed the channel, whereas the PIV-derived velocities were time-averaged over the one-minute acquisition period. In addition, the ADCP data were collected on four passes that were not exactly coincident with one another and had to be spatially aggregated. The output from the PIV algorithm, in contrast, was a regular grid of velocity vectors with a spacing dictated by the interrogation window and step size parameters we selected. This regular sampling provided a more smoothly varying pattern of cross-stream velocities as well as narrower error bars. Another advantage of PIV is that the method is image-based and produces continuous coverage (i.e., a velocity field) throughout the central portion of an image, not just a single isolated transect (Figure 3). This type of data thus provides greater flexibility in coupling velocity estimates with other data sources, such as bathymetry.

An important limitation of thermal PIV is that this method yields only a surface flow velocity that must somehow be converted to a depth-averaged velocity prior to discharge calculation; this shortcoming pertains to any surface-based velocity measurement technique (e.g., PIV of optical images, radar, GPS-based drifters). Typically, this conversion is achieved by multiplying the surface velocity by a velocity index < 1 and this approach brought the PIV-based surface velocities for Blue River cross section 1 to within the error bars of the ADCP data collected along this transect (Figure 5a). We used a value of 0.85, the most common velocity index e.g., [45], but a lower value would have provided better agreement between the PIV- and ADCP-based velocities. Using a velocity index of 0.85 overestimated the depth-averaged velocity in this case, resulting in a positive intercept term for the regression equation in Figure 5b that indicated a positive bias of the PIV-based velocities relative to the ADCP measurements. In general, the velocity index depends on the vertical structure of the flow field and thus varies among rivers and even locally within a reach. For example, Legleiter et al. [22] calculated velocity indices that ranged from 0.82 to 0.93 for five rivers in Alaska. Such variability in the velocity index will introduce a multiplicative error to any surface velocity estimate inferred from remotely sensed data.

4.1.2. Uncertainties Associated with Depths Inferred from Polarized Lidar

Depths measured while wading, by the ADCP and via lidar, were compared in Figure 6. Making this comparison involved determining the best-fit line through the horizontal coordinates of the lidar point cloud and projecting both types of field data onto this line. Although the wading measurements were fewer in number, they were in closer proximity to the lidar transect than the four more irregular passes of the ADCP, which could have contributed to the closer agreement between the lidar-based depths and the wading surveys. Because the wading survey coincided spatially with the lidar swath and consisted of a single transect that did not require spatial averaging, this data source was the most reliable independent check on the accuracy of the lidar-based depths. Relative to the wading data, the ADCP tended to underestimate the depth whereas the lidar was generally biased deep. However, the lidar appears to have captured more subtle features of the bed topography, such as the prominent bump located 12 m from the right end point in Figure 6a.

Another source of uncertainty in comparing the lidar to our field measurements was the nature of the river bed itself. Although the bed was primarily composed of sand and gravel, some submerged aquatic vegetation was also present. Whereas some of the laser pulses might have reflected off the top of the substrate surface (e.g., protruding gravel, vegetation canopy), the survey rod used while wading was more likely to come to rest at lower elevations within the sand, between gravels, or at the base of any plants. One thus might expect the unfiltered lidar point cloud to underestimate the depth relative to the wading survey. For cross section 1 on the Blue River, however, the lidar tended to overestimate the depth. The linear regression used to compare the lidar to the wading depths indicated a strong agreement ($R^2 = 0.95$), but the intercept term of 0.05 m confirmed the positive bias evident in Figure 6a over the 0.3–0.7 m range of depths sampled along this transect. For cross section 2 (Supplemental Materials Figure S2), the lidar retrieved depths up to 1.2 m but a similar positive intercept of 0.06 m was observed.

Although an explanation for this deep bias is not immediately obvious, one factor that could contribute to this surprising result is the difficulty of identifying the water surface. Lidars do not directly measure the depth but rather calculate the depth as the difference between the range to the water surface and the river bed, as determined from the time of flight of laser pulses returned to the sensor. Any error in establishing the location of the water surface will affect depth estimation because both the refraction of the laser pulse and the change in the speed at which it travels must be accounted for at the air-water interface. The ASTRALiTe polarizing lidar provides an alternative means of distinguishing water surface and bottom returns, which can prove challenging for more common waveform-based bathymetric lidars, particularly in shallow depths where the air-water interface and the stream bed are in such close proximity to one another. However, the initial results presented herein suggest that incorporating polarization information did not enable a completely reliable, binary classification of returns as either water surface or bottom. More specifically, the ASTRALiTe lidar system consists of two channels. The nominal "water-surface channel" can detect returns from the river bottom when none of the laser pulses experience a specular reflection from the water surface and return to the sensor. This situation is more likely to occur at greater off-nadir angles along the edges of the swath. Similarly, the nominal "bottom surface channel" can record reflections from the water surface when the specular return from the water surface is strong and overwhelms the optics of the sensor.

4.1.3. Implications for Discharge Calculations

The hydraulic quantities computed from the remotely sensed data and in situ ADCP measurements collected are compared in Table 3. For both cross sections, the mean velocities and cross-sectional areas determined by the remote sensing approach were greater than those calculated from the ADCP measurements. Because each of these components was overestimated when inferred from remotely sensed data, the discharges calculated from thermal PIV and bathymetric lidar exceeded those based on the ADCP data. As shown in Figure 5, even after applying a velocity index of 0.85 thermal PIV overestimated the corresponding ADCP-derived depth-averaged velocity, which in turn lead to overestimation of the discharge. Similarly, the deep bias observed for the lidar bathymetry (Figure 6) yielded greater cross-sectional areas than the field surveys and also contributed to the overestimation of discharge. In general, an overprediction of the mean velocity from remotely sensed data could be compensated for by an underprediction of the cross-sectional area or vice versa, but for both of the cross sections sampled on the Blue River overestimation of both hydraulic quantities ensured that the total discharge would be overpredicted relative to the field measurements.

Table 3. Hydraulic quantities calculated from remotely sensed data and field-based ADCP measurements for two cross sections on the Blue River.

Quantity	Cross Section	Remote sensing	ADCP
Mean velocity (m/s)	1	0.67	0.58
	2	0.42	0.36
cross sectional area (m^2)	1	6.33	5.06
	2	16.31	14.10
Total discharge (m^3/s)	1	3.71	3.05
	2	6.33	6.32

A notable feature of Table 3 is that the discharges calculated for the two cross sections, which were located in close proximity to one another, differed by 2.62 m^3/s for the remote sensing approach and by 3.27 m^3/s for the ADCP-based calculation. In addition, the agreement between the remote sensing and field-based discharges was much better for cross section 2, with only a 0.1% discrepancy at this transect, whereas the discharge was overestimated by the remotely sensed data by 22% at the upstream transect. These differences between the two cross sections are primarily a consequence of the greater coverage achieved by the lidar at cross section 2, where the sUAS was able to safely traverse the entire channel width. Vegetation along the left bank at cross section 1 prevented the sUAS from sampling the left side of the channel and restricted our discharge calculation to only a portion of the width. Because most of the flow through this reach of the Blue River was concentrated along the left side of the channel, a lack of data from this area precluded us from calculating an accurate total discharge for cross section 1. For the other transect, more complete coverage resulted in a calculated discharge closer to the value of approximately 6 m^3/s recorded at the upstream USGS gaging station (09057500 Blue River below Green Mountain Reservoir, CO).

4.2. Practical Considerations for Remotely Sensing of River Discharge

This initial pilot study revealed a number of factors that must be considered carefully in any application of an sUAS-based remote sensing approach to measuring river discharge. In this section, we discuss some of the practical lessons we learned to help guide future work based on the methodology described herein. We describe both instrumentation-related issues and operational restrictions that must be taken into account during mission planning.

Through several previous attempts to infer surface flow velocities from thermal image time series, we learned that suitable image data can only be acquired under certain environmental conditions. The thermal features tracked by the PIV algorithm are most pronounced when the temperature contrast between the water and the overlying air is greatest. This is most likely to occur at dawn when the air has cooled more overnight than the water, which has a much higher specific heat capacity. During the day, the air and water temperatures tend to converge, obscuring the water-surface thermal features of interest. In addition, reflected solar energy can become a significant fraction of the radiation recorded by the thermal camera when the sun is higher in the sky. Another way to enhance the detection of subtle differences in temperature is to use a cooled mid-wave infrared thermal camera. This type of sensor has a greater sensitivity, typically with a noise equivalent temperature difference <0.025 °C, and thus provides better clarity and contrast than more readily available uncooled thermal cameras. However, disadvantages of cooled thermal cameras include greater cost, larger size, and susceptibility to reflections from the sun [48]. Pixel size, frame acquisition rate, data capture rate, and on-board data storage capacity are also important criteria to evaluate in selecting an appropriate thermal camera.

Although this was the initial field test of the sUAS-based ASTRALiTe edge^{TM} lidar in a river, we can make some general statements about the use of this technology for measuring river bathymetry. As with airborne lidar, water clarity is an important constraint on the ability of laser pulses to penetrate through the water column to the stream bed. Substrate reflectivity can also limit the number of bottom returns. While a more powerful laser might help to overcome these obstacles, greater power would likely come at the expense of a heavier payload that would not be conducive to sUAS deployments. Regulations that ensure eye safety also constrain the amount of laser power a lidar system can use. In this study, flying the ASTRALiTe edge™ 4 m above the water surface provided a high density of bottom returns in a clear-flowing channel up to 1.2 m deep. However, this low flying height limited the coverage that could be achieved because riparian vegetation presented a hazard to the sUAS. In designing and deploying a bathymetric lidar system, a safe compromise must be reached between at least three principal factors: (1) laser power and penetration depth, (2) system weight and flight duration, and (3) flying height and obstacle avoidance.

In addition to these instrumentation-related issues, regulatory restrictions also can limit the suitability of sUAS-based remotely sensed data for characterizing streamflow. Ideally, an sUAS would be flown high enough for the imaging system to capture not only the entire channel width but also at least a small area along the banks so that targets placed to provide ground control are visible. In this study, however, a small local airport with one runway located 2 km from our study area dictated a maximum flight altitude of 100 m, per Federal Aviation Administration (FAA) regulations. As a result, the channel and both banks were only included in a single image for one of the two cross sections. For the second, wider transect, two separate hovering waypoints were required to image both banks, which was necessary because spatially distributed ground control is needed to accurately scale and geo-reference the images and the velocity vectors derived therefrom. For cross section 2, the lack of targets on both sides of the images complicated geo-referencing and we used images acquired from only one of the two waypoints for PIV and discharge calculation. At this location, the vast majority of the flow was captured by images of the left side of the channel. Images from the right side encompassed a broad shallow area dominated by submerged vegetation that conveyed a negligible portion of the total river discharge. This local configuration allowed us to derive a reasonable discharge estimate from an image sequence that did not span the entire channel width, but this will not be the case in general. More typically, restrictions on flying height could dictate that partial discharges derived from multiple image sequences distributed laterally across the channel be combined to obtain the total discharge.

4.3. Future Research Directions

This pilot study served as an initial test of the feasibility of measuring river discharge via sUAS-based remote sensing and also highlighted some important topics for further investigation. In this section, we discuss some potential refinements to instrumentation and make suggestions to improve data collection and processing.

Although the sensors used in this study fulfilled their intended purpose, both the thermal camera and polarizing lidar system will require further development and testing before this approach can become more broadly applicable. For example, the frame acquisition and data capture rate of the thermal camera was limited to a relatively low 0.5 Hz, much less than the 25–30 Hz frame rates typically used in LSPIV applications based on conventional video cameras. In this case study, the low frame rate did not compromise PIV due to the relatively low surface flow velocities observed along the Blue River at a base flow discharge in late fall but could prove to be problematic at higher discharges and/or in more energetic channels. Similarly, the sensitivity of the lidar to various environmental conditions must be examined through additional field tests. Whereas the water surface of the Blue River was very smooth at the time of this study, rougher water surface textures could affect the interaction of laser pulses with the air-water interface and thus reduce the accuracy of lidar-derived bathymetric information. The Blue River also was relatively clear at the time of our survey, but the lidar system would not be expected to perform as well in more turbid water. Our study site only included depths up to 1.2 m, so further testing in deeper rivers is needed to quantify the maximum penetration depth of the ASTRALiTe edge™. In the initial field evaluation described herein, the thermal camera and the lidar were deployed independently on separate sUAS, but in the future this instrumentation package could be deployed from a single sUAS platform. However, such consolidation would increase the weight of the payload and reduce flight duration and thus would only be beneficial if both sensors could acquire useful data from the same flying height.

Improvements to the data collection and processing we performed in this study could help remote sensing to become a viable operational tool for measuring river discharge. For example, in future field evaluations a cable (i.e., tagline) should be stretched across the channel for two reasons. First, the tagline would provide a clear linear feature that the pilot could use for guidance while operating the sUAS manually, rather than relying upon pre-programmed waypoints and automated navigation. Second, the tagline would facilitate collection of in situ velocity verification data by guiding the ADCP along a straight transect coincident with the lidar swath, rather than the irregular passes that we recorded in this study. Another, more general issue that requires further research is the selection of an appropriate velocity index to convert PIV-based surface velocities to depth-averaged velocities prior to calculating discharge. In this study we scaled and geo-referenced the thermal images based on ground control targets placed in the field, but a more efficient alternative would be direct geo-referencing of the images using position and orientation data logged on-board the sUAS; this approach would obviate the need to place ground control targets along the banks and/or in the river. If absolute spatial coordinates are not necessary for a given application, only the scale factor for converting pixels to meters is needed to perform PIV. This scale could be established by using a laser range finder or ultrasonic sensor to measure the distance from the sUAS down to the water surface and calculating the ground sampling distance based on the focal length of the camera. Finally, although the current workflow would require a hydrographer to collect data in the field and then post-process the images and analyze the lidar before performing a discharge computation, this workflow could be streamlined if some of this processing (i.e., PIV) could be performed in real-time on-board the sUAS, on the sensor directly, or processed on the ground control station system. Performing the data processing outside of the sUAS increases portability by keeping the sensor independent of the platform, thus providing the greatest flexibility for deployment. In addition, further development of the lidar processing algorithms could allow the point cloud to be visualized by an operator on the ground. This type of real-time feedback would help to ensure that remote sensing-based measurements of velocity and depth are of sufficient quality.

5. Conclusions

Remote sensing offers the potential to improve the safety of the hydrographer and increase the efficiency of on-site streamflow measurements. These advances would also allow the USGS and similar agencies worldwide to expand their hydrologic monitoring networks into remote, inaccessible locations and ungaged basins. Recent technological developments have reduced the size, weight, and power consumption of sensors to a degree that instrumentation that previously could only be carried on conventional occupied aircraft now can be deployed from sUAS platforms. In this paper, we describe a complete technique for collecting both the surface flow velocity and bathymetric data necessary to compute river discharge from an sUAS under natural conditions. This approach involved the use of a thermal infrared camera, a PIV algorithm, and a polarizing lidar. Importantly, the sensors, platforms, and software used in this study are all commercially available at this time, not just prototypes in an early stage of development.

We evaluated the performance of this approach by comparing remotely sensed velocities and depths to field measurements collected at two cross sections along the Blue River, CO, USA. The results of this investigation support the following principal conclusions:

1. Compact, sUAS-deployable thermal infrared cameras of sufficient sensitivity can detect the movement of flow features expressed at the water surface as subtle differences in temperature, thus enabling PIV without requiring visible tracer materials or artificial seeding of the flow.

2. Incorporating information on the polarization of returned laser pulses provides an alternative means of distinguishing the water surface from the channel bed even in very shallow water; these polarizing lidar systems also are small enough to be deployed from an sUAS.

3. Due to limited payload capacity and differences in flying height during data collection, the thermal camera and polarizing lidar had to be deployed from separate sUAS.

4. Surface flow velocities inferred from thermal images via PIV agreed closely (R^2 = 0.82 and 0.64) with depth-averaged flow velocities measured in situ with an ADCP when multiplied by the commonly assumed velocity index of 0.85.

5. Depths derived from the polarizing lidar closely matched those surveyed by wading in the shallower of the two cross sections (R^2 = 0.95), but the agreement was not as strong for the transect with greater depths (R^2 = 0.61), although the sensor was able to detect the bottom in depths up to 1.2 m.

6. These remotely sensed velocities and depths were combined to calculate incremental discharges that were summed laterally across the channel to obtain the total streamflow. As compared with discharges calculated in a similar manner directly from the ADCP data, the sUAS-based methodology resulted in a discharge that was 22% greater at one cross section but within 0.1% of the ADCP-based discharge at the second cross section.

Supplementary Materials: The following are available online at http://www.mdpi.com/2072-4292/11/19/2317/s1.

Author Contributions: Conceptualization, P.J.K., C.J.L.; methodology, P.J.K., C.J.L.; software, C.J.L., P.J.K.; validation, P.J.K., C.J.L.; resources, P.J.K., C.J.L.; writing—original draft preparation, P.J.K., C.J.L.; writing—review and editing, P.J.K., C.J.L.; visualization, C.J.L., P.J.K.; project administration, P.J.K., C.J.L.

Funding: This research was supported by the U.S. Geological Survey's Water Mission Area Hydrologic Remote Sensing Branch and the U.S. Geological Survey's National Unmanned Aircraft Systems Project Office.

Acknowledgments: Jonathan Nelson, Yutaka Higashi, Jeff Sloan, Jack Eggleston, and Cian Dawson (USGS); Cotton Anderson, Bill Adams, and Gerald Thompson (ASTRALiTe); Jeffery Thayer, Rory Barton-Grimley, Toby Minear, and Brian Straight (University of Colorado at Boulder); David White, Jack Davis, and Evan Menke (Juniper Unmanned). The field measurements and image data sets used in this study are available through the USGS ScienceBase Catalog data release listed in the references. Any use of trade, firm, or product names is for descriptive purposes only and does not imply endorsement by the U.S. Government.

References

1. Eberts, S.; Woodside, M.; Landers, M.; Wagner, C. Monitoring the pulse of our Nation's rivers and streams—The U.S. Geological Survey streamgaging network. *U.S. Geol. Surv. Fact Sheet 2018-3021.* **2018.** [CrossRef]

2. Gordon, R.L. Acoustic Measurement of River Discharge. *J. Hydraul. Eng.* **1989**, *115*, 925–936. [CrossRef]

3. Simpson, M.R.; Oltmann, R.N. Discharge-measurement system using an acoustic Doppler current profiler with applications to large rivers and estuaries. *USGS Water Supply Pap. 2395.* **1993**, 32. [CrossRef]

4. Oberg, K.A.; Morlock, S.E.; Caldwell, W.S. Quality assurance plan for discharge measurements using acoustic Doppler current profilers. *U.S. Geol. Surv. Sci. Investig. Rep. 5135.* **2005**, 35. [CrossRef]

5. Mueller, D.S. Extrap: Software to assist the selection of extrapolation methods for moving-boat ADCP streamflow measurements. *Comput. Geosci.* **2013**, *54*, 211–218. [CrossRef]

6. Turnipseed, D.; Sauer, V. Discharge measurements at gaging stations. *U.S. Geol. Surv. Tech. Methods Book 3* **2010**, 87. [CrossRef]

7. Fujita, I.; Komura, S. Application of Video Image Analysis for Measurements of River-Surface Flows. *Proc. Hydraul. Eng. JSCE* **1994**, *38*, 733–738. [CrossRef]

8. Lin, D.; Grundmann, J.; Eltner, A. Evaluating Image Tracking Approaches for Surface Velocimetry With Thermal Tracers. *Water Resour. Res.* **2019**, *55*, 3122–3136. [CrossRef]

9. Dugan, J.P.; Anderson, S.P.; Piotrowski, C.C.; Zuckerman, S.B. Airborne Infrared Remote Sensing of Riverine Currents. *IEEE Trans. Geosci. Remote. Sens.* **2014**, *52*, 3895–3907. [CrossRef]

10. Altenau, E.H.; Pavelsky, T.M.; Moller, D.; Lion, C.; Pitcher, L.H.; Allen, G.H.; Bates, P.D.; Calmant, S.; Durand, M.; Smith, L.C. AirSWOT measurements of river water surface elevation and slope: Tanana River, AK. *Geophys. Res. Lett.* **2017**, *44*, 181–189. [CrossRef]

11. Tauro, F.; Pagano, C.; Phamduy, P.; Grimaldi, S.; Porfiri, M. Large-Scale Particle Image Velocimetry From an Unmanned Aerial Vehicle. *IEEE/ASME Trans. Mechatronics* **2015**, *20*, 3269–3275. [CrossRef]

12. Tauro, F.; Porfiri, M.; Grimaldi, S. Surface flow measurements from drones. *J. Hydrol.* **2016**, *540*, 240–245. [CrossRef]

13. Bandini, F.; Olesen, D.; Jakobsen, J.; Kittel, C.M.M.; Wang, S.; Garcia, M.; Bauer-Gottwein, P. Technical note: Bathymetry observations of inland water bodies using a tethered single-beam sonar controlled by an unmanned aerial vehicle. *Hydrol. Earth Syst. Sci.* **2018**, *22*, 4165–4181. [CrossRef]

14. Gleason, C.J.; Smith, L.C. Toward global mapping of river discharge using satellite images and at-many-stations hydraulic geometry. *Proc. Natl. Acad. Sci. USA* **2014**, *111*, 4788–4791. [CrossRef] [PubMed]

15. Bjerklie, D.M.; Birkett, C.M.; Jones, J.W.; Carabajal, C.; Rover, J.A.; Fulton, J.W.; Garambois, P.A. Satellite remote sensing estimation of river discharge: Application to the Yukon River, Alaska. *J. Hydrol.* **2018**, *561*, 1000–1018. [CrossRef]

16. Muste, M.; Fujita, I.; Hauet, A. Large-scale particle image velocimetry for measurements in riverine environments. *Water Resour. Res.* **2008**, *44*. [CrossRef]

17. Perks, M.T.; Russell, A.J.; Large, A.R.G. Technical Note: Advances in flash flood monitoring using unmanned aerial vehicles (UAVs). *Hydrol. Earth Syst. Sci.* **2016**, *20*, 4005–4015. [CrossRef]

18. Tauro, F.; Tosi, F.; Mattoccia, S.; Toth, E.; Piscopia, R.; Grimaldi, S. Optical Tracking Velocimetry (OTV): Leveraging Optical Flow and Trajectory-Based Filtering for Surface Streamflow Observations. *Remote. Sens.* **2018**, *10*, 2010. [CrossRef]

19. Dugdale, S.J. A practitioner's guide to thermal infrared remote sensing of rivers and streams: Recent advances, precautions and considerations. *Wiley Interdiscip. Rev. Water* **2016**, *3*, 251–268. [CrossRef]

20. Chickadel, C.C.; Horner-Devine, A.R.; Talke, S.A.; Jessup, A.T. Vertical boil propagation from a submerged estuarine sill. *Geophys. Res. Lett.* **2009**, *36*. [CrossRef]

21. Puleo, J.A.; McKenna, T.E.; Holland, K.T.; Calantoni, J. Quantifying riverine surface currents from time sequences of thermal infrared imagery. *Water Resour. Res.* **2012**, *48*. [CrossRef]

22. Legleiter, C.J.; Kinzel, P.J.; Nelson, J.M. Remote measurement of river discharge using thermal particle image velocimetry (PIV) and various sources of bathymetric information. *J. Hydrol.* **2017**, *554*, 490–506. [CrossRef]

23. Gilvear, D.J.; Waters, T.M.; Milner, A.M. Image analysis of aerial photography to quantify changes in channel morphology and instream habitat following placer mining in interior Alaska. *Freshw. Biol.* **1995**, *34*, 389–398. [CrossRef]

24. Legleiter, C.J.; Roberts, D.A.; Lawrence, R.L. Spectrally based remote sensing of river bathymetry. *Earth Surf. Process. Landforms* **2009**, *34*, 1039–1059. [CrossRef]

25. Westaway, R.M.; Lane, S.N.; Hicks, D.M. The development of an automated correction procedure for digital photogrammetry for the study of wide, shallow, gravel-bed rivers. *Earth Surf. Process. Landforms* **2000**, *25*, 209–226. [CrossRef]

26. Dietrich, J.T. Bathymetric Structure-from-Motion: Extracting shallow stream bathymetry from multi-view stereo photogrammetry. *Earth Surf. Process. Landforms* **2017**, *42*, 355–364. [CrossRef]

27. Kasvi, E.; Salmela, J.; Lotsari, E.; Kumpula, T.; Lane, S. Comparison of remote sensing based approaches for mapping bathymetry of shallow, clear water rivers. *Geomorphology* **2019**, *333*, 180–197. [CrossRef]

28. Legleiter, C.J. Calibrating remotely sensed river bathymetry in the absence of field measurements: Flow REsistance Equation-Based Imaging of River Depths (FREEBIRD). *Water Resour. Res.* **2015**, *51*, 2865–2884. [CrossRef]

29. Legleiter, C.J. Inferring river bathymetry via Image-to-Depth Quantile Transformation (IDQT). *Water Resour. Res.* **2016**, *52*, 3722–3741. [CrossRef]

30. Guenther, G.; Cunningham, A.; LaRoque, P.; Reid, D. Meeting the accuracy challenge in airborne lidar bathymetry. In Proceedings of the 20th EARSeL Symposium: Workshop on LiDAR Remote Sensing of Land and Sea, Dresden, Germany, 16–17 June 2000.

31. Kinzel, P.J.; Wright, C.W.; Nelson, J.M.; Burman, A.R. Evaluation of an Experimental LiDAR for Surveying a Shallow, Braided, Sand-Bedded River. *J. Hydraul. Eng.* **2007**, *133*, 838–842. [CrossRef]

32. Hilldale, R.C.; Raff, D. Assessing the ability of airborne LiDAR to map river bathymetry. *Earth Surf. Process. Landforms* **2008**, *33*, 773–783. [CrossRef]

33. Kinzel, P.J.; Legleiter, C.J.; Nelson, J.M. Mapping river bathymetry with a small footprint green LiDAR: Applications and challenges. *J. Am. Water Resour. Assoc.* **2013**, *49*, 183–204. [CrossRef]

34. Mandlburger, G.; Pfennigbauer, M.; Wieser, M.; Riegl, U.; Pfeifer, N. Evaluation of a novel uav-borne topo-bathymetric laser profiler. *Int. Arch. Photogramm. Remote. Sens. Spat. Inf. Sci.* **2016**, *XLI-B1*, 933–939. [CrossRef]

35. Legleiter, C.J.; Overstreet, B.T.; Glennie, C.L.; Pan, Z.; Fernandez-Diaz, J.C.; Singhania, A. Evaluating the capabilities of the CASI hyperspectral imaging system and Aquarius bathymetric LiDAR for measuring channel morphology in two distinct river environments. *Earth Surf. Process. Landforms* **2016**, *41*, 344–363. [CrossRef]

36. Spicer, K.R.; Costa, J.E.; Placzek, G. Measuring flood discharge in unstable stream channels using ground-penetrating radar. *Geology* **1997**, *25*, 423. [CrossRef]

37. Costa, J.E.; Spicer, K.R.; Cheng, R.T.; Haeni, F.P.; Melcher, N.B.; Thurman, E.M.; Plant, W.J.; Keller, W.C. Measuring stream discharge by non-contact methods: A Proof-of-Concept Experiment. *Geophys. Res. Lett.* **2000**, *27*, 553–556. [CrossRef]

38. Melcher, N.B.; Thurman, E.M.; Buursink, M.; Spicer, K.R.; Hayes, E.; Plant, W.J.; Keller, W.C.; Hayes, K.; Costa, J.E.; Haeni, F.P.; et al. River discharge measurements by using helicopter-mounted radar. *Geophys. Res. Lett.* **2002**, *29*, 41–44. [CrossRef]

39. Legleiter, C.J.; Kinzel, P.J. UAS-based remotely sensed data and field measurements of flow depth and velocity from the Blue River, Colorado, October 18, 2018. *U.S. Geol. Surv. Data Release* **2019**. [CrossRef]

40. Thielicke, W.; Stamhuis, E.J. PIVlab—Towards User-friendly, Affordable and Accurate Digital Particle Image Velocimetry in MATLAB. *J. Open Res. Softw.* **2014**, *2*, e30. [CrossRef]

41. Thielicke, W. The Flapping Flight of Birds—Analysis and Application. Ph.D. Thesis, Rijksuniversiteit Groningen, Groningen, The Netherlands, 2014.

42. Thielicke, W.; Stamhuis, E.J. PIVlab—Time-Resolved Digital Particle Image Velocimetry Tool for MATLAB (version: 1.43). Available online: http://dx.doi.org/10.6084/m9.figshare.1092508 (accessed on 29 April 2019).

43. Mitchell, S.; Thayer, J.P.; Hayman, M. Polarization lidar for shallow water depth measurement. *Appl. Opt.* **2010**, *49*, 6995–7000. [CrossRef]

44. Mitchell, S.E.; Thayer, J.P. Ranging through Shallow Semitransparent Media with Polarization Lidar. *J. Atmos. Ocean. Technol.* **2014**, *31*, 681–697. [CrossRef]

45. Rantz, S.; Others, A. Measurement and Computation of Streamflow: Volume 1. Measurement of Stage and Discharge. *U.S. Geol. Surv. Water Supply Pap.* **1982**, *2175*, 284. [CrossRef]

46. Helsel, D.R.; Hirsch, R.M. Statistical Methods in Water Resources (Techniques of Water-Resources Investigations of the United States Geological Survey, Book 4, Hydrologic Analysis and Interpretation, Chapter A3). *U.S. Geol. Surv.* **2002**, 1–524. [CrossRef]

47. Dilbone, E.; Legleiter, C.J.; Alexander, J.S.; McElroy, B. Spectrally based bathymetric mapping of a dynamic, sand-bedded channel: Niobrara River, Nebraska, USA. *River Res. Appl.* **2018**, *34*, 430–441. [CrossRef]

48. Gade, R.; Moeslund, T.B. Thermal cameras and applications: A survey. *Mach. Vis. Appl.* **2014**, *25*, 245–262. [CrossRef]

 *remote sensing*

Article

Inferring Surface Flow Velocities in Sediment-Laden Alaskan Rivers from Optical Image Sequences Acquired from a Helicopter

Carl J. Legleiter * and Paul J. Kinzel

U.S. Geological Survey, Integrated Modeling and Prediction Division, Golden, CO, 80403, USA; pjkinzel@usgs.gov
* Correspondence: cjl@usgs.gov; Tel.: +1-303-271-3651

Received: 02 March 2020; Accepted: 15 April 2020; Published: 18 April 2020

Abstract: The remote, inaccessible location of many rivers in Alaska creates a compelling need for remote sensing approaches to streamflow monitoring. Motivated by this objective, we evaluated the potential to infer flow velocities from optical image sequences acquired from a helicopter deployed above two large, sediment-laden rivers. Rather than artificial seeding, we used an ensemble correlation particle image velocimetry (PIV) algorithm to track the movement of boil vortices that upwell suspended sediment and produce a visible contrast at the water surface. This study introduced a general, modular workflow for image preparation (stabilization and geo-referencing), preprocessing (filtering and contrast enhancement), analysis (PIV), and postprocessing (scaling PIV output and assessing accuracy via comparison to field measurements). Applying this method to images acquired with a digital mapping camera and an inexpensive video camera highlighted the importance of image enhancement and the need to resample the data to an appropriate, coarser pixel size and a lower frame rate. We also developed a Parameter Optimization for PIV (POP) framework to guide selection of the interrogation area (IA) and frame rate for a particular application. POP results indicated that the performance of the PIV algorithm was highly robust and that relatively large IAs (64–320 pixels) and modest frame rates (0.5–2 Hz) yielded strong agreement ($R^2 > 0.9$) between remotely sensed velocities and field measurements. Similarly, analysis of the sensitivity of PIV accuracy to image sequence duration showed that dwell times as short as 16 s would be sufficient at a frame rate of 1 Hz and could be cut in half if the frame rate were doubled. The results of this investigation indicate that helicopter-based remote sensing of velocities in sediment-laden rivers could contribute to noncontact streamgaging programs and enable reach-scale mapping of flow fields.

Keywords: flow velocity; Alaska; river; remote sensing; particle image velocimetry; PIV; large-scale particle image velocimetry; LSPIV

1. Introduction

The largest streamflow monitoring network in Alaska is maintained and operated by the U.S. Geological Survey (USGS) and currently consists of 111 continuous monitoring stations. Although these stations are distributed across the state, Alaska's vast geographic extent encompasses more than 1,200,000 km of rivers and streams; the majority of these waterways remain ungaged. As a result, the streamflow monitoring network in Alaska is sparse relative to that in the contiguous United States. Whereas the average density of streamgages in the lower 48 states is one gage per 1036 km^2, in Alaska the mean area per streamgage is over 16 times greater: 16,735 km^2 [1]. Moreover, many of the streamgages in Alaska are located in remote, roadless areas that can only be accessed by helicopter. Maintaining these gages and making periodic streamflow measurements thus places field personnel at risk. In an ongoing effort to improve safety, increase efficiency, and ultimately expand the monitoring network, the USGS is developing noncontact methods to estimate streamflow using data collected from a variety of remote platforms [1]. These remote sensing approaches comprise two broad categories: (1) techniques that rely on observations from satellites [2–4]; and (2) near-field techniques that include measurements made from bridges [5], small unmanned aircraft systems (sUAS, or drones), helicopters [6], or fixed-wing aircraft. In this study, we focus on remote sensing of surface flow velocities using optical image sequences acquired from a helicopter deployed above two Alaskan rivers. We outline a particle image velocimetry (PIV)-based workflow that could, along with information on channel bathymetry, contribute to a noncontact streamflow measurement strategy. Note that the term large-scale PIV (LSPIV) is often used to describe the application of PIV techniques, which were originally developed for examining small-scale flows under laboratory conditions, to larger natural rivers. Herein, however, we use the shorter term PIV for brevity.

Near-field remote sensing of streamflow in Alaska presents a number of unique challenges that can complicate if not preclude application of methods developed for smaller rivers and streams in more readily accessible locations. For example, although many prior investigations have estimated surface flow velocities by tracking features visible in image sequences acquired from sUAS, most of these studies seeded the flow with tracer particles to enable the use of various motion estimation algorithms including PIV, particle tracking velocimetry (PTV), and optical flow (e.g., [7–11]). However, introducing artificial tracers in a large Alaskan river like the Yukon, which is the third longest river in the United States and the second largest by flow volume [1], is simply not practical.

An alternative means of estimating surface flow velocities that does not require seeding involves using thermal rather than optical image sequences; the feasibility of this approach has been demonstrated using thermal data acquired from bridges [5], fixed-wing aircraft [12], and sUAS [13]. However, the expression of thermal features is highly sensitive to the air-water temperature contrast, implying that this approach might not be consistently effective over the course of a day. In addition, regulatory constraints and logistical considerations might preclude sUAS-based thermal image acquisition at the time when the air-water temperature contrast is most favorable: before dawn. Other, manned platforms are more expensive to operate and also are limited in several other respects. For example, helicopter exhaust can contaminate thermal images and planes cannot safely fly slow enough to provide the long dwell times most conducive to accurate velocity estimation via PIV.

Given the impracticality of seeding the flow to support large-scale PIV of optical images and the various issues associated with thermal PIV, we evaluated the potential to infer surface flow velocities from optical image sequences using a different type of tracer found in great abundance in many Alaskan rivers: sediment. We hypothesized that naturally occurring differences in the texture and color of the water surface, commonly referred to as ripples and boils, provide sufficient contrast for detection by feature tracking algorithms. This approach builds upon the work of Fujita and Komura [14], who observed "boil vortices" in Japanese rivers and used these features as natural tracers for PIV, assuming that their movement represented the surface flow velocity. Subsequently, Fujita and Hino [15] attributed differences in the color of these boil vortices to variations in the concentration of suspended sediment upwelled from within the water column by turbulence. Chickadel et al. [16] captured a similar phenomenon in a tidal estuary via thermal images acquired above a submerged rocky sill.

Although these boil vortices are not ubiquitous in all rivers and streams and might not be expressed at all discharges, when these features are present they could serve as a natural tracer and thus provide a means of inferring spatial patterns of flow velocity. This approach could be especially well suited to large rivers that transport substantial amounts of sediment in suspension or smaller rivers during periods of high runoff when elevated discharges and increased sediment supply lead to higher concentrations of suspended sediment. Glacially fed rivers, in particular, often convey substantial volumes of sediment during melting periods. For example, although only 5% of the drainage area of Alaska's Tanana River consists of mountainous glaciated terrain, material originating in these glaciated areas accounts for 76–94% of the river's sediment load [17].

In addition to stream gaging, characterizing surface velocity fields via remote sensing provides information that is useful in a number of other contexts. From a practical perspective, PIV and other similar techniques allow water surface velocities to be estimated in areas that cannot be accessed by wading or from a boat. Moreover, PIV can resolve low velocities along channel margins, in recirculation zones, and in other small-scale features of the flow field that might not be captured adequately by cross section-based field measurements. Obtaining this type of spatially distributed information is important for many applications in river management. For example, the fate and transport of floating contaminants like oil depend not only upon the main current in the river but also turbulence at the water surface and the location and extent of transient storage zones [18]. Remotely sensed surface velocity fields could be used to identify eddies that represent potential capture zones for the contaminants. Similarly, information on surface velocity patterns could complement tracer studies that use a visible dye to examine dispersion processes [19], which have important implications for the dispersal of juvenile stages of endangered fish species such as sturgeon [20]. Velocity estimates would help guide selection of the location and sampling interval of in-situ sondes used to measure tracer dye concentrations. In a biological context, eddies in rivers provide critical habitat by trapping food for fish and providing a refuge from the primary current [21]. Remote sensing of velocity fields can play a key role in characterizing these micro- to mesoscale habitat features as part of a spatially heterogeneous riverscape [22].

Our work builds upon a growing body of literature on remote sensing of surface flow velocities, summarized in the recent study by Strelnikova et al. [11]. Both Koutalakis et al. [23] and Pearce et al. [10] compared algorithms for inferring flow velocities from optical image sequences, including PIV, PTV, space-time image velocimetry (STIV), a Kanade–Lucas Tomasi optical flow technique, and several other variants. In this study, we selected a single algorithm and instead focused on extending the method to a new domain and establishing a framework to facilitate its effective application. More specifically, we used the ensemble correlation version of PIV advocated by Strelnikova et al. [11] and available within the widely used PIVlab software package [24–26] to estimate flow velocities in Alaskan rivers from optical image sequences that were acquired from a helicopter and captured the advection of boil vortices upwelling suspended sediment from within the water column. We also emphasized geo-referencing of the image sequence and derived velocity field, which might be overlooked in studies on smaller streams that only require information on image scale, because PIV output must be spatially referenced to enable comparison with field measurements of flow velocity for accuracy assessment and parameter specification. We developed a workflow that explicitly includes image stabilization and geo-referencing steps to ensure that the motion of the features tracked by the PIV algorithm is due to movement of the water and not the imaging platform.

Additional objectives of this investigation were to identify appropriate image preprocessing techniques and to facilitate selection of two key parameters of the PIV algorithm: (1) the size of the interrogation area within which features are tracked from frame to frame; and (2) the rate at which these frames are acquired. Although guidelines have been proposed [10,11,24], rigorous testing of these parameters in the context of larger rivers is lacking. In addition, we addressed the basic question of how long a river must be imaged to provide reliable velocity information by assessing the sensitivity of velocity estimates to the duration of the image sequence. Overall, we aimed to establish a general framework to guide preprocessing, parameter specification, and mission planning. This type of analysis can also help to determine which types of instrumentation might be well-suited to a particular application and provide guidance for operating these sensors. Our goal was to move beyond trial and error by introducing a more systematic strategy for data collection and PIV parameter specification. Whereas a heuristic approach fails to yield any true, general insight as to how or why a particular solution was effective to some degree, the framework introduced herein ultimately could yield a more process-based explanation as to why a particular configuration either performed well or failed to provide useful information on flow velocities.

To summarize, the objectives of this study were to:

1. Assess the feasibility of inferring flow velocities in unseeded, sediment-laden rivers via PIV of optical image sequences acquired from a helicopter.
2. Evaluate the utility of different sensor configurations (i.e., image pixel sizes and frame rates).
3. Establish a general, modular, spatially explicit workflow for performing PIV analysis.
4. Introduce a framework to optimize PIV parameter selection by performing an accuracy assessment based on field measurements of flow velocity.
5. Assess the sensitivity of PIV-derived velocity estimates to the duration of the image sequence.

2. Materials and Methods

All of the field measurements and remotely sensed data used in this study are available through a data release hosted by the USGS ScienceBase catalog [27]. The data release consists of a parent landing page with child pages for the individual data sets: field measurements of flow velocity from the Salcha and Tanana Rivers in Alaska and the image sequences acquired from each of these sites.

2.1. Study Area

The need for remotely sensed streamflow information is particularly acute in Alaska, which is the largest of the United States in terms of area but remains sparsely gaged [1]. As part of an ongoing effort to develop robust methods for remote sensing of river discharge, this study focused on two rivers near the city of Fairbanks in central Alaska: the Salcha and Tanana (Figure 1). Basic hydraulic characteristics at the time of data collection are presented in Table 1.

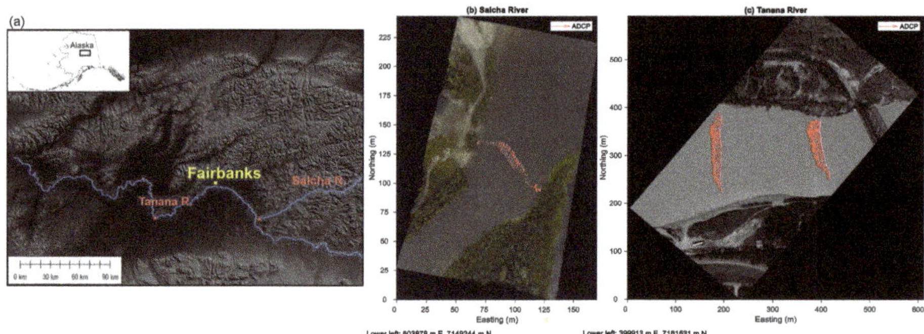

Figure 1. (**a**) Location of the Salcha and Tanana Rivers within the state of Alaska, near the city of Fairbanks. (**b**) Example image from the Salcha River. (**c**) Example video frame from the Tanana River. The red arrows in (**b,c**) represent depth-averaged velocity vectors measured directly in the field with an acoustic Doppler current profiler (ADCP).

Table 1. Basic hydraulic characteristics of the Salcha and Tanana Rivers at the time of data collection.

River	Date of Data Collection	Discharge (m^3/s)	Mean Wetted Width (m)	Mean Velocity (m/s)	Mean Depth (m)	Water Surface Slope	Drainage Area (km^2)
Salcha	8/31/2018	223	65	1.58	2.18	0.00046	5698
Tanana	7/24/2019	1722	190	1.77	5.12	0.00014	66200

The Salcha is a midsized, gravel-bed river that often flows clear but can transport large amounts of sediment in suspension during periods of high flow. Exceptionally high runoff on the Salcha occurred in 2018 and the discharge of 223 m^3/s recorded at a gaging station (USGS 15484000 Salcha River near Salchaket, AK, USA) within our study area on 31 August 2018 was in the 95th percentile of mean daily values for this date. Although no turbidity data were available, 23 suspended sediment samples were collected on the Salcha from 1953 to 1975. A power function was fit to these data to relate suspended sediment concentration in mg/L to discharge in m^3/s with an R^2 of 0.67. Based on this relationship, the suspended sediment concentration on the Salcha River on the date of image acquisition was estimated to be 55 mg/L. We deployed a mobile weather station as the remotely sensed data were being obtained and recorded a mean wind speed of 1.2 m/s from the southwest.

The Tanana is a large, glacier-fed and thus sediment-laden river [17], representative of the many glacial outwash channels that comprise a large fraction of Alaska's drainage network. Importantly, the presence of abundant suspended material within these streams could enable surface flow velocities to be inferred via PIV under natural conditions, without the need to seed the flow with artificial tracers. The discharge recorded at the USGS gaging station (15515500) on the Tanana at Nenana, AK, USA, was 1722 m^3/s during our data collection on 24 July 2019, approximately the median of the mean daily values observed on this date. Turbidity data were not available for the Tanana, but we used 139 samples of suspended sediment concentration collected between 1963 and 2005 to relate suspended sediment concentration to discharge via a power function with an R^2 of 0.92. This relation led to an estimated suspended sediment concentration of 2185 mg/L at the time remotely sensed data were collected. The mean wind speed during image acquisition on the Tanana was 1.3 m/s from the southwest.

2.2. Remotely Sensed Data

The remotely sensed data used in this study were acquired from a Robinson R44 helicopter (Figure 2), a platform that is often used in Alaska to transport field crews to gaging stations in remote, inaccessible terrain for periodic streamflow measurements and maintenance operations. The instrumentation deployed from the helicopter has evolved over time and different cameras were used on the Salcha in 2018 and on the Tanana in 2019. Details regarding the sensors used at each site are provided in the following text and in Table 2. The two instruments spanned a range of technological sophistication (and therefore cost) and collecting data with both allowed us to assess whether a relatively affordable, low-tech system might be sufficient for estimating surface flow velocities or whether a higher-tech, more expensive sensor would be required. For both systems, flight planning involved using information on focal length and the physical dimensions of the camera's detector array to calculate the flying height required to provide a swath width sufficient to see both banks of the river. Including the banks was critical because distinctive terrestrial features were needed to enable stabilization of the image sequences. In addition, we placed ground control targets along the banks of each river to facilitate image geo-referencing. The locations of these targets were established via postprocessed kinematic (PPK) global positioning system (GPS) surveys with a horizontal positional accuracy of approximately 2 cm.

High spatial resolution images of the Salcha were acquired on 31 August 2018 while the helicopter hovered at a predefined waypoint above a reach of the river where the channel widened and the flow diverged before entering a meander bend (Figure 1b). The images were obtained by a Hasselblad A6D-100c digital mapping camera [28] deployed within a pod mounted on the landing gear of the helicopter, as illustrated in Figure 2a. The instrumentation package within the pod is shown in Figure 2b and included an ATLANS GPS/Inertial Motion Unit (IMU) that provided precise, high-frequency position and orientation data during the flight. The accuracies for this sensor were 0.02° for heading, 0.01° for roll and pitch, 0.05 m for horizontal position, and 0.1 m for vertical position. The Hasselblad was programmed to acquire an image every second (i.e., at the camera's maximum frame rate of 1 Hz) as the helicopter hovered at a mean flying height of 197 m above the river. The helicopter's vertical position remained stable throughout the hovering image acquisition, with a standard deviation of flying height of 2.5 m and a range of 9.5 m; these fluctuations had a negligible impact on image scale and were accounted for by the geo-referencing software. The image data were stored on a compact flash memory card within the camera.

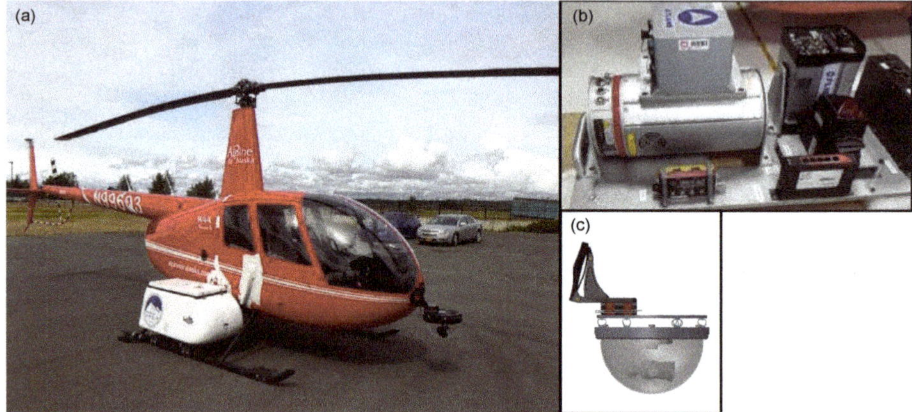

Figure 2. (**a**) The Robinson R44 helicopter used as a platform to acquire hovering image sequences. (**b**) Instrumentation housed within the pod mounted on the helicopter's landing gear. (**c**) Schematic of the mount used to attach the video camera to the nose of the helicopter. Photo in (**a**) and schematic in (**c**) courtesy of Dominic Filiano and Adam LeWinter, respectively, of the U.S. Army Corps of Engineers Cold Regions Research and Engineering Laboratory.

Table 2. Characteristics of the sensors used to collect data from the Salcha and Tanana Rivers.

Camera	Hasselblad A6D-100c	Zenmuse X5
Sensor type	CMOS	CMOS
Lens (mm)	50 mm	15 mm
Diagonal field of view (degrees)	70	72
Number of image pixels	11,600 × 8700	3840 × 2160
Pixel size on detector array (μm)	4.6	4.5
Frame rate (frames per second)	1	29.97
Image file size (megabytes)	577	1.33
Ground sampling distance/image length/swath width at the 197 m flying height used on the Salcha (m)	0.0184/160/213	0.0592/128/227
Ground sampling distance/image length/swath width at the 591 m flying height used on the Tanana (m)	0.0543/473/631	0.178/383/682
GPS/IMU integration	Yes	No
Approximate cost (US dollars)	20,000	2000
Site where deployed	Salcha	Tanana
Duration of hover sequence (s)	68	68.5
Number of images at maximum retained frame rate	68	685

The Hasselblad Phocus software package was used to import the raw images, which had a compressed, proprietary format, and export them as .tif files. We then used SimActive Correlator3D software (Version 8.3.5) [29] to geo-reference the image sequence. This process involved using the internal orientation parameters of the camera, trajectory information from the GPS/IMU, a light detection and ranging (lidar) digital elevation model (DEM), and surveyed ground control targets to perform aerial triangulation and bundle adjustment, which resulted in a series of orthorectified GeoTif images with a pixel size of 0.018 m that served as input to the PIV workflow described in Section 2.4.

Video of the Tanana River was acquired on 24 July 2019 using a Zenmuse X5 camera [30] housed within an enclosure connected to a Meeker mount that in turn was attached to the nose of the helicopter (Figure 2a,c). The Zenmuse was not integrated with the GPS/IMU within the pod, but the camera was mounted on a three-axis gimbal to enhance stability. The Zenmuse was operated via remote control from within the helicopter, which allowed the operator to trigger the camera once the aircraft was hovering above the desired waypoint and then view the live video feed. The mean flying height during image acquisition on the Tanana was 591 m above the river; the standard deviation and range of the flying height were 3.9 m and 13 m, respectively. The original video files recorded by the Zenmuse had a frame rate of approximately 30 Hz and were saved in a .mov format. We used a custom MATLAB script to parse the video into individual image frames. Relative to the original, full-frame rate video, we retained only every third frame to obtain an image sequence with a maximum frame rate of 10 Hz. The resulting images were not geo-referenced but provided suitable input for the PIV workflow described below.

2.3. Field Data

In coordination with the image acquisition at each site, we collected direct, in situ field measurements of flow velocity to assess the accuracy of PIV-derived velocity estimates and guide PIV algorithm parameterization. A TRDI RiverRay acoustic Doppler current profiler (ADCP) [31] with a Hemisphere A101 differential GPS with a horizontal precision of 0.6 m [32] was deployed from a catamaran towed behind a boat with an outboard motor. ADCP data were collected along channel-spanning cross sections oriented perpendicular to the primary flow direction. One transect was recorded within the footprint of the Salcha River image sequence (Figure 1b) whereas the greater spatial extent of the images from the Tanana River encompassed two ADCP transects (Figure 1c). The ADCP data were not intended for use as a discharge measurement and a single pass across the channel was performed at each transect. The RiverRay was controlled by an operator in the boat using the TRDI WinRiver II software package, which allowed for real-time visualization of the velocity field throughout the water column as the boat traversed the channel. The ADCP data were postprocessed in WinRiver II and exported as text files that were then imported into the USGS Velocity Mapping Toolbox (VMT) [33]. We used VMT to perform spatial averaging with horizontal and vertical grid node spacings of 1 m and 0.4 m, respectively. VMT's GIS export tool produced text files consisting of easting and northing spatial coordinates (UTM Zone 6N, NAD83) and the east and north components of the depth-averaged velocity vectors at these locations. Following the recent study by Pearce et al. [10], we used this distilled summary of the ADCP data without applying any kind of correction factor (i.e., velocity index [34]) to estimate the corresponding surface flow velocities from the depth-averaged velocity vectors. Assuming a specific velocity index would have been unnecessarily indirect and would have obfuscated the comparison of field-measured and remotely sensed velocities. Quantitative accuracy assessment of PIV output was based on velocity magnitudes.

2.4. Particle Image Velocimetry (PIV) Workflow

One of the primary objectives of this investigation was to establish a flexible, modular workflow for inferring river surface flow velocities from image sequences. In pursuit of this aim, we developed an approach implemented primarily within MATLAB, making use of the widely used PIVlab add-in as the core PIV algorithm [24–26], but also incorporating functionality from the open source FIJI (ImageJ) image processing software package [35]. The various components of the workflow are summarized in Figure 3 and described below.

1. The first step toward a map of remotely sensed surface flow velocities is to acquire an image sequence from a helicopter or unmanned aircraft system (UAS) hovering above the river reach of interest, or from any other air- or space-borne platform capable of observing the same area of the river for a duration of at least several seconds. Once the original image sequence has been acquired, the user must select the frame rate to retain for PIV.
2. The next step is to stabilize and geo-reference the image sequence, but at this point the workflow diverges depending on whether the images can be geo-referenced based on GPS/IMU data collected onboard the aerial platform and/or ground control targets placed in the field prior to the flight. In this case, a program such as SimActive Correlator 3D [29] can be used to geo-reference and thus stabilize the images. Conversely, if the imaging system is not integrated with a GPS/IMU, the image sequence must first be internally stabilized such that all of the images in the sequence are aligned with one another. We used the TrakEM2 plugin [36] to FIJI for this

purpose. The first image in the stabilized sequence is then geo-referenced to an appropriate base image by identifying surveyed ground control targets and/or selecting image-to-image tie points. The resulting set of tie points is used to define an affine transformation from row, column image coordinates to easting, northing spatial coordinates. This transformation is then applied to all frames of the stabilized image stack to place them in the same real world coordinate system.

3. The resulting stabilized, geo-referenced image stack is then output to a directory and a FIJI macro called programmatically from within MATLAB to prepare the images for PIV. This image preprocessing involves using FIJI functions to perform FFT bandpass filtering and apply a histogram equalization contrast stretch. The preprocessed images output from FIJI are then imported back into MATLAB.

4. Next, the actual spatial footprints of each image in the stack are extracted by automatically tracing the perimeter of the nonzero image pixels. These footprints thus are distinct from the nominal bounding boxes of the images, which might include large areas of no data following stabilization and geo-referencing, particularly if the images are rotated (i.e., oriented at an angle rather than being aligned directly east-west or north-south). To illustrate this step, the footprints extracted from a 1 Hz image sequence from the Tanana River are shown as polygons in Figure 4.

5. Overlaying the footprints from all of the images allows the user to identify the area of common coverage throughout the entire stack and digitize a region of interest (ROI) that can also serve as a water-only mask. For the example shown in Figure 4, the ROI for the Tanana River is represented by the blue polygon. This phase of the workflow ensures that only locations encompassed by all of the images are included in the ROI used for PIV.

6. To focus the analysis, reduce storage requirements, and improve computational efficiency, all of the images in the stack are cropped to a common bounding box based on the digitized ROI defined in the previous step.

7. As a final preprocessing step, the mask corresponding to the digitized ROI is applied to each image to obtain a stack of coregistered water-only images suitable for input to the PIV algorithm that is then exported to a folder.

8. The core PIV analysis is performed by calling the ensemble correlation algorithm available within the latest release of PIVlab (version 2.31, published 4 October 2019) programmatically from within MATLAB. Ensemble correlation was developed to facilitate PIV applications in which the seeding density is low and/or the distribution of trackable features is inhomogeneous. In contrast to standard PIV implementations that operate on a frame pair-by-frame pair basis, the ensemble correlation technique averages correlation matrices over the entire image sequence before searching for the correlation peak, which results in more robust velocity estimates, particularly for longer-duration image sequences [11]. The key input to the PIV algorithm that must be specified by the user in units of pixels is the interrogation area (IA) parameter, which sets the area within which features are tracked by computing correlations between successive images. The step size determines the spacing of the velocity vectors output from PIVlab and in this study was always set as half the IA. In our workflow, none of the optional image preprocessing steps available within PIVlab are applied and a single pass of the PIV algorithm is used. Similarly, we did not filter or interpolate the velocity vectors output from PIVlab.

9. Once the PIV analysis is complete, the workflow resumes by postprocessing the output from PIVlab, which is in units of pixels/frame. The spatial referencing information for the image stack is used to establish the image scale (i.e., the number of pixels per meter). Along with the known frame rate of the image sequence, this scaling allows the velocity vectors output from PIVlab to be converted to units of meters per second. Similarly, the spatial information associated with the images is used to geo-reference the PIV-derived velocity vectors to the same real world coordinate system as the images. In addition, note that for the newly available ensemble correlation PIV algorithm, time-averaging of the velocity vectors is not necessary. Whereas the older, image pair-based version of PIVlab produced a vector field for each successive pair of images, the ensemble correlation-based implementation generates a single vector field because

the underlying correlation analysis is done over the entire image ensemble, not on a frame pair-by-frame pair basis as in the original method.

10. The final stage of our workflow involves plotting the PIV-derived velocity vectors as a map overlain on the first image in the stack and assessing the accuracy of the PIV output via comparison with field measurements of flow velocity.

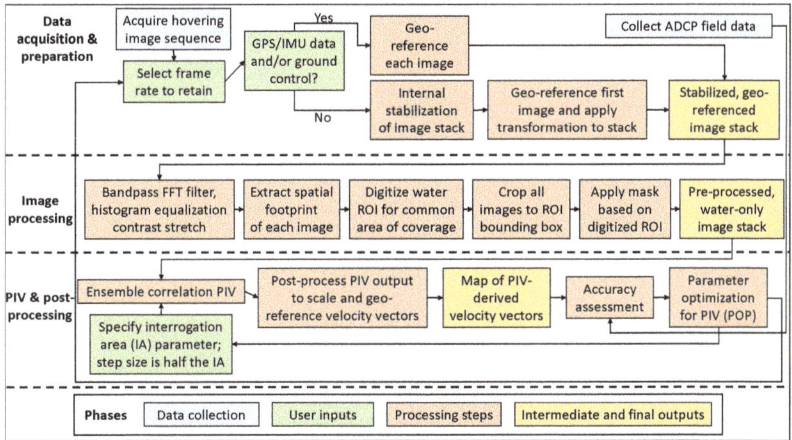

Figure 3. Flow chart outlining the data, user inputs, processing steps, and outputs of the particle image velocimetry (PIV) workflow.

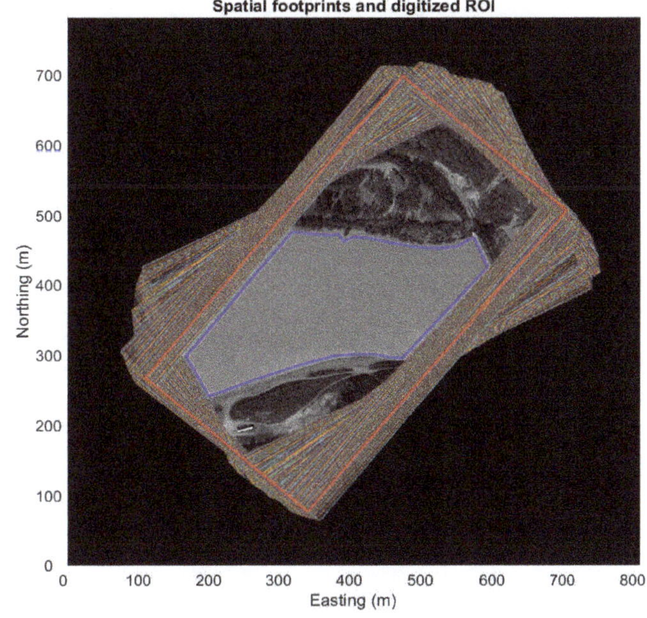

Figure 4. Actual spatial footprints of each image in the 1 Hz image sequence of the Tanana River are represented as rectangular polygons, with the footprint of the displayed image in red with a thicker line. The digitized region of interest (ROI) encompassing the water within the common area of coverage for the entire image sequence is represented by the blue polygon.

2.5. Accuracy Assessment

Accuracy assessment of PIV-derived velocity estimates was based on the depth-averaged velocity magnitudes recorded in the field by the ADCP. In comparing the PIV- and ADCP-based velocities, we retained only those ADCP measurements that were located within a specified search radius of one of the PIV output vectors. This search radius was set to half the product of the PIV step size parameter, which determined the spacing of the output vector field and was specified as half the IA parameter in units of pixels, and the image pixel size. This definition of the search window ensured that adjacent PIV interrogation areas did not overlap one another and also prevented a given ADCP measurement from being compared to more than one PIV-derived velocity. In cases where the search window included more than one observation from the ADCP, all of the field measurements within the search window were averaged to obtain a single representative value for comparison to the velocity estimated via PIV at the center of the search window.

Once the PIV-derived and ADCP-measured velocities were paired in this manner, we calculated several metrics to quantify the performance of the PIV workflow for various input data sets and parameter values. The root mean squared error (RMSE) served as an index of the precision of the PIV estimates. The absolute accuracy of these estimates was summarized in terms of the mean bias, which we calculated as the difference between the ADCP-measured and PIV-derived velocity magnitudes. A positive value of the mean bias thus indicated underprediction of the ADCP-measured velocity by the PIV algorithm. Conversely, a negative value of the mean bias implied that the velocity inferred via PIV was greater than that measured in the field by the ADCP. In addition to the dimensional RMSE and mean bias in units of m/s, we normalized these values by the mean velocity measured by the ADCP at each site (1.58 m/s for the Salcha River and 1.77 m/s for the Tanana) to express the error metrics in units of percent.

We also performed observed (ADCP) versus predicted (PIV) (OP) regressions [37] and used the OP regression R^2 value as a summary of the level of agreement between the PIV-derived velocities and those recorded by the ADCP. Moreover, the slope and intercept of the OP regression provided further information on the performance of the PIV algorithm. Whereas a slope of 1 and an intercept of zero would indicate perfect agreement, any departure from these values implied that the PIV-derived velocities were not scaled correctly (i.e., slope less than or greater than 1) and/or were biased (i.e., nonzero intercept) relative to the ADCP data.

2.6. Evaluating the Effects of Image Preprocessing and Spatial Resolution on PIV-Derived Velocities

We used the hovering image sequence of the Salcha River acquired by the Hasselblad camera to examine how image preprocessing (i.e., step #3 in the PIV workflow outlined in Section 2.4) and pixel size (i.e., ground sampling distance) influenced PIV-derived velocity estimates. To perform this analysis, we resampled the original orthorectified image sequence, which had a pixel size of 1.8 cm, to pixel sizes of 5, 10, and 20 cm. Each of these resampled stacks, as well as the original 1.8 cm-pixel size images, were used in turn as input to the PIV algorithm. We used the stack of images with a 10 cm pixel size as a base case for evaluating the importance of the image preprocessing phase of the PIV workflow. To make this assessment, we performed PIV for both the raw images and images that had been enhanced in FIJI by applying an FFT bandpass filter and a histogram equalization contrast stretch. We compared the PIV output derived from the raw and preprocessed image stacks both graphically, by inspecting maps of the PIV-derived velocity fields, and quantitatively via OP regressions based on field measurements of flow velocity from a transect across the Salcha River.

Our investigation of the effects of image spatial resolution involved performing two separate PIV analyses for each of the four image pixel sizes we considered: 1.8, 5, 10, and 20 cm. For the first type of PIV analysis, we held the IA and step size parameters fixed (in units of pixels) across the four different pixel sizes such that as the pixel size increased so did the spacing of the PIV output vectors, which is determined by the step size parameter. In the second type of PIV analysis, we held the PIV output vector spacing fixed at 6.4 m (i.e., a 64-pixel step size for the image with 10 cm pixels) and adjusted the

IA and step size parameters of the PIV algorithm to maintain a consistent vector spacing as the image pixel size varied. For example, for the 5 cm-pixel image, the IA and step size were set to 256 pixels and 128 pixels, respectively, to maintain an output vector spacing of 6.4 m. These two versions of the PIV analysis allowed us to assess whether the effect of a larger image pixel size was a reflection of sparser, less closely spaced PIV output vectors, which would reduce the sample size of the ADCP versus PIV comparison used for accuracy assessment or due to inherent differences in the ability of the PIV algorithm to detect and track features in images with a coarser spatial resolution. We compared the PIV output for different image pixel sizes for both fixed PIV parameters (i.e., variable output vector spacing) and fixed vector output spacing (i.e., variable step size parameter) by producing maps of the PIV-derived velocity fields and computing the various metrics of performance described in Section 2.5.

2.7. Parameter Optimization for PIV (POP)

Successful application of PIV techniques to infer surface flow velocities requires appropriate parameter values, but key inputs to PIV algorithms often are specified iteratively by trial-and-error or on the basis of heuristic guidelines. In this study, we explored an alternative approach, which we refer to as Parameter Optimization for PIV (POP), that involves tuning the PIV algorithm by quantifying the accuracy of velocity estimates derived using various PIV parameter combinations. The high-frame rate video we acquired from the Tanana River allowed us to explore two key dimensions of the PIV parameter space: interrogation area and frame rate. To enable rigorous, quantitative selection of parameter values, we performed PIV for IAs ranging from 16 to 320 pixels in 16-pixel increments and frame rates ranging from 0.5 to 10 Hz. We varied the frame rate by adjusting the number of frames skipped, relative to the 10 Hz maximum frame rate we retained from the original video file, from 1 to 20. For example, skipping 10 frames led to an image sequence sampled at 1 Hz. For each of the 20 IAs and 20 frame rates, we implemented the PIV workflow described in Section 2.4 and computed the RMSE between ADCP-measured and PIV-derived velocities. The combination of IA and frame rate that yielded the smallest RMSE was taken to be the optimal parameter combination. In addition, we used the PIV output for each IA/frame rate combination to compute other metrics of performance including the mean bias, normalized RMSE, normalized mean bias, and OP regression R^2. These results were summarized by producing plots, called POP charts, that display values of each metric as a function of IA and frame rate and thus provide a visual summary of the effects of these parameters on the accuracy of PIV-derived velocities. In essence, the POP framework provided a means of exploiting any available field measurements of velocity to calibrate the PIV algorithm for a particular data set.

2.8. Sensitivity to Image Sequence Duration

An important practical question in remote sensing of flow velocities is how long a river must be imaged in order to derive reliable velocity information. To address this question, we inspected the POP charts produced for the Tanana River to identify IA/frame rate combinations that yielded strong agreement between PIV estimates and field measurements of velocity. We selected an IA of 128 pixels and held this parameter fixed while varying the duration of the image sequence; six different durations were considered. To account for the potentially confounding effect of the number of images, we evaluated image sequences with frame rates of 1 Hz and 2 Hz and truncated the original, full-length sequences for both frame rates to obtain stacks with durations of 2, 4, 8, 16, 32, and 64 s. For example, the 2 s duration image sequence at a frame rate of 1 Hz consisted of a single pair of images whereas the 2 Hz sequence of the same duration included four images. For each of the six durations, the number of images was equal to the duration for the 1 Hz sequence and the number of images in the 2 Hz sequence was twice the duration. Each of these truncated image sequences was used as input to the ensemble correlation PIV algorithm and the resulting velocity estimates were compared to the ADCP field measurements by computing the various accuracy metrics described in Section 2.5.

3. Results

3.1. Image Stabilization and Geo-Referencing

The first steps in our PIV workflow involved stabilizing and geo-referencing the image sequences, but both of these processes were subject to error that introduced some uncertainty into the velocity estimates derived subsequently via PIV. For the hovering image sequence from the Salcha River, data from the GPS/IMU onboard the helicopter and integrated with the Hasselblad camera, along with surveyed ground control targets placed in the field prior to the flight, allowed us to geo-reference each individual image in the stack without an initial, internal stabilization of the stack. The Correlator 3D software we used to orthorectify each image reported an average projection error of 0.39 pixels and an average horizontal error of 1.5 cm, based on three ground control points. This error was less than the size of a single image pixel, the smallest feature displacement between successive frames that the PIV algorithm could possibly detect, and thus was not a significant source of uncertainty in the velocity fields we ultimately derived for the Salcha.

On the Tanana River, the Zenmuse camera used to record video while the helicopter hovered was not integrated with the GPS/IMU and the individual image frames we extracted could not be geo-referenced using the same approach as on the Salcha. Instead, to account for motion of the platform during the video, we used the TrakEM2 FIJI plugin to stabilize the image sequence. One frame from the middle of the stack was used as a reference to which all of the other images were aligned using a scale-invariant feature transform (SIFT) algorithm that detected stable features such as bridges and buildings with fixed spatial locations throughout the entire duration of the video. The maximum alignment error parameter in TrakEM2 was set to five pixels and an affine transformation was used to align the 685 images extracted from the original video. TrakEM2 reported minimum, mean, and maximum displacement errors of 0.626, 0.853, and 1.069 pixels, respectively. For the ground sampling distance of this image sequence, these errors represented distances of 0.094, 0.13, and 0.16 m, respectively. As for the Salcha, these stabilization errors were less than the size of a single image pixel and thus did not introduce any significant uncertainty to the PIV-derived velocity estimates.

Once internally stabilized in this manner, the image stack had to be geo-referenced to enable comparison of PIV-based velocities with field measurements from the ADCP. This phase of the geo-referencing process, performed in the Global Mapper software package (Version 19.0.2) [38] using a satellite image as a base, introduced a greater amount of spatial uncertainty. We used six tie points to establish an affine transformation between the row, column coordinates of one of the stabilized images and easting, northing spatial coordinates. Three of these tie points were ground control targets placed in the field along the south bank of the Tanana and visible in the Zenmuse image. The other three tie points were distinctive, invariant features identified in both the Zenmuse image and the geo-referenced satellite image used as a base. We used the resulting affine transformation matrix to predict the locations of the tie points on the basis of their image coordinates. Comparing these predictions to their actual locations led to a geo-referencing RMSE of 4.33 m. This error was consistent throughout the image sequence because all of the images were stabilized to one another via TrakEM2 and the same affine transformation matrix was applied to each individual frame to geo-reference the entire stack. Within our workflow, this geo-referencing error thus did not influence the PIV-derived velocities, which were based on relative displacements of features between successive images in the stabilized stack.

The absolute geo-referencing error did introduce some uncertainty into the comparison of PIV-derived and ADCP-measured velocities we used to assess accuracy and guide parameter selection. However, the PIV- and ADCP-based velocities were only paired with one another at a spatial resolution set by the spacing of the PIV output vectors (see Section 2.5), which is a function of the PIV step size parameter. In this study, the step size was always set to half the IA and so the potential error associated with image geo-referencing only would have been significant for the smallest IAs we considered. For example, for an IA of 256 pixels, the corresponding step size of 128 pixels represented a spatial distance of 19.2 m and a typical geo-referencing error of 4.33 m thus would have been only 22.5% of the spacing between PIV output vectors, implying that an incorrect pairing of an ADCP measurement with a PIV-derived velocity was highly unlikely.

3.2. Image Preprocessing

To assess the importance of preprocessing images so as to facilitate detection and tracking of features, we performed PIV for both raw and enhanced images of the Salcha River with 10 cm pixels. The results of this analysis are illustrated in Figure 5, which compares the velocity fields generated by PIVlab based on (a) raw and (b) preprocessed images. The primary effect of applying a bandpass filter and contrast stretch was to improve the coherence and consistency of the velocity vectors relative to those derived from the unmanipulated images. Whereas the velocity field in Figure 5a featured large gaps at the bottom of the image and along the left bank farther downstream and low, suspect velocities oriented toward the banks on the right side of the channel, no such gaps were present in Figure 5b and all vectors were directed primarily downstream.

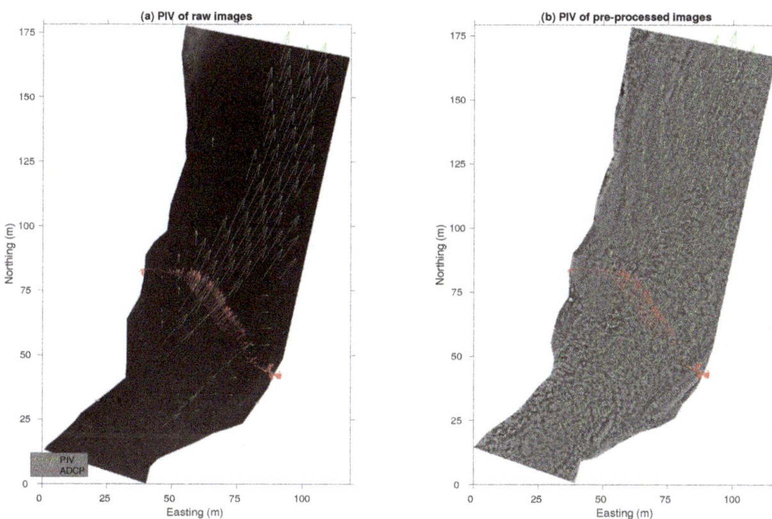

Figure 5. Comparison of PIV-derived velocity vectors derived from (**a**) raw images and (**b**) images preprocessed via FIJI. Image pixel size is 10 cm in both cases and the velocity vectors measured in the field with an ADCP are overlain in red for comparison.

Quantitative accuracy assessment yielded an OP R^2 value of 0.96 for both the raw and preprocessed image stacks but the regression for the raw images was based on a smaller sample size ($n = 6$) than that for the preprocessed images ($n = 8$) due to the gap in the velocity field inferred from the raw images. This gap occurred in the middle of the channel, where the flow field was fairly homogeneous, and so the range of velocities included in the OP regression was similar for both of the image sequences used as input to the PIV algorithm. Had the ADCP transect used for accuracy assessment been located in a different part of the reach that coincided with the spurious vectors derived from the raw image sequence, the OP R^2 value for this data set would have been lower.

In any case, these results, as well as our experience with many other raw and preprocessed image stacks, imply that enhancing the original image data leads to improved detection and tracking of features by the PIVlab ensemble correlation algorithm and thus more accurate estimates of flow velocity. Although our workflow involved preprocessing the images in FIJI, many alternative approaches to image preparation are also available. For example, PIVlab includes a contrast-limited adaptive histogram equalization (CLAHE) algorithm and several other image preprocessing options accessible via a graphical user interface. We used FIJI because this package provides a wide range of image processing tools and is freely available, open source, and amenable to scripting, batch processing, and external access (e.g., calling FIJI from within MATLAB). Irrespective of the implementation details, the objective of the preprocessing phase of the PIV workflow is to obtain an enhanced image sequence in which distinct, trackable features can be identified readily by the PIV algorithm.

3.3. Effects of Image Spatial Resolution

Another factor influencing the detectability of advecting surface features is the spatial resolution of the images. To examine this effect, we performed two sets of PIV analyses for image sequences from the Salcha River with pixel sizes of 1.8 (the native resolution of the geo-referenced images), 5, 10, and 20 cm. For the first group of PIV runs, the IA and step size PIV parameters were held fixed while the underlying image pixel size varied, leading to more widely spaced output vectors for images with larger pixels (Figure 6a–d). In the second round of PIV, these parameters were adjusted so as to maintain a constant vector spacing across the different pixel sizes (Figure 6e–h). Table 3 lists the PIV input parameters, output vector spacing, and various performance metrics for each of these iterations.

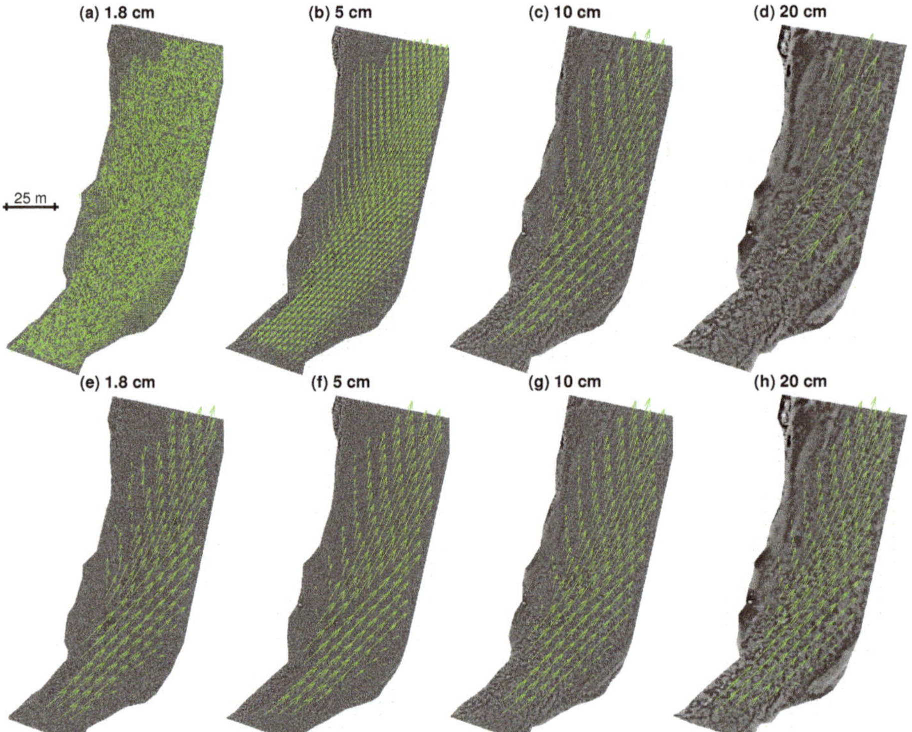

Figure 6. Comparison of PIV-derived velocity fields derived from images with varying pixel sizes for fixed PIV parameters (top row) and fixed output vector spacing (bottom row). Note that the vector length in the top row is dictated by the spacing between vector origins and does not scale with the magnitude of the velocity vectors, which is similar across the four panels.

Table 3. Summary of PIV results from the Salcha River 2018 data set. OP = Observed versus Predicted; RMSE = root mean squared error; bias is calculated as ADCP–PIV; RMSE and bias are normalized by the mean velocity recorded by the ADCP (1.58 m/s).

Pixel Size (m)	Interrogation Area (pixels)	Step Size (pixels)	Vector Spacing (m)	OP R^2	OP Intercept (m/s)	OP Slope	RMSE (m/s)	Mean Bias (m/s)	Normalized RMSE (%)	Normalized Mean Bias (%)
0.018	128	64	1.15	0.158	0.212	0.074	1.169	0.860	74.2	54.6
0.05	128	64	3.2	0.941	0.157	0.952	0.224	−0.098	14.2	−6.2
0.1	128	64	6.4	0.961	0.287	0.920	0.244	−0.184	15.5	−11.7
0.2	128	64	12.8	0.965	−0.485	1.305	0.249	0.026	15.8	1.7
0.018	712	356	6.4	0.903	0.389	0.918	0.336	−0.263	21.3	−16.7
0.05	256	128	6.4	0.922	0.210	0.966	0.281	−0.167	17.8	−10.6
0.1	128	64	6.4	0.961	0.287	0.920	0.244	−0.184	15.5	−11.7
0.2	64	32	6.4	0.924	0.323	0.912	0.266	−0.193	16.9	−12.3

For the first set of PIV analyses, the images with the highest resolution (1.8 cm pixels) provided a high density of output vectors but also yielded an erratic inferred flow field (Figure 6a) and poor agreement with field measurements of velocity (OP R^2 = 0.16, Table 3). We attributed these poor results to the noise present in the high resolution images and to the small size of the pixels relative to the features advecting at the water surface. Resampling the original, full-resolution image sequence to even a slightly coarser pixel size of 5 cm, however, led to a much more coherent velocity field (Figure 6b) and increased the OP R^2 to 0.94. Doubling the resampled pixel size to 10 cm yielded a sparser but still informative representation of the velocity field (Figure 6c) and further increased the OP R^2 to 0.96. This configuration led to RMSE and mean bias values of 0.244 and −0.184 m/s, respectively, corresponding to 15.5% and −11.7% of the mean velocity measured in situ by the ADCP. The 10 cm pixel size thus represented a reasonable compromise between image noise, spatial detail, and computational demands and served as a base case for evaluating the effects of image preprocessing described above. Further coarsening the images to 20 cm pixels while holding the PIV input parameters fixed (in units of pixels) led to output velocity vectors that were too widely spaced to provide meaningful insight regarding the flow field within the reach.

For the second round of PIV runs, the IA and step size parameters were adjusted to maintain a constant output vector spacing of 6.4 m, which also ensured that the same set of ADCP measurements were used to assess the accuracy of PIV-derived velocity estimates for the various image pixel sizes we considered. When the vector spacing was held fixed in this manner, distinctions between velocity fields inferred from images with different spatial resolutions were less pronounced (Figure 6e–h). The corresponding OP regression R^2 values were also quite similar, ranging from 0.90 for 1.8 cm pixels to 0.96 for 10 cm pixels (Table 3). The images with the smallest pixel size resulted in a slightly more irregular vector field with some voids (Figure 6e), but none of the resampled image sequences led to output that was markedly better or worse than any of the other pixel sizes, given the fixed output vector spacing. Along with the consistency of the accuracy assessment results summarized in Table 3, the similarity of the velocity fields derived from images with pixel sizes varying by an order of magnitude implied that the PIV workflow was robust to the effects of image spatial resolution. As long as the pixel size was 5 cm or greater, the noise present in the original 1.8 cm-pixel images was reduced to a sufficient degree to enable coherent, accurate velocity fields to be inferred via PIV.

Figure 7 illustrates the importance of using an appropriate pixel size (i.e., resampling the original data) and preprocessing images prior to PIV. In essence, PIV algorithms identify distinctive features expressed at the water surface and track their displacement between successive frames to infer flow velocities. We reproduced this process by visually inspecting the first frame in the 10 cm-pixel image sequence and placing a green dot to mark the location of an easily recognizable feature in the preprocessed image shown in Figure 7e, which is a zoomed-in subset of the enhanced image shown in Figure 5b. In this study, this feature represents a boil of upwelling sediment rather than the floating foam, debris, or seeded particles that have been used in most previous applications of PIV (e.g., [11]).

The location of the boil in the next frame, recorded by the Hasselblad camera 1 s later, is indicated by the red dot in Figure 7f; the length of this displacement vector is 1.99 m, leading to a plausible velocity estimate of 1.99 m/s. The same start and end points were then overlain on the full resolution (1.8 cm pixel) preprocessed images shown in Figure 7c,d, respectively, as well as the original, raw images in Figure 7a,b. The boil we selected as an example was clearly evident in the coarsened, processed image but was harder to distinguish in the preprocessed image with a smaller pixel size due to the greater noise in this sequence; this feature was not apparent at all in the original, unenhanced images. Spatial resampling and image preprocessing thus were critical steps in the PIV workflow in this case.

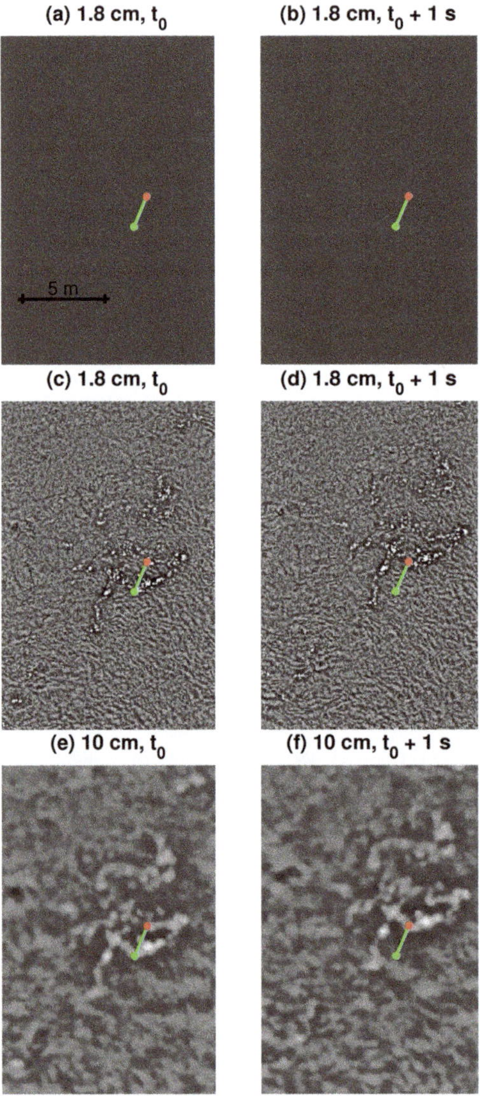

Figure 7. Importance of image preprocessing and spatial resolution. Green point represents location of the tracked feature at time 0 and red point indicates the location of the same feature in the next frame in the image sequence, acquired 1 s later. Measured displacement is 1.99 m.

3.4. Parameter Optimization for PIV (POP)

In an effort to guide selection of appropriate PIV parameters for a particular application, we used a video from the Tanana River to explore the effects of IA and frame rate on the accuracy of PIV-derived velocity estimates. Our Parameter Optimization for PIV (POP) approach was essentially a form of calibration that involved using velocity measurements collected in the field at the same time as the video to identify the PIV parameters that resulted in the closest agreement between the velocities estimated via PIV and the ADCP reference data. Although the in situ velocity measurements were depth-averaged and the PIV estimates presumably represented approximate surface flow velocities, we did not apply a velocity index to account for the difference between depth-averaged and surface velocities before comparing the PIV estimates to the ADCP measurements, consistent with the recent study by Pearce et al. [10]. In addition, note that a critical consideration in any application of the POP framework is obtaining field data that span a broad range of velocities. In this case, the velocities recorded along two ADCP transects across the Tanana River varied from near 0 to over 2.5 m/s.

Figure 8 summarizes the results of our parameter optimization analysis via matrices with different IAs on the horizontal axis and various frame rates on the vertical axis. The entries in each matrix represent values of the (a) OP regression R^2, (b) normalized RMSE, and (c) normalized mean bias. Figure 8 indicates that for any IA larger than 64 pixels and any frame rate less than 1 Hz, the OP R^2 value was greater than 0.89. Similarly, these parameter combinations lead to normalized RMSE values less than 11.7% and normalized bias values less than 4.5% of the reach-averaged mean velocity of 1.77 m/s. These results imply that the ensemble correlation PIV algorithm not only provided highly accurate velocity estimates relative to the ADCP data but also was robust to the specific combination of parameters provided as input.

For the approximately 200 m-wide Tanana River, the features detected via PIV were large enough that even for an IA of 320 pixels, equivalent to 48 m, a sufficient number of trackable features were captured within the IA to enable accurate estimation of flow velocities. Similarly, for the flow velocities on the order of 1.5–2 m/s observed on this river, the displacement of surface features was detectable for frame rates as low as 0.5 Hz (i.e., images acquired every other second). Conversely, PIV performance deteriorated for the smallest IAs and highest frame rates because too few trackable features were present within the 16- or 32-pixel (2.4 or 4.8 m) IAs and/or because those features that were detected did not move far enough between successive images when the frame rate was higher than about 5 Hz.

Although a single particular combination was identified as optimal via the POP approach, Figure 8 indicates that many other IAs and frame rates would have led to accuracies nearly as high. Selecting one of these other pairs of parameters thus could achieve comparable performance while providing more detailed spatial information in the form of more closely spaced PIV output vectors. Figure 9 compares the PIV results from the optimal parameter combination, which leads to a relatively coarse vector spacing due to the large IA, to output based on a smaller IA that yielded an OP R^2 of 0.95, a reduction of only 0.04 relative to the optimum and provided a much more detailed representation of the flow field that might be highly desirable, if not essential, for certain applications. In addition, the error metrics summarized in Table 4 indicate that the two parameter combinations yielded similar RMSE and mean bias values. In both cases, the normalized RMSE and mean bias for the Tanana River were smaller in absolute value, but opposite in sign, relative to the RMSE and mean bias observed on the Salcha River. On the Tanana, the PIV-derived velocities tended to slightly underestimate the depth-averaged velocities measured in the field with an ADCP, whereas on the Salcha the PIV estimates were larger than the ADCP measurements.

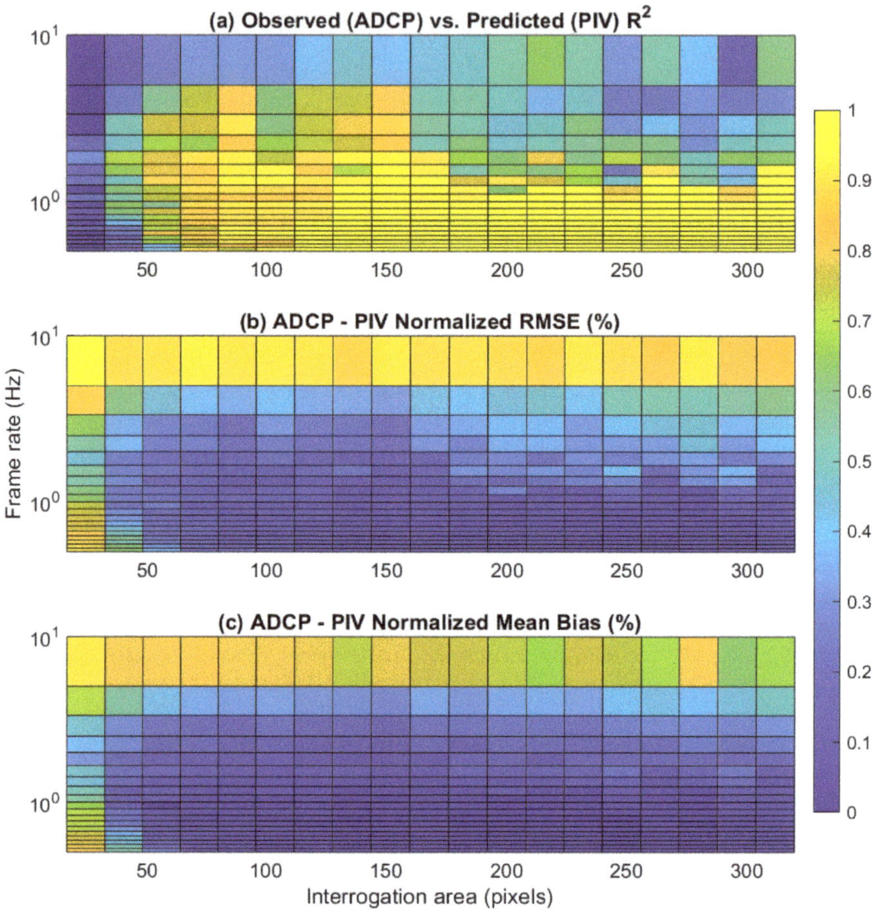

Figure 8. Parameter Optimization for PIV (POP) results for video sequences from the Tanana River. (**a**) Observed (ADCP) versus predicted (PIV) regression R^2 values. (**b**) Normalized Root Mean Squared Error (RMSE) between field-measured ADCP and PIV-derived velocities. (**c**) Normalized mean bias between field-measured ADCP and PIV-derived velocities. RMSE and mean bias are normalized by the mean of the ADCP velocity measurements. Note that for display purposes the normalized RMSE and mean bias values in panels (**b**,**c**) are expressed as proportions between 0 and 1 and not as percentages between 0 and 100.

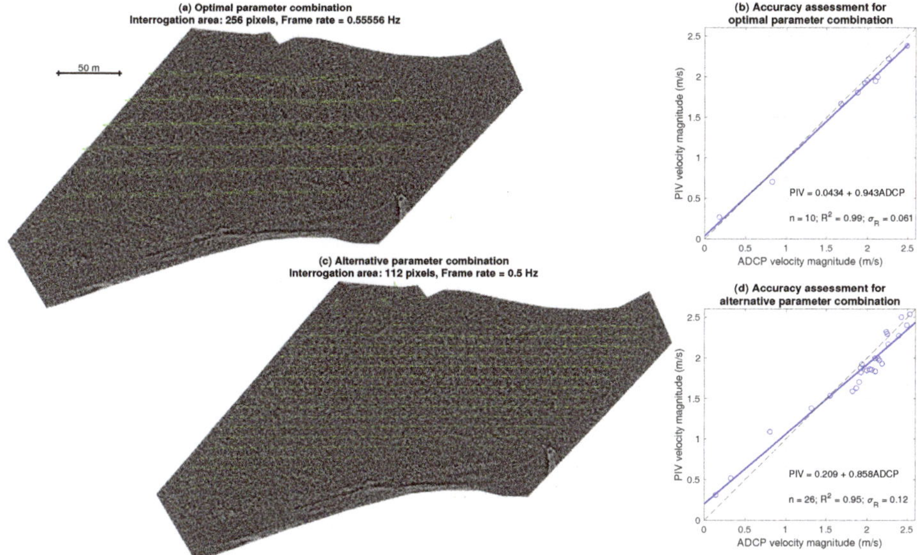

Figure 9. PIV-derived velocity vectors for the (**a**) optimal parameter combination and (**b**) an alternative parameter combination that was nearly as accurate but provided a higher density of output vectors. Corresponding observed vs. predicted regression accuracy assessments for the (**c**) optimal and (**d**) alternative parameter combinations. Note that the vector length in each panel is dictated by the spacing between vector origins and does not scale with the magnitude of the velocity vectors, which is similar for both panels.

Table 4. Summary of PIV results from the Tanana River 2019 data set for the two parameter combinations illustrated in Figure 9. Pixel size is fixed at 0.15 m. OP = Observed versus Predicted; RMSE = root mean squared error; bias is calculated as ADCP–PIV; RMSE and bias are normalized by the mean velocity recorded by the ADCP (1.78 m/s).

Frame Rate (Hz)	Interrogation Area (pixels)	Step Size (pixels)	Vector Spacing (m)	OP R^2	OP Intercept (m/s)	OP Slope	RMSE (m/s)	Mean Bias (m/s)	Normalized RMSE (%)	Normalized Mean Bias (%)
0.556	256	128	19.20	0.993	0.043	0.943	0.087	0.056	4.9	3.1
0.5	112	56	8.4	0.950	0.209	0.858	0.155	0.057	8.7	3.2

3.5. Sensitivity to Image Sequence Duration

In addition to our analysis of the effects of the IA and frame rate parameters on the accuracy of PIV-derived velocity estimates, we also used video from the Tanana River to examine the effect of image sequence duration. This sensitivity analysis involved holding the IA fixed at 128 pixels and progressively truncating the original, full-length 1 and 2 Hz image sequences. Repeating the analysis for two different frame rates accounted for the potentially confounding effect of the total number of images as well as the duration of the image sequence. This approach allowed us to address the question of not only how long the helicopter would need to hover to provide reliable velocity information but also whether the number of images, rather than the time period over which they were acquired, was an important factor.

This assessment is summarized in Figure 10 and was based on three metrics that summarized the performance of the PIV algorithm by comparing remotely sensed velocities to field measurements: the OP regression R^2, normalized RMSE, and normalized mean bias. For the image sequence with a frame rate of 1 Hz, very poor agreement between PIV-derived and ADCP-measured velocities was observed for the shortest duration, 2 s sequence, with some improvement for the 4 and 8 s sequences.

For dwell times of 16 s or longer, the 1 Hz sequence yielded OP R^2 values greater than 0.94, with little improvement for the longer 32 or 64 s durations. The corresponding normalized RMSE and mean bias stabilized around 8% and 0, respectively. These results imply that much shorter image sequences, on the order of 25% of the duration of the Tanana River video we recorded, would have been long enough to provide reliable velocity estimates.

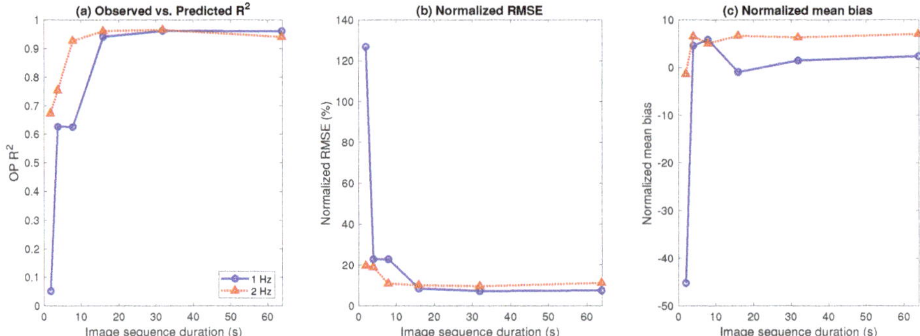

Figure 10. Sensitivity of the accuracy of PIV-derived velocities to image sequence duration for the Tanana River for a fixed interrogation area of 128 pixels and frame rates of 1 and 2 Hz. OP = Observed (ADCP) vs. Predicted (PIV). The RMSE and mean bias are normalized by the reach-averaged mean velocity recorded by the ADCP: 1.78 m/s.

Repeating this sensitivity analysis for an image sequence with a frame rate of 2 Hz enabled us to evaluate whether acquiring the same number of images but over a shorter time period would provide a similar level of agreement between the velocities inferred via PIV and those measured in situ. Figure 10 indicates that even if the dwell time was only 2 s, having as few as four images would yield a moderate OP R^2 of 0.67, much higher than the 0.05 R^2 for the 2 s, 1 Hz sequence; the normalized RMSE and mean bias also were much less for the 2 Hz sequence. The OP R^2 reached 0.93 for a duration of 8 s for the 2 Hz data set, half the time required for the OP R^2 to exceed 0.9 for the 1 Hz sequence. These results imply that dwell times as short as 8 s would be adequate if the number of images was increased by collecting data at a higher but still relatively modest frame rate of 2 Hz. The superior performance of the 2 Hz sequence relative to the 1 Hz sequence for a given duration rate provided further evidence of the ability of the ensemble correlation PIV algorithm to provide robust velocity estimates. Strelnikova et al. [11] attributed this capability to the algorithm's averaging correlation matrices over the image sequence before searching for the correlation peak, which in effect increases the density of features in an image sequence in proportion to the number of images. This characteristic of ensemble correlation PIV makes the technique well-suited to quasistationary flows with a low density of trackable features, as was the case in our study.

4. Discussion

4.1. Remote Sensing of Surface Flow Velocities in Large Rivers

Whereas previous research on remote sensing of surface flow velocities primarily has been directed toward relatively small, slowly flowing streams, this study focused on much larger rivers up to 200 m in width. In addition, rather than seeding the flow with artificial tracers, which can be problematic even in much smaller channels [11] and would be all but impossible on a large river, we inferred flow velocities by exploiting a naturally occurring tracer: boil vortices comprised of upwelling suspended sediment. The advecting boils we tracked using a PIV algorithm might not represent true surface velocities as would be the case for foam, debris, or introduced particles floating on the water surface. Another unique challenge arose in this context: the sediment boil vortices were not rigid features like a discrete floating object but rather a mobile, deforming expression of the flow field itself. These boils represent a surface manifestation, in the form of color differences that create a contrast between the feature and the background water, of turbulent mixing processes within the water column, analogous to the thermal features utilized for PIV in prior studies [5,13]. Although the physics governing sediment boil vortices might be complex and will require further research, the results presented herein demonstrate that these boils are an observable phenomenon [14,15] strongly correlated with depth-averaged velocity magnitudes.

Another innovative aspect of this investigation was that whereas many previous studies have acquired image data from bridge- or bank-mounted cameras or sUAS, we obtained image sequences from a hovering helicopter. This platform allowed us to achieve broader spatial coverage by flying higher than a typical sUAS, which tends to be limited by regulatory constraints and/or endurance (i.e., maximum flight time or battery life). Importantly, the greater extent of the helicopter-based images encompassed both banks of even the larger of the two rivers we examined. This broader field of view thus enabled image stabilization and facilitated not only establishing image scale but also geo-referencing PIV-derived velocity vectors to an established coordinate system. The resulting spatially referenced velocity estimates allowed us to perform a quantitative accuracy assessment based on field measurements of flow velocity acquired with an ADCP.

This accuracy assessment was critical to achieving two of our primary objectives: establishing a general framework for inferring flow velocities via remote sensing and introducing a more rigorous, systematic approach to specifying PIV input parameters. Although Koutalakis et al. [23] and Pearce et al. [10] recently compared different velocity estimation algorithms and Strelnikova et al. [11] examined the influence of the IA and step size parameters on ensemble correlation-based PIV, these studies focused on smaller streams; the extent to which their results might scale up to larger rivers remained untested. Moreover, in practice many applications of PIV tend to proceed on the basis of ad hoc, heuristic guidelines with little rationale for selecting a particular image pixel size, frame rate, or duration. Important factors to consider in making such decisions include the size of the features to be tracked and the rate at which they are advected by the flow. The images must be sufficiently spatially detailed to capture these features and the frame rate must be fast enough to ensure that they do not move so far between successive frames so as to not be trackable but not so fast that their movement from one frame to the next is too small to be detectable. The effects of these sensor characteristics are confounded with those of PIV input parameters, mainly the size of the IA and the number of frames skipped relative to the maximum frame rate of the image sequence. One means of setting these parameters, illustrated in Figure 7, involves visually identifying a feature in an image, measuring its displacement in the next frame, and using this information to select a reasonable IA and step size. Our Parameter Optimization for PIV (POP) framework generalizes and automates this approach by explicitly evaluating a range of IAs and frame rates and identifying that combination which yields the strongest agreement between PIV-derived and field-measured flow velocities. This technique is essentially a form of calibration that can be used to tune a PIV algorithm for application to a specific river of interest, regardless of the size of the channel or the type of features being tracked.

4.2. Implications, Applications, and Limitations

The results of this study have several important implications for operational use of remote sensing techniques for estimating flow velocities; guidelines for data collection based on our experience in Alaska are summarized in Table 5. Our systematic evaluation of various image pixel sizes, PIV IAs, frame rates, and image sequence durations indicated that the native 30 Hz frame rate of the Zenmuse video camera we used was much faster than necessary. The POP analysis summarized in Figure 8 showed that even retaining only every third frame from the original video, thereby reducing the frame rate to 10 Hz, was excessive and that accurate velocity estimates could be derived from image sequences with frame rates as low as 0.5 Hz. However, the sensitivity analysis to image sequence duration presented in Figure 10 indicated that acquiring data at 2 Hz rather than 1 Hz would allow a similar level of accuracy to be achieved in half as much time by doubling the number of images obtained during a given time period. Overall, our results implied that a standard, nonvideo camera capable of triggering from once every other second to twice per second would have been adequate for the large rivers we examined, which had mean flow velocities on the order of 1.5–2 m/s. The reason such modest frame rates would have been acceptable, even for the relatively high flow velocities we observed, was that the images were acquired from a helicopter hovering at a much greater height above the water than previous ground- or sUAS-based studies. When viewed from this elevated perspective, the displacement of features from frame to frame is less for a given frame rate. Had we collected data from a bank-mounted camera or a sUAS deployed 20 m rather than 200 m above the river, the movement of features between frames would have been greater and a higher frame rate might have been necessary. In any case, the POP framework provides a means of evaluating these trade-offs in the context of a particular field site and imaging system.

Table 5. Guidelines for data collection to support PIV on sediment-laden rivers in Alaska.

Sensor	Standard, nonvideo RGB camera capable of triggering at least once per second
Frame rate	0.5–2 frames per second
Flying height	Sufficient to include both banks in the field of view so that image sequence can be stabilized
Pixel size	Application-driven, but typically greater than 5 cm; 15 cm sufficient for most cases
Hover duration	8–16 s, depending on frame rate (duration can be shorter for higher frame rates)
Acquisition	Trackable features must be present (i.e., flows capable of transporting suspended sediment) Avoid sun glint, but sun cannot be so low as to have insufficient lighting Little or no wind and no rain or snow

Similarly, the selection of an appropriate pixel size for mapping velocity fields should be dictated by the application context. For example, a hydrologist might only be concerned with bulk flow characteristics, such as the mean velocity of a channel cross section, for calculating river discharge. For this purpose, a relatively coarse spatial resolution might be adequate. A fisheries biologist or contaminant transport specialist, in contrast, might be interested in eddies and other small-scale features of the flow field and thus would require an image sequence consisting of smaller pixels to obtain more closely spaced PIV output vectors. The ideal spatial resolution also might vary locally within a given reach of river. For the portion of the Salcha River shown in Figures 5 and 6, the ADCP transect used for accuracy assessment was located in an area of flow expansion where widening of the channel produced eddies on both banks that would only be resolved by an image of sufficient resolution. The more homogeneous flow field farther downstream would be more conducive to discharge calculations and could be characterized via coarser-resolution image data. Computational demands are another factor to consider: storage requirements and processing time will scale with pixel size, frame rate, and image sequence duration.

This work was motivated by a particular application: developing noncontact streamflow measurement techniques suitable for large, remote rivers in Alaska. Due to the state's vast geographic extent, helicopters often are used to transport field personnel to gaging stations for periodic streamflow measurements and maintenance operations. Because helicopters already are deployed to many Alaskan rivers, these platforms could easily be used to acquire hovering image data for PIV to augment or complement conventional, field-based streamgaging. Collecting both field and remotely sensed data simultaneously thus is logistically feasible and would enable POP analyses to be performed for individual streamflow monitoring stations. POP would serve to identify optimal PIV parameters and future monitoring could then be based on remotely sensed data using a site-specific, calibrated PIV algorithm. In addition, water depths recorded by the ADCP (and potentially topographic surveys of the banks and floodplain) would provide the cross-sectional area information needed to calculate discharge; structure from motion photogrammetry could also be used to obtain above-water topography, e.g., [39].

Once the transition is made to noncontact streamflow measurement, the only field activity required would be placing ground control targets for geo-referencing images, but if a GPS/IMU of sufficient accuracy was available, ground control might not be necessary. An efficient strategy might involve a pilot dropping off a hydrographer to make ADCP measurements and place ground control targets and then acquiring hovering image sequences while the field data are being collected; the results summarized in Figure 10 imply that hover durations as short as 8–16 s, depending on frame rate, would be sufficient. Many cameras, including the Zenmuse used in this study, can be controlled by a mobile phone application and thus could be operated easily and safely by the pilot. This approach would not require adding another person to the field crew. The mission could be further simplified and, in principle, performed by a single person if permanent ground control targets can be established at the monitoring station and left in place.

Although remote sensing of flow velocities has clear, demonstrated potential, a number of caveats must also be considered. Most importantly, some kind of visible, trackable features must be present in the river of interest. In this study, we exploited boil vortices in sediment-laden rivers, but this technique will not be applicable to clear-flowing rivers unless some sort of naturally occurring floating material, such as debris, foam, or ice, is available in sufficient abundance to enable PIV or another optical flow algorithm. The timing of image acquisition must be considered during mission planning because at certain sun angles sun glint can degrade image quality and preclude PIV and so can insufficient lighting if flights occur when the sun is too low. In this study, the Salcha River images were acquired around 10:30 local time and the Tanana was imaged beginning at 15:00 local time. Collecting data on overcast, cloudy days would avoid problems due to sun glint. Adverse weather conditions such as rain or snow could degrade image quality and compromise PIV results. Similarly, wind could bias PIV-derived estimates of surface flow velocity by introducing an extraneous, direction-dependent component of motion not driven by the flow of water within the channel. Wind would also make acquisition of stable hovering image sequences more difficult, particularly if lightweight sUAS were used rather than a helicopter. If the water surface is too rough, with rapids or standing waves, PIV might not be effective and this approach is best-suited to lower-gradient rivers with relatively smooth water surfaces. In addition, note that this approach assumes steady flow conditions and is not applicable to rapidly varying flows as might occur during a flash flood or glacial outburst. Site selection for a hybrid field- and remote sensing-based streamflow monitoring station could be directed by the same criteria used to establish conventional streamgages: a straight, uniform channel free of obstructions and backwater effects. The presence of exposed bar surfaces where ground control targets can be easily placed in the field and identified in the images is desirable. Ideally, the flying height and sensor field of view would be sufficient to provide coverage of at least one but preferably both banks to enable image stabilization and geo-referencing.

In addition to these environmental factors, several more practical concerns might constrain the application of remote sensing to velocity mapping. For example, helicopters are more expensive to operate than fixed wing aircraft or sUAS. On board data storage during a flight can be a limiting factor for long deployments, small pixel sizes, high frame rates, and/or long-duration image sequences. The workflow outlined herein involves a number of steps and can be computationally intensive and, for the time being, would most likely have to be implemented in a postprocessing mode, back in the office rather than in real-time in the field. Because the POP framework is essentially an empirical method, calibration of the PIV algorithm ideally would occur on a case-by-case basis for each combination of site and sensor. Moreover, because the foundation of this approach is an assessment of the accuracy of PIV-derived velocity estimates relative to in situ field measurements, having a sufficient sample size that spans a broad range of velocities is critical. In the context of discharge calculations, another important issue is how to convert PIV-derived, presumably surface flow velocities to depth-averaged velocities. Although a velocity index of 0.85 is a common default value for making this conversion, based on an assumed logarithmic velocity profile, this value is by no means universally applicable and ideally would be tuned locally using ADCP data [5,34]. For example, in this study the negative mean bias values on the order of 10%–15% observed on the Salcha (Table 3) indicated that the PIV-derived velocity estimates were larger than the ADCP-based measurements of the depth-averaged velocity, implying that a velocity index of approximately 0.85 might be appropriate for this site. On the Tanana River, however, the mean bias was positive and smaller in absolute value (3.1%, Table 4), indicating that the PIV-derived velocities tended to underestimate the depth-averaged velocity by a small amount and implying a velocity index close to or slightly greater than 1 for this channel.

4.3. Future Research Directions

This initial study points toward several potentially fruitful avenues for additional research. The underlying physics of the sediment boil vortices exploited in this study merit further investigation, building upon the work of Chickadel et al. [16]. Given the encouraging results reported herein, an exciting prospect enabled by this approach is scaling up remote sensing of flow velocities to longer reaches by collecting multiple, overlapping hovers at a series of waypoints along the river. Whereas a basic discharge measurement would only require a single hover, a multistation deployment would yield continuous, detailed, spatially distributed information on velocity patterns. This type of data could help to support contaminant mapping efforts and reach-scale habitat assessments spanning broad geographic extents. In addition, spatially distributed velocity fields, in combination with knowledge of water surface slope, could be used to invert the governing equations of fluid flow to solve for water depth. This approach thus could provide bathymetric data for turbid rivers that are not conducive to passive optical depth retrieval or green lidar.

Other extensions include assessing the potential to perform PIV based on satellite image data, provided sufficiently high spatial resolutions, frame rates, and image sequence durations can be achieved from space. A more practical alternative in the short term might involve collecting data from manned fixed wing aircraft, which would be less expensive and have a greater range than helicopters. The inability to hover could be overcome by using an advanced gimbal system with geo-location and pointing capabilities to obtain longer dwell times. This objective also could be achieved with a motorized mount that could sweep backward as the plane flies forward. Even if only small portions of successive images in a sequence overlap one another, the area of common coverage might be sufficient to extract a cross section. Our results (Figure 10) imply that as long as the same area of the river is captured in approximately 16 images over a duration of 8 s, accurate velocity estimates could be inferred. Thus, while collecting data from manned aircraft might not yield continuous maps, obtaining a series of cross sections could be feasible.

Enabling more widespread application of this approach will require progress in several other areas. Developing flexible, user-friendly software for executing the workflow outlined herein would help to grow a larger community of practitioners. Similarly, working toward a real-time capability in which PIV-derived velocity vectors could be computed and visualized by a sensor operator during flight would facilitate data collection and enable more rigorous quality control while on site. A final research objective is to provide greater process-based insight regarding the applicability and limitations of remote sensing techniques by collecting data from different sizes and types of rivers. The results of these field tests could be used to relate sensor characteristics and PIV parameters to river attributes in a nondimensional framework. The ultimate goal of this effort would be to provide robust, general guidelines so that a POP analysis would not have to be repeated at each new site. Appropriate values of the IA and frame rate could instead be based on the dimensions of the channel and an estimate of the mean flow velocity.

5. Conclusions

Alaska's vast geographic extent encompasses many large, remote rivers that can be accessed only by helicopter. To augment conventional, field-based streamgaging within the state, which can be expensive, difficult, and potentially dangerous, we evaluated the feasibility of mapping flow velocities in sediment-laden rivers using image sequences acquired from a helicopter. This type of platform can be deployed from a sufficient flying height to include the entire width of the river and provide a view of both banks, which is critical for image stabilization and geo-referencing. Moreover, the ability of a helicopter to hover above the channel facilitates acquisition of image sequences of sufficient duration for reliable velocity estimation. This approach thus can contribute to a noncontact discharge measurement program, which would also require an independent source of information on channel cross-sectional area. In addition, the ability to collect data while hovering at multiple locations along a river opens up the possibility of producing continuous, reach-scale mapping of the flow field. This type of spatially distributed velocity information would capture both bulk flow characteristics and smaller features like eddies to inform studies of dispersion, contaminant transport, and aquatic habitat.

In addition to the larger size of the rivers we considered, another important distinction between this study and most previous research on remote sensing of flow velocities was our use of a natural tracer for PIV rather than artificial, introduced particles. Moreover, instead of assuming the presence of foam or floating debris, we exploited sediment boil vortices that upwell suspended material from within the water column to produce visible, mobile features at the water surface. This approach does not require seeding the flow and thus provides a more practical solution for large, inaccessible rivers. Additional contributions of this study include an evaluation of different sensor configurations, a general workflow for performing PIV analyses, a framework for optimizing PIV input parameters, and an assessment of the sensitivity of PIV output to image sequence duration. The results reported herein support the following conclusions.

1. We evaluated the utility of an aerial mapping camera and a relatively inexpensive video camera that had different requirements in terms of ancillary data and supporting software. Although the mapping camera was capable of providing higher spatial resolution (i.e., smaller image pixels), we found that resampling the data to a coarser pixel size was critical for accurate velocity estimation. Similarly, the high frame rate (30 Hz) of the video camera was excessive; reliable velocity fields were inferred from helicopter-based image sequences with frame rates as low as 0.5 Hz in rivers with mean flow velocities on the order of 1.5–2 m/s. These results imply that relatively basic imaging systems could be used to map velocity patterns in large, sediment-laden rivers.
2. The general, modular workflow we introduced included an image preparation phase that involved stabilization and geo-referencing, an image processing phase that enhanced the data to facilitate detection and tracking of water surface features such as sediment boil vortices, a PIV analysis phase based on a robust ensemble correlation algorithm, and a postprocessing phase in which the PIV output is scaled, geo-referenced, and compared to field-based velocity measurements for accuracy assessment. Comparison of velocity fields derived from raw and preprocessed images highlighted the importance of image enhancement. This workflow was applied to data acquired from both the digital mapping camera and the video camera and can readily accommodate data from other sensors and platforms.
3. Whereas prior applications of PIV have tended to rely on ad hoc, heuristic guidelines, we developed a more systematic, rigorous approach to Parameter Optimization for PIV (POP). The POP framework identifies the combination of two key PIV input parameters, the size of the interrogation area (IA) within which features are tracked and the frame rate at which images are acquired, that yields the strongest agreement between remotely sensed velocities and those measured in the field. Applying this method to video from a 200-m wide river indicated that the ensemble correlation PIV algorithm was robust, with observed versus predicted regression $R^2 > 0.9$ across a broad range of IAs (64–320 pixels) and frame rates (0.5–2 Hz). Poor accuracies occurred only when the pixel size was smaller than the features to be tracked or when the frame rate was too fast to detect the movement of these features between successive images. These results imply that even relatively large IAs and modest frame rates can yield reliable velocity information.
4. Analysis of the sensitivity of PIV-derived velocities to image sequence duration indicated that at a frame rate of 1 Hz dwell times as short as 16 s would provide accurate velocity fields, with

negligible improvement for longer durations. If the frame rate were doubled to 2 Hz such that the same number of images could be obtained in half as much time, an 8 s duration was sufficient to provide an R^2 of 0.93. These results suggest that the number of images, in addition to the time period over which they are acquired, is an important factor and provide further evidence that the ensemble correlation PIV algorithm is highly robust. Although this conclusion was based on a particular sensor and set of input parameters for a specific field site, our results can be used to guide mission planning for subsequent helicopter-based data collection and to inform future efforts to map velocities from fixed wing aircraft and potentially satellite platforms.

5. In summary, for the two rivers we considered and the flying heights from which we acquired image data, the spatial resolution of the mapping camera and the frame rate of the video camera we used were excessive. Resampling the data to larger pixels sizes and lower frame rates led to more accurate PIV output. Similarly, sensitivity analysis indicated that the duration of the image sequences we obtained was longer than necessary. All three of these factors—larger pixels, lower frame rates, and shorter durations—would contribute to reduced data storage and computational requirements and favor more widespread application of remote sensing of flow velocities as a viable operational tool.

Author Contributions: Conceptualization, C.J.L. and P.J.K.; methodology, C.J.L. and P.J.K.; software, C.J.L. and P.J.K.; validation, C.J.L. and P.J.K.; formal analysis, C.J.L. and P.J.K.; investigation, C.J.L. and P.J.K.; resources, C.J.L. and P.J.K.; data curation, C.J.L. and P.J.K.; writing–original draft preparation, C.J.L. and P.J.K.; writing–review and editing, C.J.L. and P.J.K.; visualization, C.J.L. and P.J.K.; supervision, C.J.L. and P.J.K.; project administration, C.J.L. and P.J.K.; funding acquisition, C.J.L. and P.J.K. All authors have read and agreed to the published version of the manuscript.

Funding: Funding for this study was provided by the U.S. Geological Survey Water Mission Area.

Acknowledgments: Jon Nelson of the USGS Geomorphology and Sediment Transport Laboratory and several people from the USGS Alaska Science Center contributed to this work, most notably but not limited to Jeff Conaway, Karenth Dworsky, and Heather Best. Adam LeWinter, Peter Gadomski, Dominic Filiano, and Carl Green of the U.S. Army Corps of Engineers Cold Regions Research and Engineering Laboratory assisted helicopter- and field-based data collection; LeWinter designed the Zenmuse camera mount. Jack Eggleston of the USGS Hydrologic Remote Sensing Branch provided logistical and financial support. Any use of trade, firm, or product names is for descriptive purposes only and does not imply endorsement by the U.S. Government.

Conflicts of Interest: The authors declare no conflict of interest.

References

1. Conaway, J.; Eggleston, J.; Legleiter, C.; Jones, J.; Kinzel, P.; Fulton, J. Remote sensing of river flow in Alaska—New technology to improve safety and expand coverage of USGS streamgaging. *U.S. Geol. Surv. Fact Sheet* **2019**, *3024*. [CrossRef]

2. Durand, M.; Gleason, C.J.; Garambois, P.A.; Bjerklie, D.; Smith, L.C.; Roux, H.; Rodriguez, E.; Bates, P.D.; Pavelsky, T.M.; Monnier, J.; et al. An intercomparison of remote sensing river discharge estimation algorithms from measurements of river height, width, and slope. *Water Resour. Res.* **2016**, *52*, 4527–4549. [CrossRef]

3. Bjerklie, D.M.; Birkett, C.M.; Jones, J.W.; Carabajal, C.; Rover, J.A.; Fulton, J.W.; Garambois, P.A. Satellite remote sensing estimation of river discharge: Application to the Yukon River, Alaska. *J. Hydrol.* **2018**, *561*, 1000–1018. [CrossRef]

4. Altenau, E.H.; Pavelsky, T.M.; Moller, D.; Pitcher, L.H.; Bates, P.D.; Durand, M.T.; Smith, L.C. Temporal variations in river water surface elevation and slope captured by AirSWOT. *Remote Sens. Environ.* **2019**, *224*, 304–316. [CrossRef]

5. Legleiter, C.J.; Kinzel, P.J.; Nelson, J.M. Remote measurement of river discharge using thermal particle image velocimetry (PIV) and various sources of bathymetric information. *J. Hydrol.* **2017**, *554*, 490–506. [CrossRef]

6. Kinzel, P.; Legleiter, C.; Nelson, J.; Conaway, J.; LeWinter, A.; Gadomski, P.; Filiano, D. Near-field remote sensing of Alaska Rivers. In Proceedings of the Federal Interagency Sedimentation and Hydrologic Modeling Conference (SEDHYD), Reno, NV, USA, 24–28 June 2019.

7. Tauro, F.; Porfiri, M.; Grimaldi, S. Surface flow measurements from drones. *J. Hydrol.* **2016**, *540*, 240–245. [CrossRef]

8. Detert, M.; Johnson, E.D.; Weitbrecht, V. Proof-of-concept for low-cost and non-contact synoptic airborne river flow measurements. *Int. J. Remote Sens.* **2017**, *38*, 2780–2807. [CrossRef]

9. Tauro, F.; Tosi, F.; Mattoccia, S.; Toth, E.; Piscopia, R.; Grimaldi, S.; Tauro, F.; Tosi, F.; Mattoccia, S.; Toth, E.; et al. Optical Tracking Velocimetry (OTV): Leveraging Optical Flow and Trajectory-Based Filtering for Surface Streamflow Observations. *Remote Sens.* **2018**, *10*, 2010. [CrossRef]

10. Pearce, S.; Ljubičić, R.; Peña-Haro, S.; Perks, M.; Tauro, F.; Pizarro, A.; Dal Sasso, S.; Strelnikova, D.; Grimaldi, S.; Maddock, I.; et al. An Evaluation of Image Velocimetry Techniques under Low Flow Conditions and High Seeding Densities Using Unmanned Aerial Systems. *Remote Sens.* **2020**, *12*, 232. [CrossRef]

11. Strelnikova, D.; Paulus, G.; Käfer, S.; Anders, K.H.; Mayr, P.; Mader, H.; Scherling, U.; Schneeberger, R. Drone-Based Optical Measurements of Heterogeneous Surface Velocity Fields around Fish Passages at Hydropower Dams. *Remote Sens.* **2020**, *12*, 384. [CrossRef]

12. Dugan, J.P.; Anderson, S.P.; Piotrowski, C.C.; Zuckerman, S.B. Airborne Infrared Remote Sensing of Riverine Currents. *IEEE Trans. Geosci. Remote Sens.* **2014**, *52*, 3895–3907. [CrossRef]

13. Kinzel, P.; Legleiter, C. sUAS-Based Remote Sensing of River Discharge Using Thermal Particle Image Velocimetry and Bathymetric Lidar. *Remote Sens.* **2019**, *11*, 2317. [CrossRef]

14. Fujita, I.; Komura, S. Application of Video Image Analysis for Measurements of River-Surface Flows. *Proc. Hydraul. Eng. JSCE* **1994**, *38*, 733–738. [CrossRef]

15. Fujita, I.; Hino, T. Unseeded and Seeded PIV Measurements of River Flows Videotaped from a Helicopter. *J. Vis.* **2003**, *6*, 245–252. [CrossRef]

16. Chickadel, C.C.; Horner-Devine, A.R.; Talke, S.A.; Jessup, A.T.C.L. Vertical boil propagation from a submerged estuarine sill. *Geophys. Res. Lett.* **2009**, *36*, L10601. [CrossRef]

17. Wada, T.; Chikita, K.A.; Kim, Y.; Kudo, I. Glacial Effects on Discharge and Sediment Load in the Subarctic Tanana River Basin, Alaska. *Arct. Antarct. Alp. Res.* **2011**, *43*, 632–648. [CrossRef]

18. Park, I.; Seo, I.W.; Kim, Y.D.; Han, E.J. Turbulent Mixing of Floating Pollutants at the Surface of the River. *J. Hydraul. Eng.* **2017**, *143*. [CrossRef]

19. Nelson, J.; McDonald, R.; Legleiter, C.; Kinzel, P.; Terrell-Ramos, T.; Higashi, Y.; Seo, I.W.; Baek, D.; Lee, D.H.; Ryu, Y. New methods for predicting and measuring dispersion in rivers. *E3S Web Conf.* **2018**, *40*, 05052. [CrossRef]

20. Erwin, S.O.; Bulliner, E.A.; Fischenich, J.C.; Jacobson, R.B.; Braaten, P.J.; DeLonay, A.J. Evaluating flow management as a strategy to recover an endangered sturgeon species in the Upper Missouri River, USA. *River Res. Appl.* **2018**, *34*, 1254–1266. [CrossRef]

21. Angermeier, P.L.; Schlosser, I.J. Species-Area Relationships for Stream Fishes. *Ecology* **1989**, *70*, 1450–1462. [CrossRef]

22. Fausch, K.D.; Torgersen, C.E.; Baxter, C.V.; Li, H.W. Landscapes to riverscapes: Bridging the gap between research and conservation of stream fishes. *Bioscience* **2002**, *52*, 483–498. [CrossRef]

23. Koutalakis, P.; Tzoraki, O.; Zaimes, G. UAVs for Hydrologic Scopes: Application of a Low-Cost UAV to Estimate Surface Water Velocity by Using Three Different Image-Based Methods. *Drones* **2019**, *3*, 14. [CrossRef]

24. Thielicke, W.; Stamhuis, E.J. PIVlab—Towards User-friendly, Affordable and Accurate Digital Particle Image Velocimetry in MATLAB. *J. Open Res. Softw.* **2014**, *2*, e30. [CrossRef]

25. Thielicke, W. The Flapping Flight of Birds—Analysis and Application. Ph.D. Thesis, Rijksuniversiteit Groningen, Groningen, The Netherlands, 2014.

26. Thielicke, W.; Stamhuis, E.J. PIVlab—Time-Resolved Digital Particle Image Velocimetry Tool for MATLAB. *Figshare Dataset* **2019**, *12*. Available online: https://figshare.com/articles/PIVlab_version_1_35/1092508 (accessed on 17 April 2020).

27. Legleiter, C.J.; Kinzel, P.J. Field measurements of flow velocity and optical image sequences acquired from the Salcha and Tanana Rivers in Alaska in 2018 and 2019 and used for used for particle image velocimetry (PIV). *U.S. Geol. Surv. Data Release* **2020**. [CrossRef]

28. Hasselblad A6D 100c. Available online: http://static.hasselblad.com/2017/05/A6D-User-Manual-v1.0.3.pdf988 (accessed on 19 February 2020).

29. SimActive Correlator3D. Available online: https://www.simactive.com/correlator3d-mapping-software-features.html (accessed on 19 February 2020).

30. DJI Zenmuse X5. Available online: https://www.dji.com/zenmuse-x5 (accessed on 19 February 2020).

31. TeledyneMarine RiverRayADCP. Available online: http://www.teledynemarine.com/riverray-adcp?ProductLineID=13 (accessed on 19 February 2020).

32. HemisphereA101. Available online: https://www.hemispheregnss.com/wp-content/uploads/2019/01/hemispheregnss_a101 (accessed on 20 February 2020).

33. Parsons, D.R.; Jackson, P.R.; Czuba, J.A.; Engel, F.L.; Rhoads, B.L.; Oberg, K.A.; Best, J.L.; Mueller, D.S.; Johnson, K.K.; Riley, J.D. Velocity Mapping Toolbox (VMT): A processing and visualization suite for moving-vessel ADCP measurements. *Earth Surf. Process. Landf.* **2013**, *38*, 1244–1260. [CrossRef]

34. Mueller, D.S. extrap: Software to assist the selection of extrapolation methods for moving-boat ADCP streamflow measurements. *Comput. Geosci.* **2013**, *54*, 211–218. [CrossRef]

35. FIJI-ImageJ. Available online: https://imagej.net/Fiji (accessed on 18 February 2020).

36. Cardona, A.; Saalfeld, S.; Schindelin, J.; Arganda-Carreras, I.; Preibisch, S.; Longair, M.; Tomancak, P.; Hartenstein, V.; Douglas, R.J. TrakEM2 Software for Neural Circuit Reconstruction. *PLoS ONE* **2012**, *7*, e38011. [CrossRef]

37. Pineiro, G.; Perelman, S.; Guerschman, J.P.; Paruelo, J.M. How to evaluate models: Observed vs. predicted or predicted vs. observed? *Ecol. Model.* **2008**, *216*, 316–322. [CrossRef]

38. GlobalMapperAll-in-one GIS Software. Available online: https://www.bluemarblegeo.com/products/global-mapper.php (accessed on 26 February 2020).

39. King, T.V.; Neilson, B.T.; Rasmussen, M.T. Estimating Discharge in Low-Order Rivers With High-Resolution Aerial Imagery. *Water Resour. Res.* **2018**, *54*, 863–878. [CrossRef]

 *remote sensing*

Article

Concept and Performance Evaluation of a Novel UAV-Borne Topo-Bathymetric LiDAR Sensor

Gottfried Mandlburger [1,2,*], **Martin Pfennigbauer** [3], **Roland Schwarz** [3], **Sebastian Flöry** [1] and **Lukas Nussbaumer** [1]

1. Department of Geodesy and Geoinformation, TU Wien, Wiedner Hauptstr. 8-10, 1040 Vienna, Austria; sebastian.floery@geo.tuwien.ac.at (S.F.); lukas.nussbaumer@student.tuwien.ac.at (L.N.)
2. Institute for Photogrammetry, University of Stuttgart, Geschwister-Scholl-Str. 24D, 70174 Stuttgart, Germany
3. Riegl Research Forschungs Gesellschaft mbH, Riedenburgstr. 48, 3580 Horn, Austria; martin.pfennigbauer@riegl.co.at (M.P.); roland.schwarz@riegl.co.at (R.S.)
*   Correspondence: gottfried.mandlburger@geo.tuwien.ac.at; Tel.: +43-1-58801-12235

Received: 15 February 2020; Accepted: 13 March 2020; Published: 19 March 2020

Abstract: We present the sensor concept and first performance and accuracy assessment results of a novel lightweight topo-bathymetric laser scanner designed for integration on Unmanned Aerial Vehicles (UAVs), light aircraft, and helicopters. The instrument is particularly well suited for capturing river bathymetry in high spatial resolution as a consequence of (i) the low nominal flying altitude of 50–150 m above ground level resulting in a laser footprint diameter on the ground of typically 10–30 cm and (ii) the high pulse repetition rate of up to 200 kHz yielding a point density on the ground of approximately 20–50 points/m^2. The instrument features online waveform processing and additionally stores the full waveform within the entire range gate for waveform analysis in post-processing. The sensor was tested in a real-world environment by acquiring data from two freshwater ponds and a 500 m section of the pre-Alpine Pielach River (Lower Austria). The captured underwater points featured a maximum penetration of two times the Secchi depth. On dry land, the 3D point clouds exhibited (i) a measurement noise in the range of 1–3 mm; (ii) a fitting precision of redundantly captured flight strips of 1 cm; and (iii) an absolute accuracy of 2–3 cm compared to terrestrially surveyed checkerboard targets. A comparison of the refraction corrected LiDAR point cloud with independent underwater checkpoints exhibited a maximum deviation of 7.8 cm and revealed a systematic depth-dependent error when using a refraction coefficient of $n = 1.36$ for time-of-flight correction. The bias is attributed to multi-path effects in the turbid water column (Secchi depth: 1.1 m) caused by forward scattering of the laser signal at suspended particles. Due to the high spatial resolution, good depth performance, and accuracy, the sensor shows a high potential for applications in hydrology, fluvial morphology, and hydraulic engineering, including flood simulation, sediment transport modeling, and habitat mapping.

Keywords: UAV LiDAR; airborne laser bathymetry; full waveform processing; performance assessment; high resolution hydro-mapping

---

1. Introduction

The introduction of Unmanned Aerial Vehicles (UAV), also referred to as Unmanned Aerial Systems (UAS), as platforms for the acquisition of high-resolution 3D data has fundamentally changed the mapping sector in recent years. UAVs flying at low altitudes combine the advantages of small sensor-to-object distances (i.e., close range) and agility. The latter is achieved via remotely controlling the unmanned sensor platform either manually by a pilot within the Visual Line-Of-Sight (VLOS) or by autonomously flying predefined routes based on waypoints, allowing both within and beyond VLOS operation.

UAV-based 3D data acquisition was first accomplished using light-weight camera systems, where advancements in digital photogrammetry and computer vision-enabled automatic data processing workflows for the derivation of dense 3D point clouds based on Structure-from-Motion (SfM) and Dense Image Matching (DIM). Due to advancements in UAV-platform technology and ongoing sensor miniaturization, today compact LiDAR sensors are increasingly integrated on both multi-copter and fixed-wing UAVs, enabling 3D mapping with unprecedented spatial resolution and accuracy. The tackled applications include topographic mapping, geomorphology, infrastructure inspection, environmental monitoring, forestry, and precision farming. While UAV-borne laser scanning (ULS) can already be considered state-of-the-art for mapping tasks above the water table, UAV-based bathymetric LiDAR still lacks behind, mainly due to payload restrictions.

The established techniques for mapping bathymetry are single- or multi-beam echo sounding (SBES/MBES), including Acoustic Doppler Current Profiling (ADCP). This holds for both coastal and inland water mapping. MBES is the prime acquisition method for area-based charting of water bodies deeper than approximately two times the Secchi depth (SD), and the productivity of this technology is best in deeper water because the width of the scanned swath increases with water depth. For mapping shallow water bathymetry including shallow running waters, boat-based echo sounding is less productive and furthermore, safety issues arise. For relatively clear and shallow water bodies, airborne laser bathymetry (ALB) constitutes a well-proven alternative. This active, optical, polar measurement technique features the following advantages: (i) areal coverage only depending on the flying height but not on the water depth; (ii) homogeneous point density; (iii) simultaneous mapping of water bottom, water surface, and dry land area in a single campaign, referred to as topo-bathymetric LiDAR; and (iv) non-contact remote sensing approach with clear benefits for mapping nature conservation areas.

ALB performed from manned platforms [1] has a long history in under-water object detection and charting of coastal areas including characterization of benthic habitats [2]. In the recent past, the technology is increasingly used for inland waters [3–5]. In order to ensure eye safety, the employed laser beam divergence is typically in the range of 1–10 mrad. For a typical flying altitude of 500 m or higher the resulting laser footprint diameter on the ground measures at least 50–500 cm. Thus, the spatial resolution of ALB from manned aerial platforms is limited. Due to the shorter measurement ranges, ULS promises a better resolution and allows users to capture and reconstruct smaller structures like boulders, submerged driftwood, and the like, with potential applications for the characterization of flow resistance in hydrodynamic numerical (HN) models.

To date, only a few UAV-borne LiDAR sensors are available. The *RIEGL* BathyDepthFinder is a bathymetric laser range finder with a fixed laser beam axis (i.e., no scanning). Like most of the bathymetric laser sensors, the system utilizes a green laser ($\lambda = 532$ nm) and allows for the capture of river cross-sections constituting an appropriate input for 1D-HN models. The ASTRALiTe sensor is a scanning polarizing LiDAR [6]. The employed 30 mW laser only allows low flying altitudes and therefore provides limited areal measurement performance and the depth performance is moderate (1.2 -fold SD). The RAMMS (Rapid Airborne Multibeam Mapping System) was recently developed by Fugro Inc. and Areté Associates [7]. It is a relatively lightweight, 14 kg push-broom ALB sensor featuring a measurement rate of 25 kHz and promising a penetration performance of 3 SD according to the company's web site [8]. The sensor is operated by Fugro for delivering high-resolution bathymetric services, but independent performance and accuracy assessments are not available.

In 2018, *RIEGL* introduced the VQ-840-G, a lightweight topo-bathymetric full-waveform scanner [9] with a maximum pulse repetition rate (PRR) of 200 kHz. The scanner is designed for installation on various platforms including UAVs and carries out laser range measurements for high resolution, survey-grade mapping of underwater topography with a narrow, visible green laser beam. The system is optionally offered with an integrated and factory-calibrated IMU/GNSS system, and with an optional camera or IR rangefinder. As such, it constitutes a compact, comprehensive, and full-featured ALB system for mapping coastal and inland waters from low flying altitudes.

The inherent advantage of such a design is twofold: (i) the short measurement ranges and high scan rates enable surface description in high spatial resolution as a consequence of high point density and relatively small laser footprint; and (ii) integration on UAV platforms or lightweight manned aircraft reduces mobilization costs and enables agile data acquisition. In the context of ALB, the prior holds out the prospect of reconstructing submerged surface details below the size of boulder (i.e., dm-level) which is not feasible from flying altitudes of 500 m or higher due to the large laser footprint diameter, and the latter is especially beneficial for capturing medium-sized, meandering running waters with moderate widths (<50 m) and depth (<5 m) which would require a complex flight plan with many potentially short flight strips and a multitude of turning maneuvers for a conventional manned ALB survey. Thus, UAV-borne bathymetric LiDAR is considered a major leap forward for river mapping with an expected positive stimulus on a variety of hydrologic and hydraulic applications like flood risk simulations, sediment transport modeling, monitoring of fluvial geomorphology, eco-hydraulics, and the like.

In this article, we present the first rigorous accuracy and performance assessment of a UAV-borne ALB system by evaluating the 3D point clouds acquired at the pre-Alpine Pielach River and an adjacent freshwater pond. We compare the LiDAR measurements with ground truth data obtained from terrestrial survey with a total station in cm-accuracy and provide objective measures describing the precision, accuracy, and depth performance of the system. Furthermore, we discuss the fields of application, limits, and challenges of UAV-borne LiDAR bathymetry including potential fusion with complementary sensors (RGBI cameras, thermal IR cameras), widening the scope beyond pure bathymetric mapping towards discharge estimation, flow resistance characterization, etc.

Our main contribution is providing an in-depth evaluation of topo-bathymetric UAV-LiDAR including a discussion of limitations on the one side and benefits compared to manned ALB on the other side. Next to accuracy and performance assessment, our focus is on an objective evaluation of achievable spatial resolution, especially addressing the question to which extent the shorter measurement ranges and smaller nominal laser footprints directly induce higher resolution. Furthermore, we put the presented full-waveform scanner in context with existing line scanning technologies.

The remainder of the manuscript is structured as follows. In Section 2, we review the state-of-the-art in UAV-based bathymetry followed by a description of the senor concept in Section 3. Section 4 introduces the study area and existing datasets, and details the employed data processing and assessment methods. The evaluation results are presented in Section 5 along with critical discussion in Section 6. The paper concludes with a summary of the main findings in Section 7.

2. Related Work

In this Section, we provide a summary of related work in the context of UAV-based optical remote sensing of coastal and inland water areas, focusing on the derivation of bathymetry. For the sake of completeness, we first briefly discuss passive image-based techniques, both via multimedia photogrammetry [10] and spectrally based depth estimation [11], but mainly concentrate on bathymetric LiDAR [1] in general and small-footprint topo-bathymetric LiDAR for inland water applications in particular [12,13]. Furthermore, we present recent validation and comparison studies on subject matters as our contribution also focuses on sensor evaluation.

2.1. UAV-Borne Multimedia Photogrammetry

The advent of UAVs as carrier platforms of active and passive mapping sensors had a huge impact in the field of photogrammetry and remote sensing [14]. The introduction of Structure-from-Motion (SfM) and Dense Image Matching (DIM) techniques providing automatic orientation of entire image blocks and height estimates for every image pixel [15,16] has democratized image-based 3D mapping of topography and dramatically increased the achievable point densities [17,18]. The applicability of UAV-photogrammetry is further facilitated due to the existence of easy-to-use software solutions providing an automated processing chain from captured images to digital surface models, 3D meshes, and orthophoto maps, respectively [19–21].

While the application of SfM is directly applicable for the dry part of alluvial and coastal areas [22,23], mapping of underwater topography requires consideration of beam bending at the air-water medium boundary. Ref. [24] describes the basic principles of multimedia photogrammetry and the authors of [25] provide an early evaluation of the technique for mapping a clear and shallow gravel-bed river. More recently, Ref. [26,27] focused on fluvial and aquatic applications of SfM based on UAV imagery emphasising the unprecedented spatio-temporal coverage [27] and the necessity of automated procedures for refraction correction [26]. The latter is crucial for obtaining accurate depth measurements but is difficult to achieve because image-based reconstruction of natural water surfaces poses problems due to the general prerequisite of transparent water conditions. In this context, Refs. [28–30] propose a deep-learning-based framework to automatically correct refraction effects in SfM and multi-view stereo processing pipelines.

As for every optical method aiming at mapping bathymetry, clear water is a precondition for the successful application of multimedia photogrammetry. In addition, Ref. [31] discusses further prerequisites for exploiting the potentially high spatial resolution of mapping shallow water bathymetry via through-water dense image matching. It is concluded that sufficient bottom texture and favorable calm water surfaces are prerequisites for achieving accurate results, as image matching is heavily distorted in wavy surface conditions and spatial aggregation (i.e., low pass filtering) is necessary to suppress high-frequency noise. This, however, limits the spatial resolution to the 1m-level, even for UAV-imagery featuring a ground sampling distance (GSD) in the cm-range.

2.2. UAV-Borne Spectrally Based Depth Estimation

The foundations of mapping bathymetry from multi-spectral images are described in [11,32–35]. The main application of this technique, which uses either physical models or regression techniques to relate the spectral information of one or more color channels to water depth, is Satellite-Derived Bathymetry (SDB) based on multispectral remote sensing sensors like Landsat 8, WorldView-2, etc. [36,37]. While the resolution of SDB bathymetry is inherently limited by the GSD of satellite images, high-resolution applications of this technique based on hyperspectral UAV imagery was recently reported by [38]. The authors demonstrate the potential for bathymetric mapping at cm-resolution and approximately dm-accuracy. However, transient water surfaces negatively affect the achievable accuracy as discussed in Section 2.1.

2.3. UAV-Borne Bathymetric LiDAR

Scanning bathymetric LiDAR from manned airborne platforms, commonly referred to as Airborne Laser Bathymetry (ALB), suffers from low spatial resolution. In order to ensure eye-safe operation, the employed visible green laser beams are generally broader compared to lasers using infrared laser radiation for mapping topography. This applies to both so-called deep bathy sensors [1] with a typical beam divergence of 5 mrad as well as for relatively small-footprint topo-bathymetric laser scanners [3,12,39] employing a beam divergence of around 1 mrad. For the latter, the footprint diameter at a typical flying altitude of 600 m still features a size of 60 cm, thus limiting the achievable planimetric resolution to the m-level. One of the main advantages of UAV-borne LiDAR bathymetry is, therefore, the potentially higher planimetric resolution. In the following, we review available sensors that are lightweight enough to be carried by multi-copter and even fixed-wing UAV platforms. Most of the instruments were just recently introduced, thus only a few scientific articles are published yet. This is the reason why our review is mainly based on company websites, sensor spec sheets, and recent publications in technical journals reporting about emerging trends [40].

- *ASTRALite edge™*: This sensor features a weight of 5 kg and a PRR of 20 kH [41]. The manufacturer states a precision/accuracy of 5/10 mm and a depth penetration of >1.5 SD at a typical flying altitude for mapping bathymetry of 20 m. To overcome the known problem of separating signals from the surface and the bottom beneath, signal processing is based on a technique known as INtrapulse Phase Modification Induced by Scattering (INPHAMIS) considering the polarization of the returned laser pulses [6,42]. Show-cases of this sensor, which can be mounted on standard multi-copter UAV platforms like the DJI Matrice 600, are reported in [43–45]. The authors of [46] use the instrument for UAV-based discharge estimation requiring flow velocity but also precise river bed geometry data. A current drawback of the sensor is its relatively low flying altitude resulting in a poor areal coverage performance.
- *Fugro RAMMS*: Fugro's topo-bathymetric UAV laser scanner avoids moving parts by employing a pushbroom scanning technique [47]. RAMMS (Rapid Airborne Multibeam Mapping System) is a successor of PILLS (Pushbroom Imaging LIDAR for Littoral Surveillance) developed by Areté Associates [48], which was first introduced in 2012. Instead of single laser pulses deflected by a rotating or oscillating mirror, a laser line is emitted and the backscattered signal is captured by a camera receiver. The compact sensor (weight: 14 kg) is designed for integration in small aircraft or larger UAVs, delivers 25K range observations per second, and achieves 3 SD penetration [8,49]. Successful completion of large-area bathymetric mapping projects (area: 10K km^2) carried out in very clear coastal waters are reported in [7]. The survey was conducted from a flying altitude of 325 m resulting in an across-track point spacing of 33 cm. The mission parameters are optimized for large area coverage, but similar point densities can also be achieved with standard topo-bathymetric sensors [4]. Higher spatial resolution is generally feasible by integrating the sensor on UAV platforms [47], but to the best of our knowledge, respective reports have not been published so far.
- *TDOT GREEN*: The Japanese company Amuse Oneself Inc. developed a drone-mounted compact green laser scanner. The general concept and the sensor specifications are presented at the product's website together with exemplary uses cases for mapping coastal and river environments [50]. The scanner operates with a PRR of 60 kHz, a scan speed of 30 scans/s, and provides up to 4 echoes per laser pulse. The large total Field of Fov (FoV) of 90° is rather uncommon for bathymetric scanners as nominal laser incidence angles of 15–20° are optimal for mapping bathymetry [1]. The laser output is automatically controlled depending on the flying altitude to ensure eye-safe operation (<40 m: Class 1; >40 m: Class 3R). The laser beam divergence of 0.3 mrad results in a very small footprint diameter on the ground (e.g., 3 cm at a flying altitude of 100 m above ground level). However, no specific information is available on (i) the exact definition of the reported beam divergence and (ii) the actual laser power. [50] also reports data for validating sensor performance. Within these internal tests, the system achieved a maximum penetration of 9 m in clear seawater conditions and around 2 m in a more turbid river environment. However, no information on the actual turbidity is available and independent performance and accuracy evaluations are missing.

2.4. Comparison and Accuracy Evaluation Studies

As our contribution focuses on accuracy evaluation and performance assessment, we briefly list existing comparison studies related to optical remote sensing for mapping bathymetry in general, and ALB in particular. However, a comprehensive review is beyond the scope of this paper.

The authors of [51] provide a comparison of remote sensing methods for mapping bathymetry of shallow, clear water rivers. Their work mainly focuses on echo sounding and SfM-based techniques. Through-water photogrammetry and spectral depth approaches for water depth extraction based on UAV imagery are addressed in [52]. The authors state that a 1.5 cm DEM is achievable via the photogrammetric approach yielding an unbiased depth estimation with a standard deviation of 17 cm compared to RTK GNSS reference measurements, outperforming the spectral approach w.r.t. accuracy in their experiment at a shallow gravel-bed river.

In the area of bathymetric LiDAR, [53] introduces a comprehensive approach for the derivation of Total Vertical Uncertainty depending on flight mission and environmental parameters. In contrast, [54] summarize empirical quality control measures for the evaluation of ALB data and emphasize the necessity of collecting independent reference measurements for both bottom geometry (GNSS and/or Sonar) and water transparency. Actual geometric accuracy assessments of bathymetric LiDAR data captured from manned airborne platforms compared to independent reference data are reported in [4,55–60]. These studies discuss systematic biases as well as random errors. Depending on the used sensors (deep bathy, topo-bathy) and study areas (water depth, turbidity) the reported errors are in the cm- to dm-range compared to reference data obtained from sonar or GNSS. The systematic depth biases reported in [55,56,59,60] are of special interest as they indicate insufficient compensation of refraction-induced effects. This topic will further be discussed in Section 6.

In the context of UAV-borne LiDAR hydro-mapping, [61] evaluated a bathymetric laser range finder. The instrument features a constant laser beam axis and provides section data with an along-track point spacing of around 1 cm. The authors report a bias of 4 cm and a standard deviation of 3 cm comprising all involved error sources (sensor position and attitude, ranging, lever arms, boresight angles, etc.) compared to GNSS reference measurements of the gravel-bed river bottom. To the best of our knowledge, no accuracy assessments have been published for UAV-borne bathymetric laser scanners, which is the main focus of our contribution.

3. Sensor Concept

The RIEGL VQ-840-G is a fully integrated compact airborne laser scanner for combined topographic and bathymetric surveying. The instrument can be equipped with an integrated and factory-calibrated IMU/GNSS system and with an integrated industrial camera, thereby implementing a full airborne laser scanning system. The VQ-840-G LiDAR has a compact volume of 20.52 L with a weight of 12 kg and, thus, can be installed on various platforms including UAVs.

The laser scanner comprises a frequency-doubled IR laser, emitting pulses with about 1.5 ns pulse duration at a wavelength of 532 nm and at a PRR of 50–200 kHz. At the receiver side, the incoming optical echo signals are converted into an electrical signal, they are amplified and digitized at a digitization rate of close to 2G samples/s. The laser beam divergence can be selected between 1 mrad and 6 mrad in order to be able to maintain a constant energy density on the ground for different flying altitudes and therefore balancing eye-safe operation with spatial resolution. The receiver's iFOV (instantaneous Field of View) can be chosen between 3 mrad and 18 mrad. For topographic measurements and very clear or shallow water, a smaller setting is suitable while for turbid water it is better to increase the receiver's iFOV in order to collect a larger amount of light scattered by the water body.

The VQ-840-G employs a Palmer scanner generating an elliptical scan pattern on the ground. The scan range is ±20° across and ±14° along the intended flight direction and consequently, the laser beam hits the water surface at an incidence angle with low variation. The scan speed can be set between 10–100 lines/s to generate an even point distribution in the center of the swath. Towards the edge of the swath, where the consecutive lines overlap, an extremely high point density with strongly overlapping footprints is produced.

The onboard distance measurement is based on time-of-flight measurement through online waveform processing of the digitized echo signal [62]. A real-time detection algorithm identifies potential targets within the stream of the digitized waveform and feeds the corresponding sampling values to the signal processing unit, which is capable of performing system response fitting (SRF) in real-time at a rate of up to 2.5M targets per second. These targets are represented by the basic attributes of range, amplitude, reflectance, and pulse shape deviation and are saved to the storage device, which can be the internal SSD a removable CFast© card, or an external data recorder via an optical data transmission cable.

Besides being fed to target detection and online waveform processing, the digitized echo waveforms can also be stored on disc for subsequent off-line full waveform analysis. For every laser shot, sample blocks with a length of up to 75 m are stored unconditionally, i.e., without employing any target detection. This opens up a range of possibilities including pre-detection averaging (waveform stacking), variation of the target detection parameters or algorithm, and employing different sorts of full-waveform analysis. For the work presented here, offline SRF with modified target detection parameters was employed for optimized results. The depth performance of the instrument has been demonstrated to be in the range of 2 Secchi depths for single measurements, i.e., without waveform stacking.

Instead of or in addition to the internal camera, a high-resolution camera can be externally attached and triggered by the instrument. This option was chosen for the experiments in this study using a nadir looking Sony Alpha 7RM3. Figures 1 and 2 show conceptual drawings of the sensor and the general data acquisition concept, respectively. respectively. Table 1 summarizes the main sensor parameters.

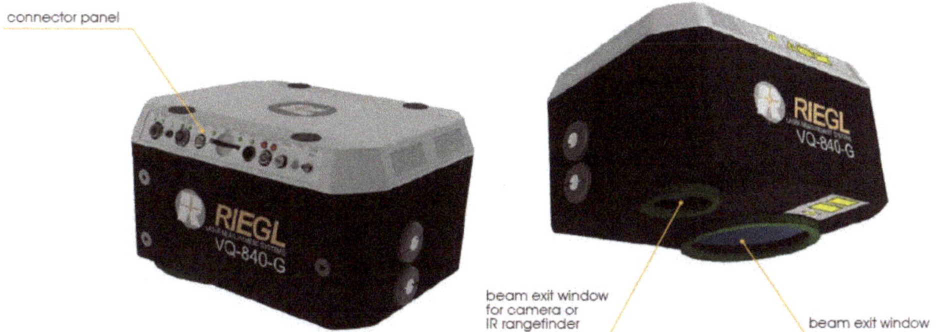

Figure 1. Computer Aided Design rendering of the VQ-840-G laser scanning system in basic configuration. The electrical and data interfaces are shown on the left hand side while the apertures for the green laser scanner and the camera at the bottom of the unit are shown on the right hand side.

Table 1. Sensor key specifications.

PRR	50–200 kHz
Scan rate	10–100 lines/s
Scan swath	±20 deg across flight path, ±14 deg along flight path
Beam divergence	1–6 mrad
Receiver's iFOV	3–18 mrad
Size	360 mm × 285 mm × 200 mm
Weight	12 kg
Power consumption	160 W

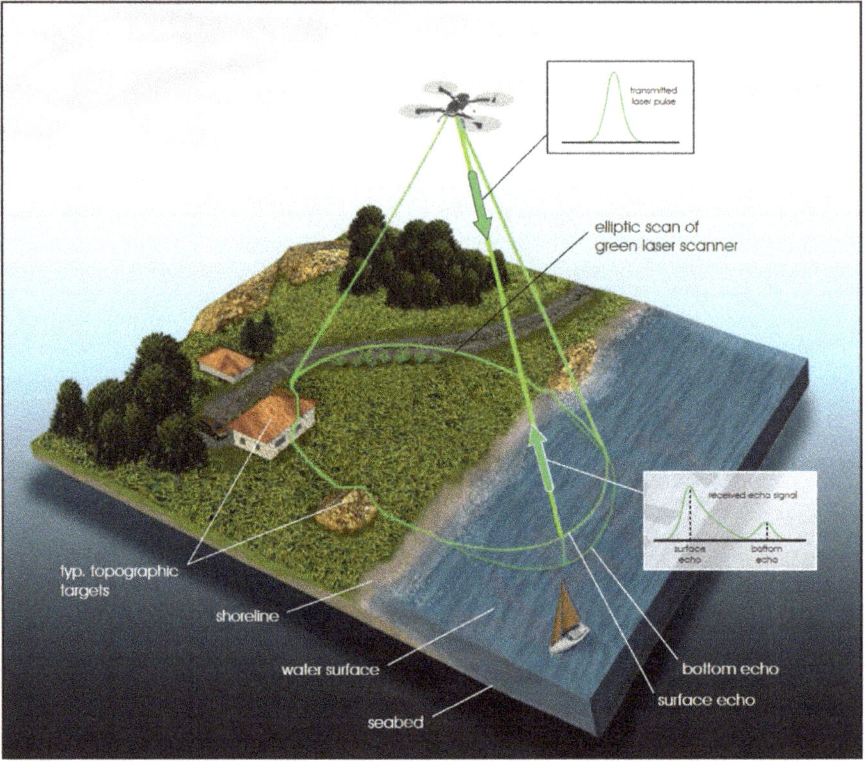

Figure 2. Visualisation of the VQ-840-G mounted on a drone in operation. The scan pattern on the ground is indicated by the green ellipse. The inserts show typical pulse shapes of the outgoing laser pulse and the echo return when measuring into the water.

4. Materials And Methods

In this Section, the study area is introduced in Section 4.1, the existing datasets are presented in Section 4.2, and the employed data processing methods are detailed in Section 4.3.

4.1. Study Area

The study area Neubacher Au (N 48°12′50″, E 15°22′30″; WGS 84) is located in the eastern part of Austria at the tail-water of the pre-Alpine Pielach River, a right-side tributary of the Danube (cf. Figure 3). The study area is part of the conservation area Niederösterreichische Alpenvorlandflüsse (Area code: AT1219000) of the European Union's Natura2000 program. The river is classified as riffle-pool type [63] with a maximum depth of approximately 3 m and exhibits a pluvio-nival regime with expected discharge peaks during snowmelt in winter/spring and torrential rain in summer. Within the investigated reach, the mean annual discharge is about 7 m^3/s, the bed-load sediment is dominated by coarse gravel (2–6.3 cm), and the average gradient is about 0.4% [64]. The entire catchment area measures 590 km^2 and the mean channel width is approximately 20 m. Although the longitudinal continuum of the Pielach River is disrupted by weirs built for hydropower use and engineering measures, the river has retained some of its natural self-forming morphological characteristics like periodically inundated sidearms, dynamic gravel bars, large woody debris, small oxbows, etc., within the study area [65].

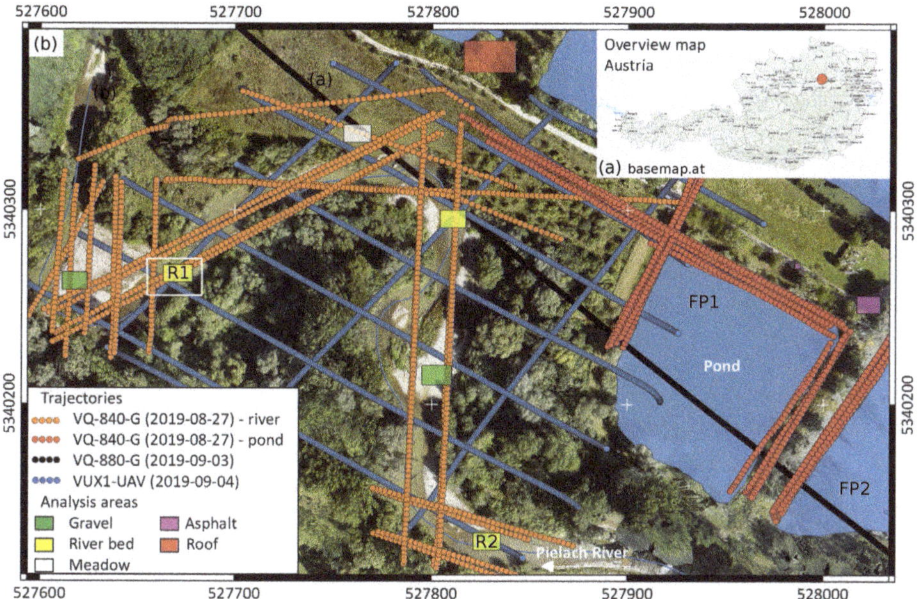

Figure 3. Study area Neubacher Au; (**a**) overview map of Austria with study area marked as a red circle; (**b**) orthophoto of the study area overlaid with flight trajectories (circles) and detailed analysis units (rectangles).

The floodplain to the north of the river features around a dozen groundwater supplied ponds (cf. right side of Figure 3). The bottom of the ponds is covered with a layer of mud, which is constantly stirred up by the fish population (e.g., *Cyprinidae*). This causes turbidity resulting in a Secchi depth of generally less than 2 m. With an overall depth of around 5–6 m, the ponds are ideal water bodies for testing the depth performance of bathymetric LiDAR sensors with a nominal depth performance of 1.5–2 SD.

Since 2013, the study area has been repeatedly captured with various topo-bathymetric laser scanners mounted on manned aircraft for monitoring river morphodynamics and instream habitats [4] as well as with UAV-borne topographic sensors for detailed analysis of the riparian area [66,67]. At the same site, first performance evaluations of a UAV-borne bathymetric laser range finder were carried out [61]. All these articles, as well as the cited literature therein, provide further details about the study area.

4.2. Datasets

In order to assess the accuracy potential and depth performance of the novel topo-bathymetric full-waveform laser scanner, two field campaigns were conducted on 28 August 2019, and 4 September 2019. The two dates are referred to as day 1 and day 2 in the following. On both days, the topo-bathymetric UAV-scanner was operated on a RiCOPTER-M octocopter platform and collected data of the ponds (flight lines marked with red dots in Figure 3) and of an approximately 650 m section of the Pielach River (orange dots).

For comparison and evaluation purposes, two additional LiDAR datasets were captured with the topographic UAV laser scanner VUX1-UAV mounted on a RiCOPTER UAV platform on day 2 and with the VQ-880-G sensor installed on a Diamond DA42 aircraft on 3 September 2019. The respective flight lines are plotted in Figure 3 (VUX1-UAV=blue dots, VQ-880-G=black dots). Both datasets served for cross-validation of the topo-bathymetric UAV-LiDAR data.

Both UAV-based sensor systems operate RGB cameras along with the laser scanners. While the VQ-840-G carries a Sony α7R III (ILCE-7RM3) full-frame mirror-less interchangeable lens camera for capturing 42 MPix RGB nadir images, the VUX1-UAV is equipped with two oblique looking 24 MPix Sony α6000 (ILCE-6000) APS-C format cameras matching the 230°-FoV of the VUX1 sensor. The RGB images of the latter served as basis for the orthophoto shown in Figure 3. The α7R images, in turn, were used for evaluating the visual depth of the different water bodies during LiDAR data acquisition. Due to technical problems, α7R images were only available for the pond flight lines on day 1 but not for the river flight lines.

The turbidity of the Pielach river depends on the flow velocity and discharge within the entire 590 km^2 catchment area. Figure 4a shows the annual mean daily discharge together with the lines for mean discharge (MQ = 7.09 m^{3}s^{-1}) and the one-year flood peak (HQ1 = 106.5 m^{3}s^{-1}), respectively. Discharges beyond the MQ-level only occurred during winter and spring, while summer 2019 was extremely dry with discharges around 2.5 m^{3}s^{-1} throughout the entire summer. In the days before the data acquisitions, typical summer thunderstorms entailed contamination with suspended debris (cf. Figure 4b) and, thus, a high turbidity level of the river water. While this is generally sub-optimal for laser bathymetry, these conditions also provided a chance to study the depth performance of the scanner and the accuracy implications within the available small range of water depths (Pielach river: ≈3 m, ponds: ≈5–6 m).

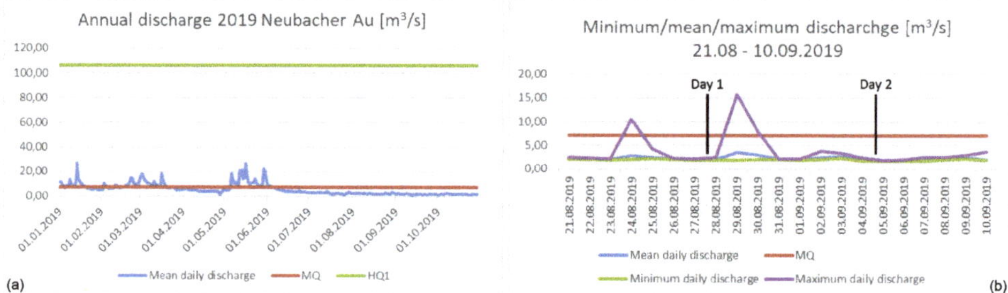

Figure 4. Discharge curves for study area Neubacher AU; (**a**) mean daily discharge for 2019 (blue) together with nominal mean discharge (MQ, red) and one-year flood peak (HQ1, green); (**b**) minimum, mean, and maximum daily discharge within the data acquisition period (flight dates marked with black ticks).

Table 2 summarizes the in-situ SD and turbidimeter measurements. Secchi depths were measured with a 30 cm black-and-white disk and for measurement of the concentration of suspended particles, a Hach 2100Q portable turbidimeter was employed. From both the SD- and turbidimeter values documented in Table 2, it can be stated that (i) pond FP2 tends to be more turbid than FP1; and (ii) turbidity was generally higher for day 2 compared to day 1. This also relates to the higher discharge peak of 15.7 m^{3}s^{-1} caused by the thunderstorm event on 29 August before the second data acquisition day compared to the lower peak of 10.5 m^{3}s^{-1} resulting from the rain event on 24 August before day 1 (cf. Figure 4b).

Table 2. Secchi depth and turbidity measurements.

Measuring Site	Day	Secchi Depth [m]	Turbidimeter [NTU]
FP1	1	1.40	4.35
FP2	1	1.00	4.52
R1	1	1.10	7.90
FP1	2	0.90	9.74
R1	2	0.90	13.4
R2	2	0.70	-

During data acquisition, different scanner settings were tested w.r.t. their implications on depth performance, accuracy potential, and spatial resolution. As stated in Section 3, the bathymetric UAV-laser scanner features both a user-definable laser beam divergence and receiver iFoV. The prior is responsible for eye-safe operation and was set to 2 mrad for all tests, while the PRR, flying altitude, and receiver iFoV were varied. Table 3 summarizes all parameter combinations. In total, the entire test procedure entailed six flights on day 1 and three flights on day 2. The overall duration of each flight amounted to 13 minutes including data acquisition, take-off, initialization of the GNSS/IMU navigation system, and landing. The highest possible PRR of 200 kHz could not be used due to the high power consumption of this scan mode.

Table 3. Flight mission parameters.

Sensor	Area	Day	Altitude [m]	PRR [kHz]	Beam div. [mrad]	Rec. iFoV [mrad]	Camera
VQ-840-G	Pond	1	50	50	2.0	6.0	α7R
–||–	Pond	1	65	50	2.0	6.0	α7R
–||–	Pond	1	75	50	2.0	6.0	α7R
–||–	Pond	1	55	100	2.0	6.0	α7R
–||–	Pond	1	75	100	2.0	6.0	α7R
–||–	Pond	1	50	50	2.0	18.0	α7R
–||–	Pond	1	75	50	2.0	18.0	α7R
VQ-840-G	River	1	50	50	2.0	6.0	—
–||–	River	1	75	50	2.0	6.0	—
VQ-840-G	River	2	55	50	2.0	18.0	α7R
–||–	River	2	75	50	2.0	18.0	α7R
VUX1-UAV	River	2	75	550	0.5	4.5	α600
VQ-840-G	River	2	650	550	1.3	7.0	—

For validation of the scan data, several reference targets were positioned in the pond and river. To evaluate the effective penetration depth of the bathymetric UAV-scanner, four white-colored metal plates of 1×1 m^2 size were lowered into pond FP1 on day 1. Each plate was floating in an individual constant water depth, with depths varying from 1.1 m to 3.1 m, corresponding to 0.8–2.2 SD. The floating of the plates was realized by attaching ropes to the four corners of each plate and mounting styrofoam buoyancy bodies to the other end of the ropes. Figure 5 shows the four plates in one of the α7R images during data acquisition on day 1.

Figure 5. The 1×1 m^2 plates (P1–P4) floating in pond FP1 for evaluation of depth performance; camera: α7R; acquisition: day 1; depth of plate from left to right: 1.1 m, 1.9 m, 2.6 m, 3.1 m.

To assess the bathymetric accuracy, an array of three parallel profiles with 6–7 checkerboard targets (i.e., black-and-white disks of 20 cm or 30 cm diameter, respectively) were anchored in the gravel riverbed and measured with a Leica TS16 total station on day 1. The approximate position of the reference target field is marked as R1 in Figure 3. The 19 reference targets exhibited a minimum, mean, and maximum depth of 0.33 m, 1.44 m, and 2.15 m, respectively. Snorkeling was employed for fixing the checkerboard targets in the riverbed as well as positioning the measurement pole in the target's center hole. This was especially necessary for all targets featuring a depth of more than 1 m. As a guide for both installation and measurement, a rope was fixed on both sides of the riverbank. The photographs in Figure 6c–e illustrate the setup of the reference target field.

Furthermore, Figure 6a shows one of the black-and-white checkerboard targets used as ground control points (GCP) for absolute orientation of the image block as well as the laser scans. In total, 25 of theses quadratic targets (edge length: 30 cm) were distributed within the entire study area and measured with a Spectra Precision SP80 GNSS receiver in real-time kinematic (RTK) mode based on correction data obtained from the local GNSS service provider EPOSA. The GCPs feature a total horizontal and total vertical uncertainty, respectively, of less than 2 cm (1σ). While GCPs are the first choice for absolute orientation of an image block, laser scanning requires planar control patches. Two A-shaped saddle roofs composed of wooden formwork panels, one on each side of the river in close vicinity to the reference target field, served as the basis for the absolute orientation of the laser flight block (cf. Figure 6d,e). The data captured on day 2 were only used for orthophoto generation and cross-validation on dry land due to the very high turbidity on this day. The topo-bathymetric LiDAR data of day 1, in turn, serve as a basis of all subsequently described data processing steps and the presented results.

Figure 6. Photo documentation of data acquisition on 27 August (day 1); (**a**) Pielach river with quadratic black-and-white checkerboard ground control point target in the foreground; (**b**) topo-bathymetric UAV laser scanner RIEGL VQ-840-G during data capture; (**c**) UAV-image of field of submerged circular black-and-white checkerboard targets aligned with rope fixed between both riverbanks; (**d**) terrestrial photo of reference target field, most of the targets are not visible with the naked eye due to the limited Secchi depth of 1.1 m, yellow wooden plates in fore- and background: saddle roof control patches for absolute orientation of laser scans; (**e**) total station Leica TS16 in front of reference field and control patches.

4.3. Data Processing Methods

In the following, the general workflow including preprocessing, refraction correction of the raw LiDAR point clouds, assessment of accuracy and spatial resolution, and evaluation of depth performance is summarized. For most of the processing steps, the scientific laser scanning software OPALS [68] was employed.

4.3.1. Preprocessing

Data preprocessing comprised the following steps:

- Post-processing of raw GNSS and total station data and calculation of 3D coordinates of ground control points and checkpoints using Leica software
- Aero-triangulation of all captured images and derivation of digital orthophotos using Pix4DMapper [20]
- Post-processing of flight trajectory data and derivation of a Smoothed Best Estimate Trajectory (SBET) [69] using observations from the installed permanent reference station
- Direct georeferencing of the laser echoes originating from online waveform processing using the manufacturer's software RiPROCESS
- Geometric calibration of flight block via strip adjustment [70,71]. The ICP-based calibration procedure included estimation of mounting parameters (boresight angles, lever arm) as well as additional trajectory corrections per strip. For each strip, a constant offset (X, Y, Z, roll, pitch, yaw) was estimated to further improve the strip fitting precision. Consideration of the terrestrially measured ground control points and saddle roof models (cf. Figure 6a,d) in the strip adjustment ensured optimal absolute orientation of the flight block and consistency to the (submerged) checkpoints.
- Quality check of the resulting 3D point cloud, calculation of strip-wise DEMs, and derivation of pairwise DEM of Difference models including statistical analysis of the resulting strip height differences in smooth areas [72]
- Derivation of strip-wise water surface models. First, approximate water levels were derived interactively with the editing tool qpals (i.e., QGIS plugin for OPALS software) based on the approach described in [73]. This initial water surface model served as the basis for selecting potential water surface points as seed points for region growing segmentation. In the last step, all classified water surface points were used to derive the final strip-wise water surface model as a regular 50 cm grid using moving least squares interpolation. The resulting water surface models capture the low-frequency contributions of the wave-induced water surface shape while smoothing out higher frequency parts.
- Refraction correction of the raw laser echoes according to the Snell's law [74]. Based on a recent publication [75], a refraction coefficients of $n = 1.33$ was employed for beam deflection whereas coefficients of $n >= 1.36$ were used for round-trip-time correction.

4.3.2. Evaluation

1. *Precision:* Assessment of precision comprised evaluation of measurement noise and geometric consistency of overlapping strips.

    - *Measurement noise:* Assessment of the sensor's noise level was carried out at smooth surfaces (e.g., asphalt) within the individual strips by estimating a best-fitting plane for each point of an appropriate surface patch via least-squares adjustment based on the k nearest neighbor points (k = 12) within a maximum search radius of 50 cm. For surfaces with known planar geometry, the residuals (i.e., point-to-plane normal distances) provide an estimate of the sensor's ranging precision (cf. Figure 7a).

- *Strip fitting precision:* The height deviations in smooth parts (cf. Figure 7b) of overlapping flight strips provide an estimate of the overall quality of the sensor system including GNSS/IMU navigation device, LiDAR unit, and scanning mechanism. First, strip-wise DEMs were interpolated using moving least squares interpolation, and subsequently, the height discrepancies in smooth surface parts were visualized in color-coded maps and statistically analyzed (histograms, standard statistics). For the quantification of bias and dispersion, robust statistics (bias: median, dispersion: σ_{MAD}= Median of Absolute Differences) are employed. Separate analysis of dry land and submerged areas enable a distinction between effects stemming from sensor calibration and flight block orientation (land) and effects connected to water surface estimation and refraction correction.

2. Accuracy: estimation of absolute accuracy relies on a comparison of the laser derived point cloud with the independently captured reference points measured as point-to-plane distances (cf. Figure 7c). For each reference point, the k-nearest laser points (k = 8) within a maximum search radius of 15 cm served as input for fitting a tilted plane via least-squares adjustment. The normal distance between the reference point and the plane constitutes an objective measure of overall accuracy as it comprises all possible error sources.

3. *Depth performance:* Quantifying the achieved maximum depth penetration is accomplished via a comparison of the known depths of the metal plates with the depths derived from the corresponding laser returns. In case the plate was discernible, the points in the center of the plate were selected and served as the basis for the median-based estimation of the plate's representative height. The process was repeated for all strips containing laser echoes of the plate also considering the individual sensor settings of the respective strip (beam divergence, iFoV, flying altitude).

4. Spatial resolution: the theoretical resolution derived from the flight mission and sensor parameters (beam divergence, pulse repetition and scan rate, flying altitude and speed), is confronted with the actual resolution derived from the captured laser points. The studied objects are a rope spanned across the river and the right-angle shaped wooden plates (cf. Figure 6d). For the linear feature, all neighboring laser points are projected to the rope axis allowing analysis of the lateral distances (cf. Figure 7d) and for the roof-shaped feature, the shape of the point cloud is analyzed in the vicinity of the crest line, where the point cloud shows substantial rounding of the sharp edge.

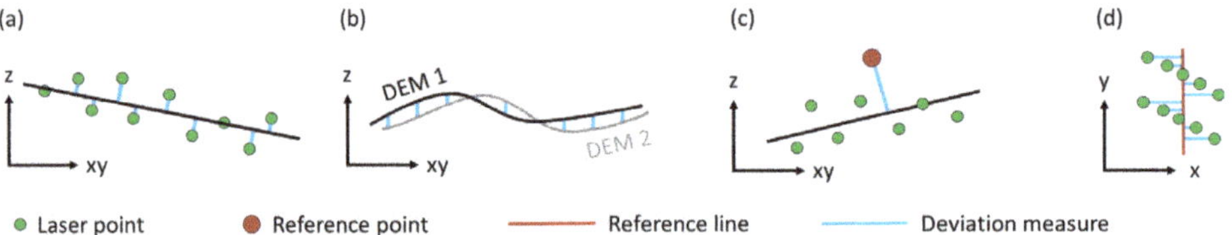

Figure 7. Conceptual sketch of employed deviation measures (blue lines). (**a**) measurement noise; (**b**) strip fitting precision; (**c**) accuracy w.r.t. reference points; (**d**) spatial resolution.

5. Results

In this section, we present the quality evaluation results separated into precision (Section 5.1), accuracy (Section 5.2), depth performance (Section 5.3), and spatial resolution (Section 5.4).

5.1. Precision

5.1.1. Measurement Noise

Figure 8 shows five selected surface patches (asphalt: A1, A2, gravel: G1, G2, and meadow: M1). For each patch we provide the orthophoto and DEM (color-coded height map superimposed with 1 cm contour lines) in the first two rows. The last row exhibits a plot of the local interpolation error (σ_0).

In case of a sufficiently planar surface, σ_0 quantifies the measurement noise (reproducibility). This is especially the case for A1. All other patches exhibit small scale topographic variations (gravel, G1 and G1) or undulations due to grass vegetation (A2, M1).

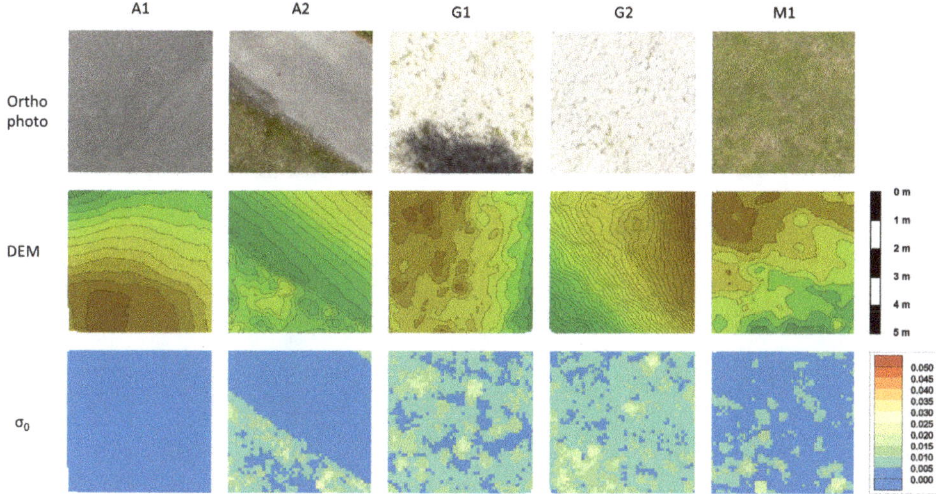

Figure 8. Visualization of measurement noise for selected 5×5 m^2 surface patches; A1: asphalt, A2: asphalt and grass, G1/G2: gravel bank, M1: meadow; first row: digital orthophoto, second row: color coded height map superimposed with 1 cm contour lines, third row: color coded map of local surface roughness (i.e., σ_0 [m]).

Table 4 summarizes the quantitative results. For the asphalt dominated patches Table 4 also documents the noise level variations depending on the mission parameters (flying altitude, iFoV). For A1, σ_0 is below 2 mm for a narrow iFoV of 6 mrad and slightly higher (3.5 mm) for the wider iFoV of 18 mrad. For the mixed asphalt/grass patch A2, increasing the flying altitude from 50 m to 70 m does not effect the precision (mean: $\approx$4 mm, median: $\approx$3 mm). In both cases, the higher mean value reflects the higher variations within the grass vegetation, while the more robust median can be seen as representative for the dominant asphalt area. For all other patches, the micro-relief leads to increased σ_0 values, which are in the range of 1 cm for the gravel patches (G1, G2) and around 5 mm for the meadow patch M1. All reported values are below 15 mm, the precision value reported in the sensor's spec sheet [9].

Table 4. Evaluation of measurement noise reported as local σ_0 (mean, median, and interquartile range (IQR)).

Patch	Altitude [m]	iFoV [mrad]	Mean [mm]	Median [mm]	IQR [mm]
A1	65	6	1.9	1.8	0.7
A1	50	18	3.4	3.3	1.4
A2	50	6	4.4	2.3	4.7
A2	50	18	4.2	3.1	3.6
A2	70	18	3.9	2.9	3.1
G1	50	6	7.6	6.9	4.5
G2	50	6	8.2	7.4	3.9
M1	50	6	4.6	3.2	2.1

5.1.2. Strip Fitting Precision

Figure 9 shows a color-coded map of strip height differences in smooth surface areas on top of a DSM map of the entire UAV-LiDAR flight block (color-coded elevation map superimposed with shaded relief map). The dominant white color tones of the strip differences indicate deviations of less than 2 cm and confirm the good fitting precision of the flight strips. The histogram of the strip height differences plotted in the lower-left corner of Figure 9 exhibits a symmetric shape. The distribution is unbiased (mean: 1 mm, median: 0 mm) with a dispersion of 9 mm (σ_{MAD}). Occasional strip differences in vegetated areas, which have passed the automatic smoothness filter, are responsible for the comparably high standard deviation of 10 cm. 96% of the strip height differences are within a range of -24 mm to $+25$ mm, which further underlines the fidelity of the entire sensor system including navigation device, laser ranging unit, and scanner.

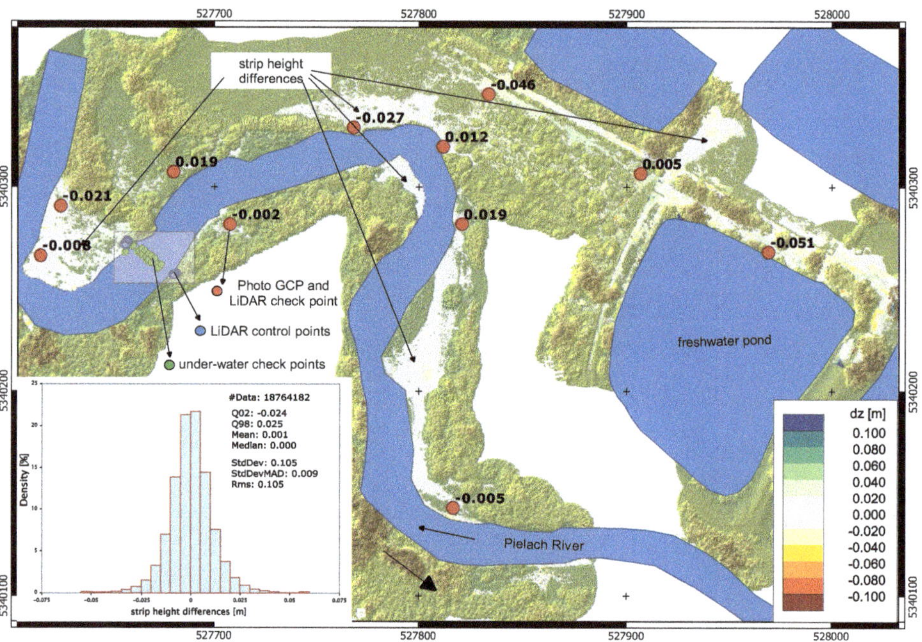

Figure 9. Color coded mosaic of strip height differences, background: LiDAR DSM (color coded elevation map superimposed with shaded relief map), lower left: histogram of height differences in smooth surface areas, lower right: color bar for strip differences.

5.2. Accuracy Assessment

The absolute accuracy, measured as signed point-to-plane distances between terrestrially measured reference points and LiDAR-derived planes are plotted in Figure 9 (labeled red dots) and Figure 10 for land and underwater, respectively.

On the land side, the deviations are mainly below 3 cm, with only two checkpoints at the western end of the flight block exhibiting deviations of $\approx$5 cm. The poorer absolute fitting accuracy in this part of the block is due to the unavailability of appropriate laser control patches in this area of the block, where absolute orientation was of minor concern for the conducted evaluation. Control patches (wooden saddle roof formwork constructions) were located in the vicinity of the river section featuring the submerged reference targets (cf. Figure 6c–e). In this area utmost accuracy of the absolute block orientation was crucial, and the deviations in this region are 2 cm or better. Table 5 summarize the deviations on dry land for selected reference points on the saddle roof shaped control patches

(CP, cf. blue dots in Figure 9) and on the water-land-boundary (WLB, cf. black dots in Figure 9). The deviations are generally below 1 cm (cf. points 1001–1020, 2001–2010, 108–110) and only the points on the ridge of the control patches (1022–1024, 2010–2012) exhibit larger, systematically negative deviations of more than 3 cm. This will be further discussed in Sections 5.4 and 6.

Table 5. Accuracy assessment for reference targets on land; selected point-to-plane distances [m] between reference points and LiDAR derived planes.

ID	Area	Dev.	ID	Area	Dev.	ID	Area	Dev.
1001	CP(W)	−0.006	2001	CP(E)	0.013	108	WLB	−0.005
1005	CP(W)	−0.007	2003	CP(E)	0.001	109	WLB	−0.008
1010	CP(W)	0.008	2005	CP(E)	−0.006	110	WLB	0.006
1016	CP(W)	−0.004	2008	CP(E)	−0.002			
1020	CP(W)	−0.006	2010	CP(E)	−0.035			
1021	CP(W)	−0.031	2011	CP(E)	−0.045			
1022	CP(W)	−0.034	2012	CP(E)	−0.079			
1023	CP(W)	−0.033	2013	CP(E)	0.003			
1024	CP(W)	−0.036	2014	CP(E)	−0.001			

Figure 10 shows the results of the bathymetric accuracy assessment for all 20 submerged reference targets. The upper left plot shows the locations of the reference targets, arranged in three cross-sections, together with the point labels of the middle section, for which the deviations are presented numerically in Table 6 together with the respective water depth. For refraction correction of the raw laser echoes, we strictly used a refractive index of $n = 1.33$ for beam deflection (i.e., angular component). For run-time correction, we varied the coefficient from $n = 1.33$ to $n = 1.36$ according to literature [75,76] and also beyond that till $n = 1.42$. Table 6 shows that the use of standard refraction coefficients (1.33, 1.36) leads to a systematic, depth-dependent, negative bias indicated by the red color tones. The respective columns in Table 6 also clearly show the depth-dependent increase of the deviations reaching 7.8 cm for the deepest point 206 (water depth = 2.1 m) for refractive index of 1.36. Due to this unexpected systematic effect, we first double-checked the accuracy of the water surface, but did not find any evidence that the modeled water surface model is responsible for the bias (cf. deviations water-land-boundary points in Table 5 and the cross-sectional plot in Figure 11). In addition, we performed refraction correction with different software tools (RiPROCESS, OPALS) leading to consistent results. We, therefore, performed tests with refraction coefficients beyond $n = 1.36$ and found that the bias entirely disappears for $n = 1.42$. This can be seen from the white color tones of the respective plots in Figure 10 as well as from the underlined values in Table 6.

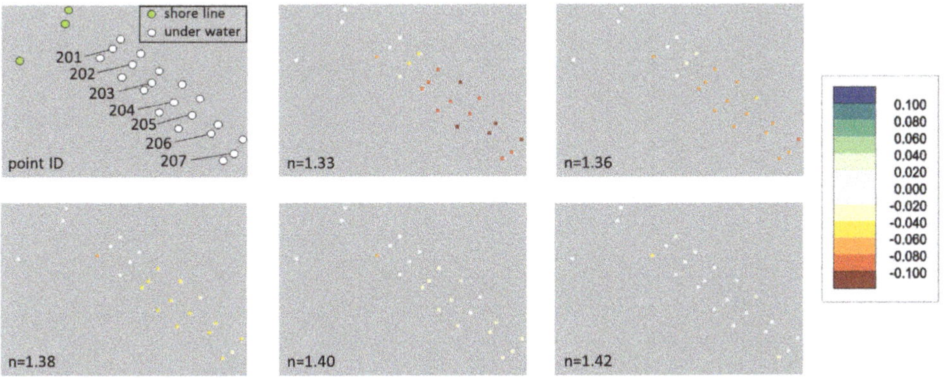

Figure 10. Color coded point-to-plane distances [m] between check points and LiDAR derived planes, run-time correction based on refraction coefficients ranging from $n = 1.33$ to $n = 1.42$.

Table 6. Accuracy assessment for underwater reference targets; selected point-to-plane distances [m] between reference points and LiDAR derived planes using different refraction coefficients for run-time correction.

ID	Depth	n = 1.33	n = 1.36	n = 1.38	n = 1.40	n = 1.42
201	0.395	−0.006	**0.004**	0.009	0.014	**0.016**
202	0.934	−0.048	**−0.027**	−0.016	−0.001	**0.011**
203	1.260	−0.088	**−0.061**	−0.043	−0.026	**−0.009**
204	1.425	−0.094	**−0.066**	−0.045	−0.026	**−0.006**
205	1.881	−0.098	**−0.069**	−0.041	−0.019	**0.000**
206	2.097	−0.123	**−0.078**	−0.050	−0.029	**0.001**
207	1.443	−0.094	**−0.063**	−0.039	−0.022	**−0.006**

Figure 11 shows a vertical section of the refraction corrected LiDAR point cloud ($n = 1.36$) colored by signal amplitude overlaid with (i) the reference points (magenta) and (ii) the modelled water surface (black). The saddle roof on the right (i.e., western) embankment shows very good agreement with the reference points whereas the submerged reference targets lie systematically above the LiDAR point cloud of the river bottom.

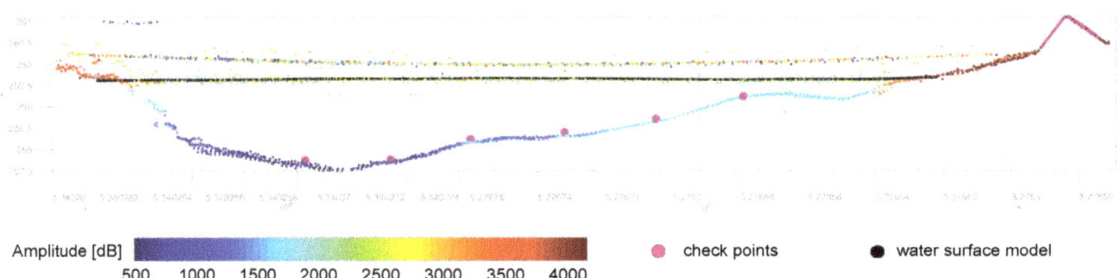

Figure 11. Vertical section of refraction corrected 3D point cloud ($n = 1.36$) of strip 105334 colored by signal amplitude; larger magenta dots: checkpoints, black: modelled water surface.

5.3. Depth Performance

To evaluate the sensor's depth performance, four white metal plates were lowered into pond FP1 at the eastern border of the flight block (cf. Figure 5) in depths of 1.1–3.1 m corresponding to 0.8–2.2 SD. The plates were captured multiple times with different sensor settings (PRR, altitude, iFoV). Figure 12 depicts the 3D point clouds of two selected strips (063116, 093054), both flown 50 m above ground level with a PRR of 50 kHz, but with a different iFoV of 6 mrad and 18 mrad, respectively. Figure 12 reveals that the wider iFoV generally delivers more points on the detectable plates P1-P3 and the plates appear larger than the nominal size of 1 m^2. The signal amplitude is higher for the wider 18 mrad iFoV, indicated by the more reddish color tones, as more backscattered laser light reaches the detector.

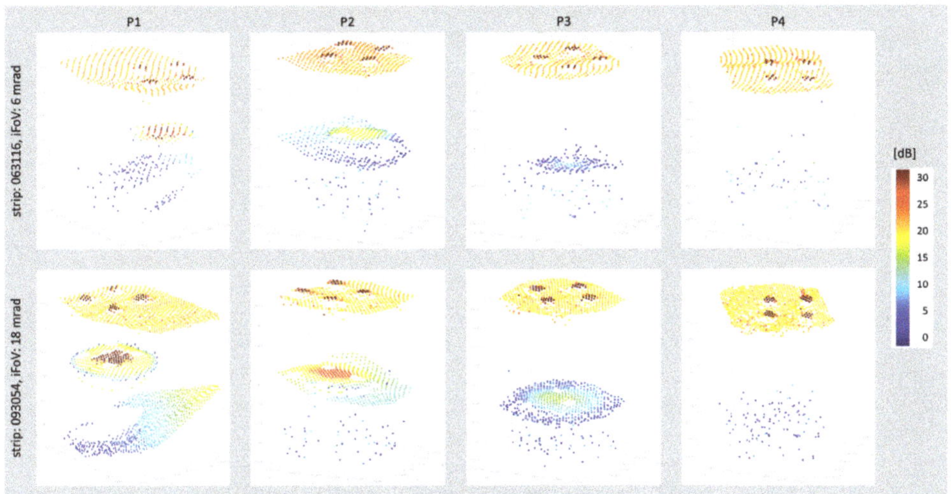

Figure 12. Perspective view of 3D point points of lowered metal plates P1-P4 colored by signal amplitude [dB]. upper row: strip 063116, iFoV: 6 mrad, lower row: strip 093054, iFoV: 18 mrad.

For the same strips, Table 7 provides a quantitative summary. Table 7 documents that plates P1–P3 could be detected in the point clouds, but not P4 at a depth of 3.11 m or 2.2 SD, respectively. This holds for the strip flown with a 6 mrad iFoV as well as for the strip with the wider 18 mrad iFoV. As expected, the signal amplitude level of the 18 mrad iFoV is generally higher for all plates, and at P3 the receiver still measures 14.5 dB. This, however, was not enough to identify P4 located just beyond the 2 SD limit. The relative depth of P3 amounts to 1.94 SD, thus, slightly below the sensor specification of 2 SD. Furthermore, the nominal-actual depth comparison reveals the same systematic depth overestimation as already reported in Section 5.2.

Table 7. Depth performance assessment.

Plate	Strip	iFoV	Altitude [m]	Nominal Depth [m]	Actual Depth [m]	Deviation [m]	rel. Depth [SD]	Amplitude [dB]
P1	063116	6	50	1.16	1.18 ± 0.01	0.02	0.84	26.8
P2	063116	6	50	1.99	2.03 ± 0.02	0.04	1.45	17.6
P3	063116	6	50	2.64	2.71 ± 0.01	0.07	1.94	7.8
P4	063116	6	50	3.11	-	-	-	-
P1	093054	18	50	1.16	1.18 ± 0.01	0.01	0.84	29.7
P2	093054	18	50	1.99	2.03 ± 0.02	0.04	1.45	22.7
P3	093054	18	50	2.64	2.74 ± 0.01	0.10	1.94	14.5
P4	093054	18	50	3.11	—	—	—	—

Figure 13 further illustrates the impact of the user-definable receiver iFoV in terms of depth penetration at a near shore transect of pond FP1. In both cases, the point clouds were captured from a flying altitude of 50 m but with a different iFOV (strip 063419: brown, 6 mrad; strip 093410: blue, 18 mrad). The larger iFoV leads to a higher depth penetration. In the depicted case of a freshwater pond with an SD of 1.4 m at the time of data acquisition, the maximum depth is about 2.67 m with the 6 mrad iFoV and 3.07 m using the larger 18 mrad iFOV, corresponding to 1.9 SD and 2.2 SD, respectively. The gain in the presented case is 40 cm (0.3 SD) or 15%, respectively. Thus, the manufacturer's specification of 2 SD was verified in the experiment. The wider iFoV also marginally increases the measurement noise, which can be seen from the higher amount of clutter points in Figure 13 as well as from Table 4 (A1, median). This aspect is further discussed in Section 5.4.

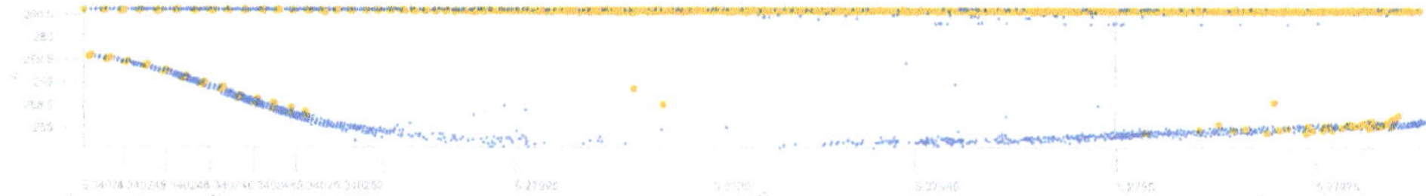

Figure 13. Vertical section of refraction corrected 3D point clouds of strip 063419 (brown, iFoV: 6 mrad) and 093410 (blue, iFoV: 18 mrad); flying altitude: 50 m; refraction coefficient used for run-time correction: ($n = 1.36$).

5.4. Spatial Resolution

The spatial resolution of any LiDAR system is limited by the size of footprint diameter, which, in turn, is a function of the beam divergence γ and the measurement range R. In this experiment, a constant beam divergence of 2 mrad was employed to ensure eye safety at the standard flying altitude of 50 m. The footprint diameter D calculates to $D = R\gamma$, i.e., 10 cm for the standard flight setup in our experiment.

To evaluate the actual footprint diameter, two suitable objects have been chosen and analyzed in detail, namely (i) a linear object and (ii) a sharp angle. Figure 14 shows the 3D point clouds of the study objects in a perspective view (a-c) and a sectional view (d) colorized by signal amplitude. The linear object (rope) does not appear as a confined line of points but rather as a band featuring a width of 30 cm, thus, 3 times the theoretical footprint size.

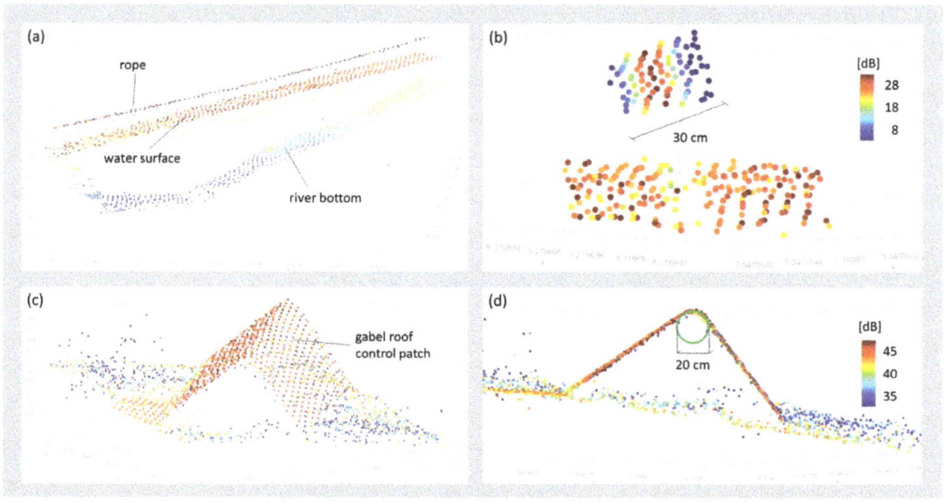

Figure 14. Empirical assessment of spatial resolution; top row: rope spanned across river, bottom row: gable roof shaped control patch; (**a–c**) perspective view of 3D point cloud; (**d**) vertical section; points colored by signal amplitude [dB].

At the ridgeline of the gable roof-shaped control patch, the points do not form a sharp edge but appear as a circular cylinder featuring a diameter of 20 cm as the finite laser footprint illuminates both roof sides. This effect is responsible for the larger deviations for all control points along the ridgeline reported in Section 5.2 (cf. Table 5, e.g., point 1022) while all control points, for which the laser footprint is entirely on one side of the wooden panel, exhibit deviations below 1 cm.

It is noted that, as expected, also the receiver iFoV substantially influences the spatial resolution, even for constant beam divergence settings. The plate experiment presented in Section 5.3 shows this clearly. While all plates exhibit an area of 1 m^2, increasing the iFoV has a big impact on the apparent size of the plate as depicted in the point cloud. For plate P1, for instance, the area covered by laser returns reflecting from the plate with the 18 mrad iFoV setting measures 3.5 m^2. This is caused by beam spreading due to volume scattering in the water column. With a larger iFOV, more signal from the periphery of the conical laser beam reaches the detector enabling echo detection even if the plate is only partially within the illuminated footprint. However, full and partial hits of the plates can be distinguished via analysis of the amplitude values. Amplitudes are nearly constant if the entire laser footprint impinges on the plate, and rapidly drop outwards with a decreasing percentage of illuminated plate area (cf. bottom left plot of Figure 12).

6. Discussion

The presented compact topo-bathymetric laser scanner exhibits a weight of 12 kg and a power consumption of 160 W. This makes the instrument suitable for integration on light manned aircraft, gyrocpters, and helicopters as well as on larger UAV platforms, as was the case for our data acquisition. The considerable payload and power consumption, however, limit UAV flight endurance with the selected UAV. The longest net acquisition airtime in our experiment was about 10 min excluding starting, landing, and IMU/GNSS initialization. This might be sufficient endurance for corridor mapping in a visual line of sight (VLOS) context, but poses hard constraints for larger project areas and beyond VLOS operation. Thus, our UAV-borne experiment is regarded as a successful proof-of-concept, but further sensor miniaturization is needed for integration on mainstream UAV platforms like the *DJI Matrice 600 Pro* or comparable platforms. No matter if the sensor is integrated on a light manned or unmanned aerial platform, in both cases the mobilization costs are considerably lower compared to traditional ALB surveys and smaller aircraft feature a higher maneuverability. The sensor, therefore, constitutes a cost-effective alternative, especially for mapping bathymetry of narrow river channels, as a supplement to area-wide topographic laser scanning which is today often available on a country-wide level.

A major benefit of the sensor design is the adjustable pulse repetition rate, laser beam divergence, and receiver iFoV, making eye-safe operation possible for low flying altitudes like the standard 50 m altitude in our experiment. Due to the smaller sensor-to-target distances and the narrow laser cone, the resulting nominal laser footprint diameter in the dm-range is much smaller compared to existing topo-bathymetric sensors operated on manned aircraft featuring footprint sizes of about half a meter. This gain in spatial resolution opens new applications in the context of hydraulic engineering (roughness estimation) for reliable flood simulations, eco-hydraulics (habitat modelling), fluvial geomorphology, and many more.

As stated before, the limiting factor for spatial resolution in the lateral direction of the laser beam is the *laser footprint*, i.e., the illuminated area on the target. As reported in Section 5.4, this is not the same as the area resulting from the vendor specifications. The power distribution in the laser footprint has Gaussian shape and it is common practice to specify the beam diameter by twice the radius where the optical power has decreased by a factor of $1/e^2$. For a beam diameter w, the power distribution $I(r)$ as a function of off-axis distance r is given by

$$I(r) = I_0 e^{-\left(\frac{2\sqrt{2}}{w}r\right)^2},$$ (1)

with I_0 the power at the center. The power level at $r = w/2$ is 0.14% of the power level at the maximum. Expressing the power level at $r = w/2$ in Decibels gives

$$10\log_{10}\left(\frac{I(0.5w)}{I_0}\right) \approx -8.7\text{dB}.$$ (2)

A signal causing a return strength in the center of the beam that is more than 8.7 dB above the receiver detection threshold will cause a detectable signal in an adjacent measurement even when further apart than the nominal beam diameter radius. The effective size of the footprint, thus, is determined not only by the properties of the laser beam but also by the sensitivity of the receiver circuitry and the iFOV. A small isolated object will be seen in adjacent laser points until it is too weak to be detected. However, the object will appear with diminishing amplitudes towards the edges (cf. Figure 14b).

The finite footprint and high receiver sensitivity also cause smoothing effects if the target's shape is not flat within the illuminated area. This applies to discontinuities in height (cliff, wall) or slope (sharp edge), or in case of high curvature (e.g., small boulders). In any case, this leads to rounding artifacts as exemplified in Figure 14d for a sharp edge. In the context of mapping river channel bathymetry, this limits the ability to precisely reconstruct objects smaller than ≈20 cm. While the instrument's spatial resolution is too low for roughness estimation on grain size level, it is suitable for the detection of blocks, boulders, and other flow resistance relevant objects. As an example, the circular underwater reference targets clearly stood out from the surrounding river bottom although they were firmly anchored in the ground and, thus, only protruded from the ground surface by a few cm.

For smooth surfaces, the sensor showed a measurement noise below 1 cm verified at horizontal asphalt surfaces (cf. surface patch A1 in Figure 8 and Table 4) as well as on tilted surfaces (cf. Table 5, e.g., points 1001–1020). Beyond the local level, this also holds for the strip fitting precision of the entire flight block, which even outperformed comparable data acquisitions with a state-of-the-art topographic UAV laser scanner employed in the same study area in the previous work [66,71]. High relative accuracy is of special importance, e.g., for detailed hydrodynamic-numerical modeling in flat areas where a few cm can determine whether or not certain areas are flooded.

While the absolute accuracy of the flight block, measured as point-to-plane distances between terrestrially surveyed reference targets and the LiDAR point cloud, is generally better than 3 cm on land (cf. red dots in Figure 9), systematic deviations were observed for submerged checkpoints. In addition to geometric calibration of the sensor system and (ranging, scanning, lever arms, boresight angles, trajectory), the overall accuracy of LiDAR-derived bathymetry depends on the accuracy of (i) the modeled water surface and (ii) the refraction coefficients for run-time correction. Data analysis has revealed that laser echoes from the water surface are dense (i.e., practically every laser shot returned an echo from the surface) and the signal amplitude is sufficiently strong to enable reliable range measurement. Figures 11 and 14a,b illustrate that (i) the recorded signal amplitudes are around 20 dB; (ii) the point density is high; and (iii) the vertical spread is remarkably low. It is well known from literature including our own work [1,57,77,78] that the laser signal reflected from water surfaces is a mixture of reflections from the surface and volume backscattering from the water column beneath the surface. Surface data obtained from conventional green-only ALB sensors operated from manned platforms in flying altitudes of around 600 m often show a fuzzy appearance. In contrast, surface data acquired in this experiment are concise and the point clouds show a vertical spread of only a few cm. Independent validation of the modeled water surface with terrestrially surveyed checkpoints at the water-land-boundary (cf. Table 5) also confirmed the accuracy of the water surface model. Thus, we rule out that inaccuracies of the modeled water surface are responsible for the detected depth-dependent bias.

Concerning run-time correction, we first started with standard values documented in the literature. [76] reports a valid range of $n = 1.329$–1.367, but the upper bound is only valid for extreme salinity and temperature conditions. In recent work, [75] reported that group velocity is relevant for run-time correction of pulsed lasers rather than the phase velocity of the underlying laser wavelength ($\lambda = 532$ nm). With a respective experiment, they showed that a value of around 1.36 is valid for freshwater conditions. Therefore, this value was also used for run-time correction in this work, but comparison with the reference points still showed a systematic, depth-dependent error. The bias only disappeared when using a hypothetical refraction coefficient of $n = 1.42$. As there is no physical explanation for such a high refractive index, the question for the reason of the systematic deviation still remains. Multi-path effects due to complex forward scattering within the water columns are the most likely explanation for the observed phenomenon. As documented in Table 2 and as can be seen from the on-site photographs in Figure 6, the turbidity level was high (Secchi depth in the river: 1.1 m) due to a thunderstorm event in the days before data acquisition. This resulted in a high load of suspended sediment. Scattering at the sediment particles leads to a widening of the laser footprint, which entails non-linearity of the ray paths leading to elongated path lengths in water, which might result in the observed underestimation of the river bottom height. Following this line of argument, a repeat survey is planned in the winter season 2020 in clear water conditions to verify and quantify the influence of turbidity on the UAV-LiDAR derived depth measurements.

The depth performance of 2 SD stated by the manufacturer was confirmed in the conducted evaluation. All results relied on online waveform processing [62] and did not make use of sophisticated waveform processing [79,80]. Processing of data acquired with different iFoV settings confirmed an increased depth performance using the widest possible setting of 18 mrad compared to the default setting of 6 mrad. The gain measures 0.3 SD or 15% but comes at the price of a reduced spatial resolution as the receiver's perceivable target area is larger, exemplary illustrated in Figure 12 at plate P1. In a practical context, this clearly shows that a trade-off between high spatial resolution and maximum depth performance needs to be found. A potential best practice procedure may include a repeat survey of the same flight lines first with scanner settings optimized for spatial resolution (high PRR, narrow beam divergence and iFoV) and second with parameters optimized for maximum depth penetration. With the studied instrument, this can be achieved in a single flight mission as individual sensor settings are feasible on a per flight line basis.

The conducted depth performance assessment based on the four lowered metal plates yielded that plate P4 at a depth of 2.2 SD could not be identified in any of the flight strips including the ones with 18 mrad iFoV. This is astonishing, as the natural bottom of the pond could be measured at approximately the same depth (cf. blue points in Figure 13). By extrapolating the signal amplitude decrease from plate P3 (14.5 dB), enough signal should have been available to also detect P4. One of the plausible reasons is an abrupt turbidity increase within the freshwater pond at a water depth of around 3 m. We consider this the most likely explanation as the pond is mainly used for fishing purposes and the dominating species (*Cyprinus carpio*) tends to dig into the mud at the bottom of the pond potentially swirling up mud particles.

Beyond the pure sensor settings, an additional increase of depth performance is expected via sophisticated waveform processing. [81] reported an increased areal coverage when employing a surface-volume-bottom based exponential decomposition approach, in both very shallow and deeper water areas. [5,82] use waveform stacking, i.e., averaging of neighboring waveforms, to enhance ALB depth performance reporting an increase of 30%. The presented sensor is specifically well suited for the application of waveform stacking, also referred to as pre-detection averaging, as waveform recording does not depend on a trigger event (i.e., echo detection) during data acquisition but waveforms are continuously stored within a certain range gate. While this is not possible for typical ALB flying altitudes of 600 m due to storage capacity limits, it is well possible for the shorter ranges in UAV-borne laser bathymetry. By combining the possibilities offered by the sensor (PRR, beam divergence, iFoV) with data processing based on the cited strategies, a maximum depth penetration of about 2.9 SD can be expected, but further experiments are required to verify this value.

Further advantages arise from the integration of active and passive sensors in a single hybrid system. Examples for applications benefiting from the concurrent acquisition of laser and image data are water body detection [83] and discharge estimation [46].

7. Conclusions

In this paper, we introduced the concept of a novel compact topo-bathymetric laser scanner and presented a comprehensive assessment evaluating measurement noise, strip fitting precision, accuracy compared to terrestrially measured reference data, depth performance, and spatial resolution. With a weight of 12 kg the sensor is suitable for integration on light manned as well as powerful unmanned platforms, reducing the mobilization costs compared to conventional ALB. The scanner features an adjustable pulse repetition rate (50–200 kHz), scan rate (10–100 lines/s), beam divergence (1–6 mrad), and receiver's iFoV (of 3–18 mrad). With nominal flying altitudes from 50–150 m, this enables flexible flight planning and full user control concerning the properties of the resulting 3D point clouds w.r.t. point density, spatial resolution, achievable depth performance, and aerial coverage.

For assessing the overall quality, a flight mission was designed and carried out on 28 August 2019, at the pre-Alpine Pielach River (Austria) with the sensor mounted on an octocopter UAV platform. After geometric calibration and refraction correction, the resulting 3D point clouds were compared to reference points obtained by terrestrial surveys with RTK GNSS and total station. The evaluations exhibited a local measurement noise at smooth asphalt surfaces of 1–3 mm, a relative strip fitting precision of about 1 cm, and an absolute flight block accuracy of 2–3 cm compared to check points on dry land.

Assessment of the bathymetric accuracy yielded a depth-dependent bias when employing a representative refraction coefficient of $n = 1.36$ for run-time correction in water. The maximum deviation of 7.8 cm at a water depth of 2.1 m only disappeared when using a hypothetical refraction coefficient of $n = 1.42$, which is physically implausible. We hypothesize that the bias results from multi-path effects caused by forward scattering at dissolved sediment particles, as turbidity was high (SD: $\approx$1.1 m) due to thunderstorm events in the days before data acquisition.

Depth performance evaluation, based on four metal plates lowered into a freshwater pond in different depths, confirmed a maximum water penetration depth of 2 SD for the laser echoes derived with online waveform processing. In addition to that, the entire echo waveform is also stored for off-line waveform analysis. Waveform recording does not depend on a triggering event but the entire waveform is rather captured within a user-definable range gate. This enables waveform stacking by summing up the waveforms of neighboring laser shots, which potentially increases the depth performance to 2.5–2.8 SD. Respective experiments addressing these open questions are currently in preparation.

To sum up, the presented compact topo-bathymetric laser scanner is well suited for mapping river channel bathymetry. The sensor system poses an alternative to conventional ALB for mapping smaller rivers and shallow lakes when mounted on flexible UAV-platforms and also for larger coastal environments when integrated on light manned aircraft. One of the main benefits compared to other UAV-based bathymetric laser sensors is the full adjustability of the sensor parameters enabling the end-user to balance accuracy, depth performance, spatial resolution, and aerial coverage.

Author Contributions: G.M. designed the data acquisition (selection of study area, flight mission parameters, reference measurements) and was responsible for data processing (geo-referencing, performance assessment), writing main parts of the text, and supervision the overall manuscript preparation. M.P. provided the text for sensor concept section. R.S. has performed the waveform analysis and other data processing tasks. S.F. and L.N. designed and installed the measurement field for accuracy assessment, and conducted the terrestrial surveying. All authors contributed to the interpretation of the results and to revision of the manuscript. All authors have read and agreed to the published version of the manuscript.

Funding: The presented investigation was partially conducted within the project "Bathymetry by Fusion of Airborne Laser Scanning and Multi-Spectral Aerial Imagery" (SO 935/6-2) funded by the German Research Foundation (DFG). The work of Martin Pfennigbauer and Roland Schwarz was funded by the Austrian Research Funding Agency (FFG) within the project "Aerial search & Rescue support and supErvision of inAccessible terrainS" (AREAS) 86702.

Acknowledgments: The authors acknowledge TU Wien Bibliothek for financial support through its Open Access Funding Program.

Conflicts of Interest: The authors declare no conflict of interest.

References

1.	Guenther, G.C.; Cunningham, A.G.; Laroque, P.E.; Reid, D.J. Meeting the accuracy challenge in airborne lidar bathymetry. In Proceedings of the 20th EARSeL Symposium: Workshop on Lidar Remote Sensing of Land and Sea, Dresden, Germany, 16–17 June 2000.

2.   Parrish, C.E.; Dijkstra, J.A.; O'Neil-Dunne, J.P.M.; McKenna, L.; Pe'eri, S. Post-Sandy Benthic Habitat Mapping Using New Topobathymetric Lidar Technology and Object-Based Image Classification. *J. Coast. Res.* **2016**, *76*, 200–208. [CrossRef]

3.   Kinzel, P.J.; Legleiter, C.J.; Nelson, J.M. Mapping River Bathymetry With a Small Footprint Green LiDAR: Applications and Challenges. *JAWRA J. Am. Water Resour. Assoc.* **2013**, *49*, 183–204. [CrossRef]

4.   Mandlburger, G.; Hauer, C.; Wieser, M.; Pfeifer, N. Topo-bathymetric LiDAR for monitoring river morphodynamics and instream habitats-A case study at the Pielach River. *Remote Sens.* **2015**, *7*, 6160–6195. [CrossRef]

5.   Maas, H.G.; Mader, D.; Richter, K.; Westfeld, P. Improvements in lidar bathymetry data analysis. *ISPRS Int. Arch. Photogramm. Remote Sens. Spat. Inf. Sci.* **2019**, *XLII-2/W10*, 113–117. [CrossRef]

6.   Mitchell, S.E.; Thayer, J.P. Ranging through Shallow Semitransparent Media with Polarization Lidar. *J. Atmos. Ocean. Technol.* **2014**, *31*, 681–697. [CrossRef]

7.   Goosen, R. This Is How Airborne Multibeam Lidar Coastal Mapping in Paradise is Done. 2019. Available online: https://www.hydro-international.com/content/article/this-is-how-airborne-multibeam-lidar-coastal-mapping-in-paradise-is-done (accessed on 8 February 2020).

8.   Fugro. Rapid Airborne Multibeam Mapping System, ALB. 2020. Available online: https://www.fugro.com/about-fugro/our-expertise/innovations/rapid-airborne-multibeam-mapping-system (accessed on 8 February 2020).

9.   Riegl. VQ-840-G topo-hydrographic full waveform scanner data sheet. 2020. Available online: http://www.riegl.com/uploads/tx_pxpriegldownloads/RIEGL_VQ-840-G_Preliminary-Datasheet_2019-09-02.pdf (accessed on 17 March 2020).

10.   Murase, T.; Tanaka, M.; Tani, T.; Miyashita, Y.; Ohkawa, N.; Ishiguro, S.; Suzuki, Y.; Kayanne, H.; Yamano, H. A photogrammetric correction procedure for light refraction effects at a two-medium boundary. *Photogramm. Eng. Remote Sens.* **2008**, *74*, 1129–1136. [CrossRef]

11.   Lyzenga, D.R. Passive remote sensing techniques for mapping water depth and bottom features. *Appl. Opt.* **1978**, *17*, 379–383. [CrossRef]

12.   Pfennigbauer, M.; Ullrich, A.; Steinbacher, F.; Aufleger, M. High-resolution hydrographic airborne laser scanner for surveying inland waters and shallow coastal zones. *Laser Radar Technol. Appl.* **2011**, *8037*, 6. [CrossRef]

13.   Tonina, D.; McKean, J.A.; Benjankar, R.M.; Wright, C.W.; Goode, J.R.; Chen, Q.; Reeder, W.J.; Carmichael, R.A.; Edmondson, M.R. Mapping river bathymetries: Evaluating topobathymetric LiDAR survey. *Earth Surf. Process. Landforms* **2019**, *44*, 507–520. [CrossRef]

14.   Colomina, I.; Molina, P. Unmanned aerial systems for photogrammetry and remote sensing: A review. *ISPRS J. Photogramm. Remote Sens.* **2014**, *92*, 79–97. [CrossRef]

15.   Schönberger, J.L.; Frahm, J. Structure-from-Motion Revisited. In Proceedings of the 2016 IEEE Conference on Computer Vision and Pattern Recognition (CVPR), Las Vegas, NV, USA, 27–30 June 2016; pp. 4104–4113. [CrossRef]

16.   Hirschmuller, H. Stereo Processing by Semiglobal Matching and Mutual Information. *IEEE Trans. Pattern Anal. Mach. Intell.* **2008**, *30*, 328–341. [CrossRef]

17.   Anderson, K.; Westoby, M.J.; James, M.R. Low-budget topographic surveying comes of age: Structure from motion photogrammetry in geography and the geosciences. *Prog. Phys. Geogr. Earth Environ.* **2019**, *43*, 163–173. [CrossRef]

18.   Nesbit, P.R.; Hugenholtz, C.H. Enhancing UAV–SfM 3D Model Accuracy in High-Relief Landscapes by Incorporating Oblique Images. *Remote Sens.* **2019**, *11*, 239. [CrossRef]

19.   Agisoft. Metashape—Photogrammetric Processing of Digital Images and 3D Spatial Data Generation. 2020. Available online: http://www.agisoft.com/ (accessed on 8 February 2020).

20.   Pix4D. Pix4Dmapper: Professional Drone Mapping and Photogrammetry Software. 2020. Available online: https://www.pix4d.com/product/pix4dmapper-photogrammetry-software (accessed on 8 February 2020).

21.   Rothermel, M.; Wenzel, K.; Fritsch, D.; Haala, N. SURE: Photogrammetric surface reconstruction from imagery. In Proceedings of the Low Cost 3D Workshop, Berlin, Germany, 6–7 December 2012.

22.   Mancini, F.; Dubbini, M.; Gattelli, M.; Stecchi, F.; Fabbri, S.; Gabbianelli, G. Using Unmanned Aerial Vehicles (UAV) for High-Resolution Reconstruction of Topography: The Structure from Motion Approach on Coastal Environments. *Remote Sens.* **2013**, *5*, 6880–6898. [CrossRef]

23. Templin, T.; Popielarczyk, D.; Kosecki, R. Application of Low-Cost Fixed-Wing UAV for Inland Lakes Shoreline Investigation. *Pure Appl. Geophys.* **2018**, *175*, 3263–3283. [CrossRef]
24. Maas, H.G. On the Accuracy Potential in Underwater/Multimedia Photogrammetry. *Sensors* **2015**, *15*, 18140–18152. [CrossRef]
25. Westaway, R.M.; Lane, S.N.; Hicks, D.M. Remote sensing of clear-water, shallow, gravel-bed rivers using digital photogrammetry. *Photogramm. Eng. Remote Sens.* **2001**, *67*, 1271–1281.
26. Dietrich, J.T. Bathymetric Structure-from-Motion: extracting shallow stream bathymetry from multi-view stereo photogrammetry. *Earth Surf. Process. Landforms* **2016**, *42*, 355–364. [CrossRef]
27. Carrivick, J.L.; Smith, M.W. Fluvial and aquatic applications of Structure from Motion photogrammetry and unmanned aerial vehicle/drone technology. *WIREs Water* **2019**, *6*, e1328. [CrossRef]
28. Agrafiotis, P.; Skarlatos, D.; Georgopoulos, A.; Karantzalos, K. Shallow Water Bathymetry Mapping From Uav Imagery Based on Machine Learning. *ISPRS Int. Arch. Photogramm. Remote Sens. Spat. Inf. Sci.* **2019**, *XLII-2/W10*, 9–16. [CrossRef]
29. Agrafiotis, P.; Karantzalos, K.; Georgopoulos, A.; Skarlatos, D. Correcting Image Refraction: Towards Accurate Aerial Image-Based Bathymetry Mapping in Shallow Waters. *Remote Sens.* **2020**, *12*, 322. [CrossRef]
30. Agrafiotis, P.; Skarlatos, D.; Georgopoulos, A.; Karantzalos, K. DepthLearn: Learning to Correct the Refraction on Point Clouds Derived from Aerial Imagery for Accurate Dense Shallow Water Bathymetry Based on SVMs-Fusion with LiDAR Point Clouds. *Remote Sens.* **2019**, *11*, 2225. [CrossRef]
31. Mandlburger, G.; Lehner, H.; Pfeifer, N. A Comparison of Single Photon and Full Waveform Lidar. *ISPRS Ann. Photogramm. Remote Sens. Spat. Inf. Sci.* **2019**, *4*, 397–404. [CrossRef]
32. Lyzenga, D.R.; Malinas, N.P.; Tanis, F.J. Multispectral bathymetry using a simple physically based algorithm. *IEEE Trans. Geosci. Remote Sens.* **2006**, *44*, 2251–2259. [CrossRef]
33. Legleiter, C.J.; Dar, A.R.; Rick, L.L. Spectrally based remote sensing of river bathymetry. *Earth Surf. Process. Landforms* **2009**, *34*, 1039–1059. [CrossRef]
34. Legleiter, C.J. Inferring river bathymetry via Image to Depth Quantile Transformation (IDQT). *Water Resour. Res.* **2016**, *52*, 3722–3741. [CrossRef]
35. Legleiter, C.J.; Fosness, R.L. Defining the Limits of Spectrally Based Bathymetric Mapping on a Large River. *Remote Sens.* **2019**, *11*, 665. [CrossRef]
36. Hernandez, W.J.; Armstrong, R.A. Deriving Bathymetry from Multispectral Remote Sensing Data. *J. Mar. Sci. Eng.* **2016**, *4*, 8. [CrossRef]
37. Sagawa, T.; Yamashita, Y.; Okumura, T.; Yamanokuchi, T. Satellite Derived Bathymetry Using Machine Learning and Multi-Temporal Satellite Images. *Remote Sens.* **2019**, *11*, 1155. [CrossRef]
38. Gentile, V.; Mróz, M.; Spitoni, M.; Lejot, J.; Piégay, H.; Demarchi, L. Bathymetric Mapping of Shallow Rivers with UAV Hyperspectral Data. In Proceedings of the Fifth International Conference on Telecommunications and Remote Sensing, Milan, Italy, 10–11 October 2016; pp. 43–49. [CrossRef]
39. Birkebak, M.; Eren, F.; Pe'eri, S.; Weston, N. The Effect of Surface Waves on Airborne Lidar Bathymetry (ALB) Measurement Uncertainties. *Remote Sens.* **2018**, *10*, 453. [CrossRef]
40. Quadros, N.; Keysers, J. Emerging Trends in Bathymetric Lidar Technology. 2018. Available online: https://www.hydro-international.com/content/article/emerging-trends-in-bathymetric-lidar-technology (accessed on 8 February 2020).
41. ASTRALiTe. Website of ASTRALiTe, Inc. 2020. Available online: https://www.astralite.net/ (accessed on 8 February 2020).
42. Mitchell, S.; Thayer, J.P.; Hayman, M. Polarization lidar for shallow water depth measurement. *Appl. Opt.* **2010**, *49*, 6995–7000. [CrossRef] [PubMed]
43. Wilder Young, J. Little Topo-Bathy Lidar. 2017. Available online: https://lidarmag.com/2017/09/17/little-topo-bathy-lidar/= (accessed on 8 February 2020).
44. ASTRALite. Press Release: ASTRALiTe Demonstrates Scanning Topo–Bathy LiDAR System on DJI Matrice 600 Pro. 2018. Available online: https://www.businesswire.com/news/home/20181119005609/en/ASTRALiTe-Demonstrates-Scanning-Topo%E2%80%93Bathy-LiDAR-System-DJI (accessed on 8 February 2020).
45. SBG Systems. UAV-Based LiDAR Can Measure Shallow Water Depth. 2019. Available online: https://spectrum.ieee.org/robotics/drones/uavbased-lidar-can-measure-shallow-water-depth (accessed on 8 February 2020).

46. Kinzel, P.J.; Legleiter, C.J. sUAS-Based Remote Sensing of River Discharge Using Thermal Particle Image Velocimetry and Bathymetric Lidar. *Remote Sens.* **2019**, *11*, 2317. [CrossRef]

47. Mitchell, T. From PILLS To RAMMS. In Proceedings of the 20th Annual JALBTCX Airborne Coastal Mapping and Charting Technical Workshop, South Bend, Indiana, 4–6 June 2019.

48. Zuckerman, S. PILLS 2.5: From Design to Operations. In Proceedings of the 20th Annual JALBTCX Airborne Coastal Mapping and Charting Technical Workshop, South Bend, Indiana, 4–6 June 2019.

49. Fugro. 2020. Available online: https://lidarmag.com/2019/11/13/fugro-ramms-technology-benefits-us-navy-mapping-system/ (accessed on 8 February 2020).

50. Amuse Oneself Inc. Product website and spec sheet of TDOT GREEN. 2020. Available online: https://amuse-oneself.com/en/service/tdotgreen (accessed on 8 February 2020).

51. Kasvi, E.; Salmela, J.; Lotsari, E.; Kumpula, T.; Lane, S. Comparison of remote sensing based approaches for mapping bathymetry of shallow, clear water rivers. *Geomorphology* **2019**, *333*, 180–197. [CrossRef]

52. Shintani, C.; Fonstad, M.A. Comparing remote-sensing techniques collecting bathymetric data from a gravel-bed river. *Int. J. Remote Sens.* **2017**, *38*, 2883–2902. [CrossRef]

53. Eren, F.; Jung, J.; Parrish, C.E.; Sarkozi-Forfinski, N.; Calder, B.R. Total Vertical Uncertainty (TVU) Modeling for Topo-Bathymetric LIDAR Systems. *Photogramm. Eng. Remote Sens.* **2019**, *85*, 585–596. [CrossRef]

54. Saylam, K.; Hupp, J.R.; Andrews, J.R.; Averett, A.R.; Knudby, A.J. Quantifying Airborne Lidar Bathymetry Quality-Control Measures: A Case Study in Frio River, Texas. *Sensors* **2018**, *18*, 4153. [CrossRef]

55. Steinvall, O.K.; Koppari, K.R.; Karlsson, U.C.M. Experimental evaluation of an airborne depth-sounding lidar. *Opt. Eng.* **1993**, *32*, 1307. [CrossRef]

56. Hilldale, R.; Raff, D. Assessing the ability of airborne LiDAR to map river bathymetry. *Earth Surf. Process. Landforms* **2008**, *33*, 773–783. [CrossRef]

57. Fernandez-Diaz, J.; Glennie, C.; Carter, W.; Shrestha, R.; Sartori, M.; Singhania, A.; Legleiter, C.; Overstreet, B. Early Results of Simultaneous Terrain and Shallow Water Bathymetry Mapping Using a Single-Wavelength Airborne LiDAR Sensor. *IEEE J. Sel. Top. Appl. Earth Obs. Remote Sens.* **2014**, *7*, 623–635. [CrossRef]

58. Fernandez-Diaz, J.C.; Carter, W.E.; Glennie, C.; Shrestha, R.L.; Pan, Z.; Ekhtari, N.; Singhania, A.; Hauser, D.; Sartori, M. Capability Assessment and Performance Metrics for the Titan Multispectral Mapping Lidar. *Remote Sens.* **2016**, *8*, 936. [CrossRef]

59. Legleiter, C.J.; Overstreet, B.T.; Glennie, C.L.; Pan, Z.; Fernandez-Diaz, J.C.; Singhania, A. Evaluating the capabilities of the CASI hyperspectral imaging system and Aquarius bathymetric LiDAR for measuring channel morphology in two distinct river environments. *Earth Surf. Process. Landforms* **2016**, *41*, 344–363. [CrossRef]

60. Wright, C.W.; Kranenburg, C.; Battista, T.A.; Parrish, C. Depth Calibration and Validation of the Experimental Advanced Airborne Research Lidar, EAARL-B. *J. Coast. Res.* **2016**, *76*, 4–17. [CrossRef]

61. Mandlburger, G.; Pfennigbauer, M.; Wieser, M.; Riegl, U.; Pfeifer, N. Evaluation of a novel uav-borne topo-bathymetric laser profiler. *Int. Arch. Photogramm. Remote Sens. Spat. Inf. Sci.* **2016**, *XLI-B1*, 933–939. [CrossRef]

62. Pfennigbauer, M.; Wolf, C.; Weinkopf, J.; Ullrich, A. Online waveform processing for demanding target situations. *Proc. SPIE* **2014**, 90800J. [CrossRef]

63. Montgomery, D.R.; Buffington, J.M. Channel reach morphology in mountain drainage basins. *GSA Bull.* **1997**, *109*, 596–611. [CrossRef]

64. Melcher, A.H.; Schmutz, S. The importance of structural features for spawning habitat of nase Chondrostoma nasus (L.) and barbel Barbus barbus (L.) in a pre-Alpine river. *River Syst.* **2010**, *19*, 33–42. [CrossRef]

65. Zitek, A.; Schmutz, S.; Jungwirth, M. Assessing the efficiency of connectivity measures with regard to the EU-Water Framework Directive in a Danube-tributary system. *Hydrobiologia* **2008**, *609*, 139–161. [CrossRef]

66. Mandlburger, G.; Pfennigbauer, M.; Riegl, U.; Haring, A.; Wieser, M.; Glira, P.; Winiwarter, L. Complementing airborne laser bathymetry with UAV-based lidar for capturing alluvial landscapes. *Proc. SPIE* **2015**, 9637. [CrossRef]

67. Wieser, M.; Mandlburger, G.; Hollaus, M.; Otepka, J.; Glira, P.; Pfeifer, N. A Case Study of UAS Borne Laser Scanning for Measurement of Tree Stem Diameter. *Remote Sens.* **2017**, *9*, 1154. [CrossRef]

68. Pfeifer, N.; Mandlburger, G.; Otepka, J.; Karel, W. OPALS—A framework for Airborne Laser Scanning data analysis. *Comput. Environ. Urban Syst.* **2014**, *45*, 2. [CrossRef]

69. Hutton, J.J.; Gopaul, N.; Zhang, X.; Wang, J.; Menon, V.; Rieck, D.; Kipka, A.; Pastor, F. Centimeter-Level, Robust Gnss-Aided Inertial Post-Processing for Mobile Mapping Without Local Reference Stations. *ISPRS Int. Arch. Photogramm. Remote Sens. Spat. Inf. Sci.* **2016**, *XLI-B3*, 819–826. [CrossRef]
70. Glira, P.; Pfeifer, N.; Briese, C.; Ressl, C. A Correspondence Framework for ALS Strip Adjustments based on Variants of the ICP Algorithm. *PFG Photogramm. Fernerkundung, Geoinf.* **2015**, *2015*, 275–289. [CrossRef]
71. Glira, P.; Pfeifer, N.; Mandlburger, G. Rigorous Strip Adjustment of UAV-based Laserscanning Data Including Time-Dependent Correction of Trajectory Errors. *Photogramm. Eng. Remote Sens.* **2016**, *82*, 945–954. [CrossRef]
72. Ressl, C.; Kager, H.; Mandlburger, G. Quality checking of als projects using statistics of strip differences. *Int. Arch. Photogramm. Remote Sens.* **2008**, *37*, 253–260.
73. Mandlburger, G.; Hauer, C.; Höfle, B.; Habersack, H.; Pfeifer, N. Optimisation of LiDAR derived terrain models for river flow modelling. *Hydrol. Earth Syst. Sci.* **2009**, *13*, 1453–1466. [CrossRef]
74. Westfeld, P.; Maas, H.G.; Richter, K.; Weiß, R. Analysis and correction of ocean wave pattern induced systematic coordinate errors in airborne LiDAR bathymetry. *ISPRS J. Photogramm. Remote Sens.* **2017**, *128*, 314–325. [CrossRef]
75. Schwarz, R.; Pfeifer, N.; Pfennigbauer, M.; Mandlburger, G. Depth Measurement Bias in Pulsed Airborne Laser Hydrography Induced by Chromatic Dispersion. *IEEE Geosci. Remote Sens. Lett.* **2020**. submitted.
76. Mobley, C.D. *Light and Water: Radiative Transfer in Natural Waters*; Academic Press: Cambridge, MA, USA, 1994.
77. Mandlburger, G.; Pfennigbauer, M.; Pfeifer, N. Analyzing near water surface penetration in laser bathymetry—A case study at the River Pielach. *ISPRS Ann. Photogramm. Remote Sens. Spat. Inf. Sci.* **2013**, *2*. [CrossRef]
78. Mandlburger, G.; Jutzi, B. On the feasibility of water surface mapping with single photon lidar. *ISPRS Int. J. -Geo-Inf.* **2019**, *8*, 188. [CrossRef]
79. Allouis, T.; Bailly, J.S.; Pastol, Y.; Le Roux, C. Comparison of LiDAR waveform processing methods for very shallow water bathymetry using Raman, near-infrared and green signals. *Earth Surf. Process. Landforms* **2010**, *35*, 640–650. [CrossRef]
80. Kogut, T.; Bakuła, K. Improvement of Full Waveform Airborne Laser Bathymetry Data Processing based on Waves of Neighborhood Points. *Remote Sens.* **2019**, *11*, 1255. [CrossRef]
81. Schwarz, R.; Mandlburger, G.; Pfennigbauer, M.; Pfeifer, N. Design and evaluation of a full-wave surface and bottom-detection algorithm for LiDAR bathymetry of very shallow waters. *ISPRS J. Photogramm. Remote Sens.* **2019**, *150*. [CrossRef]
82. Mader, D.; Richter, K.; Westfeld, P.; Weiß, R.; Maas, H.G. Detection and Extraction of Water Bottom Topography From Laserbathymetry Data by Using Full-Waveform-Stacking Techniques. *ISPRS Int. Arch. Photogramm. Remote Sens. Spat. Inf. Sci.* **2019**, *XLII-2/W13*, 1053–1059. [CrossRef]
83. Tymków, P.; Jóźków, G.; Walicka, A.; Karpina, M.; Borkowski, A. Identification of Water Body Extent Based on Remote Sensing Data Collected with Unmanned Aerial Vehicle. *Water* **2019**, *11*, 338. [CrossRef]

 *remote sensing*

Article

Quantifying Below-Water Fluvial Geomorphic Change: The Implications of Refraction Correction, Water Surface Elevations, and Spatially Variable Error

Amy S. Woodget [1,*], **James T. Dietrich** [2] **and Robin T. Wilson** [3]

[1] Department of Geography and Environment, School of Social Sciences and Humanities, Loughborough University, Loughborough LE11 3TU, UK

[2] Department of Geography, University of Northern Iowa, Cedar Falls, IA 50614, USA; james.dietrich@uni.edu

[3] Independent Scholar, Southampton SO16 6DB, UK; robin@rtwilson.com

[*] Correspondence: a.woodget@lboro.ac.uk; Tel.: +44-1509-222791

Received: 12 September 2019; Accepted: 15 October 2019; Published: 18 October 2019

Abstract: Much of the geomorphic work of rivers occurs underwater. As a result, high resolution quantification of geomorphic change in these submerged areas is important. Currently, to quantify this change, multiple methods are required to get high resolution data for both the exposed and submerged areas. Remote sensing methods are often limited to the exposed areas due to the challenges imposed by the water, and those remote sensing methods for below the water surface require the collection of extensive calibration data in-channel, which is time-consuming, labour-intensive, and sometimes prohibitive in difficult-to-access areas. Within this paper, we pioneer a novel approach for quantifying above- and below-water geomorphic change using Structure-from-Motion photogrammetry and investigate the implications of water surface elevations, refraction correction measures, and the spatial variability of topographic errors. We use two epochs of imagery from a site on the River Teme, Herefordshire, UK, collected using a remotely piloted aircraft system (RPAS) and processed using Structure-from-Motion (SfM) photogrammetry. For the first time, we show that: (1) Quantification of submerged geomorphic change to levels of accuracy commensurate with exposed areas is possible without the need for calibration data or a different method from exposed areas; (2) there is minimal difference in results produced by different refraction correction procedures using predominantly nadir imagery (small angle vs. multi-view), allowing users a choice of software packages/processing complexity; (3) improvements to our estimations of water surface elevations are critical for accurate topographic estimation in submerged areas and can reduce mean elevation error by up to 73%; and (4) we can use machine learning, in the form of multiple linear regressions, and a Gaussian Naïve Bayes classifier, based on the relationship between error and 11 independent variables, to generate a high resolution, spatially continuous model of geomorphic change in submerged areas, constrained by spatially variable error estimates. Our multiple regression model is capable of explaining up to 54% of magnitude and direction of topographic error, with accuracies of less than 0.04 m. With on-going testing and improvements, this machine learning approach has potential for routine application in spatially variable error estimation within the RPAS–SfM workflow.

Keywords: fluvial; geomorphology; change detection; remotely piloted aircraft system; refraction correction; structure-from-motion photogrammetry; water surface elevation; topographic error; machine learning

---

1. Introduction

High spatial resolution, spatially continuous, accurate, and precise methods for quantifying topographic change in fluvial environments are important for a range of river science and management applications. For instance, such data can provide valuable input to sediment routing and budgeting models (e.g., [1,2]), help predict channel dynamics through the provision of channel texture and roughness data [3–6], assist in the monitoring of patterns of geomorphic adjustment [7,8] to significant external drivers such as climate change [9], and reveal trends in channel capacity, which in turn feed flood risk models and insurance forecasting [10,11]. Such data are also essential in assessments of habitat quality and availability [12,13]. Significant early work on quantifying topography and geomorphic change, some of which has been undertaken in fluvial environments, made use of photogrammetry [14,15], terrestrial and airborne laser altimetry [8,16,17], and extensive GPS- or total station-based topographic surveys [7,18]. Such studies tend to use either multiple approaches for characterising the exposed and submerged parts of the system separately, which are subsequently fused together, or dense networks of irregularly spaced individual point measurements, which are exceptionally time-consuming to collect and must then be interpolated to form an elevation model. The resulting models typically have spatial resolutions no finer than 1 m/pixel. Ideally, a single method of quantifying topographic change in all parts of the river system could be used to enable more rapid and consistent data acquisition for monitoring and management of river habitats. Such a method should have high accuracy and precision, a very high spatial resolution (i.e., at the grain scale, <0.1 m/pixel), and provide spatially continuous data (typically as rasters or point clouds) rather than interpolations between sparse point samples (such as traditional topographic surveys, total station surveys, global navigation satellite system (GNSS) surveys, and cross sections or longitudinal profiles). This ideal method would be in alignment with the popular 'riverscape' paradigm [19], which conceptualises fluvial systems as highly heterogeneous rather than gradually varying and would provide physical habitat data at the grain scale, which is of most relevance to aquatic biota (i.e., <0.1 m spatial resolution over reach-scale extents).

Recently, we have seen a proliferation of studies offering high spatial resolution topographic data over continuous spatial extents in a variety of geomorphological settings, without the need for interpolation between data points (e.g., [4,8,20–26]). Such studies have demonstrated the use of airborne and terrestrial LiDAR and Structure-from-Motion photogrammetry using imagery from remotely piloted aircraft systems (RPAS). In fluvial environments specifically, the acquisition of survey data is complicated in submerged areas where the effects of refraction, turbidity, and water surface roughness represent well-known challenges [27,28]. However, these submerged spaces are where a majority of fluvial 'work' occurs (i.e., erosion and deposition resulting from variations in flow magnitude and shear stress). To date, most fluvial studies aimed at measuring spatially-continuous topography and/or topographic change at high spatial resolutions offer just one of the following:

(a) Coverage within submerged areas only, for a single point in time. This is particularly applicable in the case of new and emerging technologies, which have not yet progressed much beyond proof of concept. For example, Legleiter et al. [29] evaluates a range of sensors for deriving fluvial bathymetry data, including a RPAS-mounted hyperspectral sensor with a moderately high spatial resolution of 0.18 m/pixel.

(b) Coverage for multiple points in time (thereby permitting change assessments) but only for exposed areas, with a specific focus on bank erosion or floodplain dynamics, using one or more survey methods (e.g., [26,30–34]).

(c) Complete fluvial coverage (i.e., both exposed and submerged areas) but only for a single point in time and requiring the use of a combination of survey methods. For example, Javemick et al. [35] uses a combination of helicopter-acquired imagery, processed using Structure-from-Motion (SfM) photogrammetry, for exposed areas and bathymetric echo-sounding in submerged areas. In other settings, such as coastal and shallow marine environments, a combination of approaches is also common (e.g., [36]).

(d) Complete fluvial coverage for multiple points in time, but with the need for extensive fieldwork to collect calibration data within the submerged parts of the channel, which can be dangerous or prohibitively time-consuming in some settings. For example, Flener et al. [37] uses an optical bathymetric approach on images acquired by RPAS (remotely piloted aircraft system) to obtain data in submerged areas. This approach requires the associated acquisition of bathymetry elevations using a GNSS to calibrate the relationship between the spectral image data and the water depth (from which elevation can be inferred). Similar workflows are employed by [38].

As such, no study has yet demonstrated and rigorously assessed an approach which uses a single method capable of quantifying spatially continuous fluvial geomorphic change at grain scales, both above and below the water surface, without the need to collect extensive calibration data. It is here that this paper aims to contribute.

1.1. Refraction Correction

Since 2015, a method which uses high resolution RPAS (or UAS/unmanned aerial system) imagery, processed using SfM, has been shown as a promising tool for measuring both above-water (bank and floodplain) and below-water (bathymetric) topography at the grain scale in fluvial environments, when paired with refraction correction post-processing [27]. Above-water use of the RPAS–SfM approach has already provided an important step-change in our ability to quantify exposed topography without the greater expense of a terrestrial laser scanning system (e.g., [34,38–40]), but its use in submerged areas is less well established, due to the additional challenges posed by the presence of water and, in particular, the effects of refraction. Recently, two key RPAS–SfM methods have been proposed by the authors for use in submerged areas [27,28].

In 2015, Woodget et al. [27] published details of a simple approach for refraction correction of bathymetric elevation models using a small angle approximation. This RPAS–SfM approach generated high resolution orthophotos and digital elevation models (DEMs) of submerged areas. These datasets were analysed in a GIS environment to extract the position and elevation of the water's edge and create an elevation model for the position of the water surface. Snell's Law was then applied to this surface to increase the apparent water depths by a multiplier of 1.34 (i.e., the refractive index of clear water) to account for the effects of refraction (which cause the channel bed to appear closer to the camera than it really is). This workflow provides a relatively simple method for use in shallow (<1 m), non-turbid settings and is easily implemented by other researchers and practitioners (e.g., [41,42]). Furthermore, this method was shown to reduce the magnitude of mean errors associated with bathymetric topography estimations by up to 50% (i.e., accuracy of 0.02–0.05 m and precision of 0.06–0.09 m). It was not able, however, to eliminate completely the systematic increase in error with increasing water depth, and difficulties in accurately establishing the position of the water's edge were noted.

In 2017, Dietrich [28] further developed this method by proposing a more sophisticated multi-angle refraction correction approach. Instead of using the orthophotos and DEMs generated by the RPAS–SfM process, Dietrich used the dense three-dimensional (3D) point clouds (which describe the 3D topography of the scene) and applied an iterative refraction correction method based on the positions and view angles of all the cameras used within the photogrammetric reconstruction of the position of each point in the cloud. As found by [27], the challenges of accurately defining the water's edge and the compounding effect on bathymetric accuracy were noted. Dietrich's method requires more inputs than the small angle approach. Guidance on its implementation is available online and is accompanied by a dedicated python script ([43]: pyBathySfM v4.0). Furthermore, Dietrich reports improved error statistics associated with this method of refraction correction, with typical accuracy values of ± 0.01 m and precision values of 0.06–0.08 m. However, a direct comparison of these two approaches using the same input data has not yet been attempted, and nor has either method been tested for quantifying topographic change in the submerged parts of river channels. Therefore, our first objective is to investigate how our RPAS–SfM bathymetric refraction correction methods can be used to accurately measure change in submerged parts of rivers. We include in this a direct comparison of the methods of Woodget et al. [27] and Dietrich [28].

1.2. Water Surface Elevation

Recognising that one of the key impediments to accurate and reliable refraction correction relates to the challenges of defining accurately the position of the water's edge and the elevation of the water surface, our second objective is to explore different methods for estimating water surface elevations and their influence on the accuracy and precision of below-water topography estimates and estimates of change over time. Knowledge of the water surface elevation is paramount within both refraction correction approaches [27,28], as it forms the lynchpin for computing the apparent water depth, and subsequently the true water depths and refraction-corrected elevations using the geometry of refraction described by Snell's Law.

1.3. Spatially Variable Elevation Error

The quantification of topographic elevation change in any setting, and particularly within this initial assessment of refraction-corrected bathymetric change, requires the establishment of appropriate error thresholds for each epoch of data and propagation of this error into models of change, in order to distinguish real change from noise [1,16,23,44]. In the past, geomorphic change detection studies have either (a) recognised the limitations of change quantification but not defined or implemented a specific error threshold (e.g., [45]), or (b) have employed spatially uniform error thresholds based on surface topography, data acquisition method, and/or post-processing (known as a 'level of detection' or LoD, e.g., [8,46,47]). However, such broad approaches run the risk of masking real change in some places and detecting change which is not real in others [32,48]. Therefore, as others have done before us, we highlight the need for spatially variable error thresholds [24,32,48–50] and implement them within our work to optimise the accuracy of our change measurements. Previously, spatially variable error thresholds have been based on measures associated with broad polygons of wet or dry areas or of different survey methods (e.g., [8,16,38]), or on spatially-continuous measures of local topographic variability such as roughness, grain size, or slope angle or on sampling point density [1,33,48,51]. However, very few studies have tested the combined impact and significance of multiple proxies on elevation error (with the exception of the fuzzy inference systems used by [1]), and none have yet quantified spatially variable error of RPAS–SfM-derived DEMs in submerged areas. Thus, we take this a step further with our final objective, by utilising the full potential of the exceptionally high resolution RPAS imagery (<0.02 m pixel size) to investigate how this detailed information (specifically concerning topographic complexity, landscape composition, and survey quality/conditions) acquired by each RPAS–SfM survey can help to elucidate the direction and magnitude of elevation errors in submerged and exposed areas. Specifically, we explore how this information might be exploited using artificial intelligence approaches (including multiple regression and a machine learning classifier) to inform the development of an appropriate spatially variable error threshold.

To this end, we study specifically the influence of the following variables on elevation error:

1. Topographic complexity: Including slope angle and point cloud roughness.
2. Landscape composition: Including presence/absence of dense vegetation and water depth.
3. Survey quality: Including image quality, SfM point cloud density, and the precision of SfM tie points. We note the latter two SfM variables here are influenced by image texture, and therefore might also be classified as dependent on the nature of the landscape composition.
4. Survey conditions: Including the presence/absence of water surface reflections and roughness and the presence/absence of dark shadows.

These variables can be obtained for each pixel within the RPAS–SfM-derived datasets and many of them have been shown previously to have some bearing on topographic error [1,26,27,48,49,51]. Elevation error is determined by computing the elevation difference between point cloud elevations and the elevation data acquired from >3500 validation points measured using a combination of GNSS and total station surveys. Given that the existing evidence suggests that simple monotonic relationships between these variables and elevation error is unlikely [1,48], we investigate the use of multiple regression and machine learning classification algorithms for predicting elevation error. Machine learning (ML) is a form of artificial intelligence (AI), whereby a 'machine' is trained to recognise statistical patterns and trends within data and use them to predict new data outputs. As such, it includes a broad range of algorithms, encompassing everything from simple linear regressions to deep learning-based neural networks [52]. The application of AI/ML for geomorphic error thresholding purposes is novel, with existing studies within freshwater settings applying it predominantly for classification of land cover types from image-derived parameters (e.g., [53–56]) or for the identification of other specific features of interest such as buildings (e.g., [57]) and invasive species (e.g., [58]). As such, our ultimate aim is to create the first high resolution, spatially continuous SfM-derived topographic change models in submerged fluvial environments constrained by spatially variable error estimates.

2. Materials and Methods

2.1. Site Set-Up

Our field setting is a 600-m-long reach of the River Teme in Herefordshire, near the England/Wales border, where we collected RPAS imagery in August 2016 and August 2017 (Figure 1). This is a lowland, meandering, dynamic, gravel-bed river system with channel widths of 3–13 m and a total catchment drainage area of 160,000 hectares [59]. In-channel hydromorphic diversity is good, palaeochannels are common (often still connected to the main water course), and median grain size (D50) of exposed surfaces is 21.5 mm. Our section of the River Teme has previously experienced significant channel change following a flood event in the winter of 2014–15 (as evidenced by before and after Google Earth imagery).

Prior to the first RPAS image acquisition survey at the River Teme in August 2016, we established six permanent ground survey markers around the outer boundary of our area of interest (PM1–6, Figure 1). These markers comprised a 50-cm-long steel stake, topped with a flat, plastic yellow survey head and sealing cap which features a plumb point recess for accurate alignment of survey poles. We surveyed the position of these markers in August 2016 using a Trimble R8 RTK GNSS (with a manufacturer-reported accuracy 10 mm in horizontal and 20 mm in the vertical) and collected all subsequent survey data relative to these marker positions. Given the sturdy nature of these markers and light grazing land-use in this area, we were confident that the markers would not have moved between surveys and therefore could be used to ensure the accurate alignment of multiple epochs of data. Despite this, we re-surveyed the markers in August 2017 using a Leica VIVA GS10 GNSS and found that the difference from 2016 marker positions was, on average, less than ±0.008 m, ±0.007 m, and ±0.0185 m in XY and Z, respectively. This difference is very minimal and is within the bounds of uncertainty we would expect as a result of random errors, from pole tilt for example. As a result, we assume no movement has occurred within the intervening time. Three of the six markers could not be found on return to the site in 2017, so an additional two markers were added in new locations and surveyed using the Leica VIVA GS10 GNSS (PM7–8, Figure 1). This allows us to maintain the precision of subsequent internal surveys undertaken using a total station.

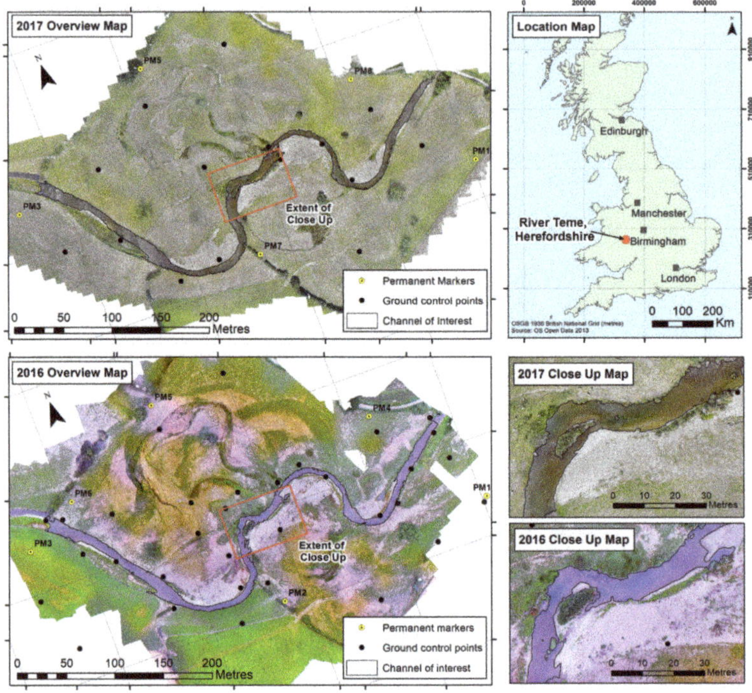

Figure 1. Location map and site overviews (orthoimagery) from 2017 and 2016 surveys.

2.2. RPAS Surveys

In August 2016, we flew manually using a rotary-winged DJI Inspire 1 RPAS over the site at a height of ca. 35 m above ground level. High resolution imagery (<0.02 m ground sample distance (GSD)) was acquired using the associated Zenmuse X3 FC350 camera, predominantly at a nadir viewing angle, with a 60–80% level of overlap. The survey was carried out in dry and calm conditions, with sunny intervals. In August 2017, we acquired our second epoch of high-resolution imagery (<0.2 m GSD) using a DJI Phantom 4 Pro RPAS, which was pre-programmed using DJI's Ground Station Pro application, to collect imagery at nadir with an 80% overlap from an altitude of ca. 35 m. Prevailing weather conditions were comparable to those in August 2016. We also collected a number of oblique images from higher altitudes with each survey to incorporate into the SfM processing, to avoid the risk of systematic doming errors which can otherwise result from inadequate camera lens calibration models [27,60–62].

Prior to each survey, we had distributed a number of ground control points (GCPs) throughout the site (n = 19 in 2016, n = 14 in 2017), ensuring they were positioned to represent the range in elevation distributions across the site (Figure 1). The position of all GCPs was surveyed using a Leica Builder 500 total station in relation to the known positions of the permanent markers.

In 2016, a total of 773 images were acquired, of which 220 images were excluded due to redundancy, blurring, or other visual distortions. In 2017, a total of 1512 images were acquired, of which 471 images were excluded due to redundancy or blurring.

2.3. SfM Processing

Each epoch of imagery was imported into Agisoft's PhotoScan Pro software for processing using SfM photogrammetry (v.1.4.0 build 5650). Image alignment was undertaken with high accuracy and was optimised following the addition of GCP locations within the model. Exports included a ca. 0.01 m resolution orthophoto, ca. 0.04 m resolution digital elevation model, and a dense 3D point cloud for each survey date. All outputs are referenced to British National Grid (OSGB 1936) using the GCPs and permanent markers established at the site.

2.4. Validation Data

In both 2016 and 2017, independent topographic validation data were acquired shortly after the RPAS flights in both exposed and submerged parts of the site. Elevation (Z) and XY positions of each validation point were recorded using the total station and tied into the same coordinate system using the permanent markers. Efforts were made to collect validation data using a random distribution, whilst maintaining coverage of all key parts of the area of interest. Particular attention was paid to acquiring validation data over a range of water depths and along breaks of slope, to ensure the validation was not biased to the more easily accessible, flatter parts of the site. This sampling strategy was prompted by knowledge that breaks of slope tend to be areas of greatest error within elevation models [1]. A total of 1522 validation points were collected in 2016, and 2091 points in 2017 (Figure 2).

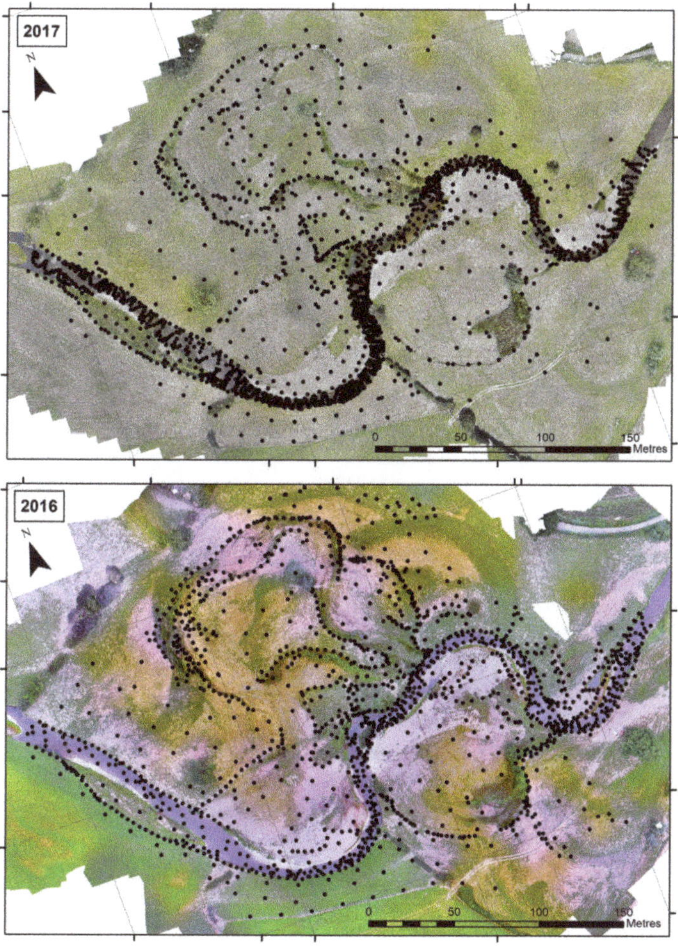

Figure 2. Validation datasets for 2017 and 2016.

2.5. Water Surface Elevations

Water surface elevations are a critical variable in any refraction correction, since the equations derived from Snell's Law are in terms of water depth rather than absolute elevations. As discussed in [28], any errors in water surface elevations translate directly as an error into the refraction-corrected depths and, ultimately, the refraction corrected elevations. As in other studies, we choose to approximate the water surface as planar in the cross-stream direction (bank-to-bank) and spatially variable in the downstream direction.

For each dataset, we digitised polyline features at the approximate edges of the water surface. The digitised lines tended to be along areas of exposed sediment where the water's edges were clearly visible. In the riffles, where the flows were more complex, the polylines followed the main channel (majority of the flow) and in the side channels, additional point features were digitised at regular intervals (~1–2 m) along the water's edge. The side channel points were not used in the smoothing operation below but were used in the final water surface interpolation.

To extract the elevations along the polyline features, we created a 1 cm buffer polygon on each line segment. The buffer polygons were used to select and extract data points directly from the SfM point clouds in CloudCompare. For the side channels, for all point features, an elevation value was assigned by taking an average of the five nearest point cloud points.

The nature of the extracted point cloud points, with variable elevations at the scale of individual sediment clasts, necessitated that the data be statistically smoothed to generate an average downstream elevation profile that approximated the water's edge. Downstream distances were assigned to all of the extracted points using a stream normal coordinate transformation [63], where the edge points are transformed into coordinates with downstream (along a centreline) and cross-stream (orthogonal to the centreline) components while also retaining their original X, Y, and Z coordinates.

We tested seven different statistical smoothing filters, mostly in the "local polynomial" class of filters. These included Loess and Lowess, along with their robust counterparts, Savitzky–Golay, Gaussian, and Moving Average. The filtering combined both right and left bank points to best estimate the planar cross-stream elevations. All these filters operate via a smoothing window. The optimal window size for each was determined by trial and error, qualitatively judging the fit quality, e.g., preserving known downstream variability, while suppressing high-frequency "noise" and outliers. Once the windows were optimised, the downstream water surface elevation profiles from the seven different filters were compared to determine the best fit for the given set of input data.

To generate the final water surface for each dataset, we resampled the smoothed profiles at 2 m intervals downstream. The 2 m intervals were along the centreline of the stream and we constructed orthogonal cross-section lines that extended beyond the banks at each interval (Figure 3). The attributes of each 2 m interval point were copied to new point features at the both ends of the cross-section lines. We combined the 2 m interval points with the original side channel points and used a Delaunay triangulation to interpolate all the points into a coherent water surface mesh. For the multi-view refraction correction processing, the mesh was used to calculate apparent depths in the point cloud via a Point-to-Mesh distance calculation. For the small angle correction, the mesh was "rasterised" into a 5 cm cell size GeoTIFF, and the apparent depths calculated by raster subtraction. We assessed the performance of this smoothed water surface by direct comparison with a manually digitised water surface (generated as per [27]), through validation of the derived submerged elevations obtained by the small angle refraction correction procedure outlined below.

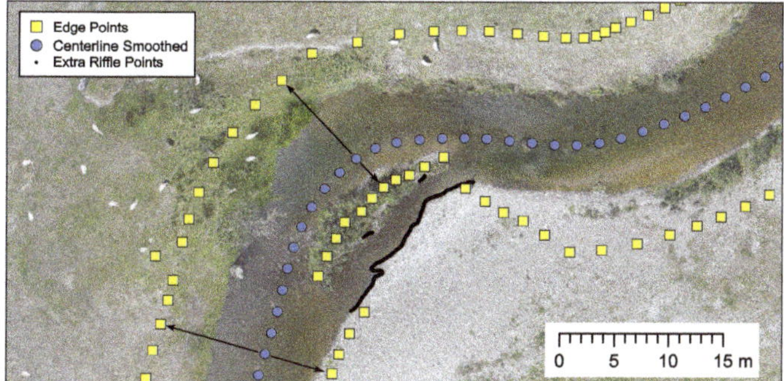

Figure 3. Water surface interpolation points. Centreline points (circles) are generated from the statistical smoothing operations. Edge points (squares) are copies of the centreline shifted orthogonally (black arrows) to create a solid interpolation surface that extends beyond the wetted channel. Additional points are added along the riffles to handle the more complex topography in the side channels.

2.6. Refraction Correction

In our SfM data, the submerged portions of the point cloud are impacted by the refraction of light at the air–water interface. The result is that the submerged point elevations are too shallow (Figure 4). These points therefore represent the apparent depth (h_a) based on the water surfaces described above. To correct the point cloud, we used algebraic manipulations of Snell's Law ($n_1 \sin i = n_2 \sin r$) to adjust the apparent depth to the "true" depth (h).

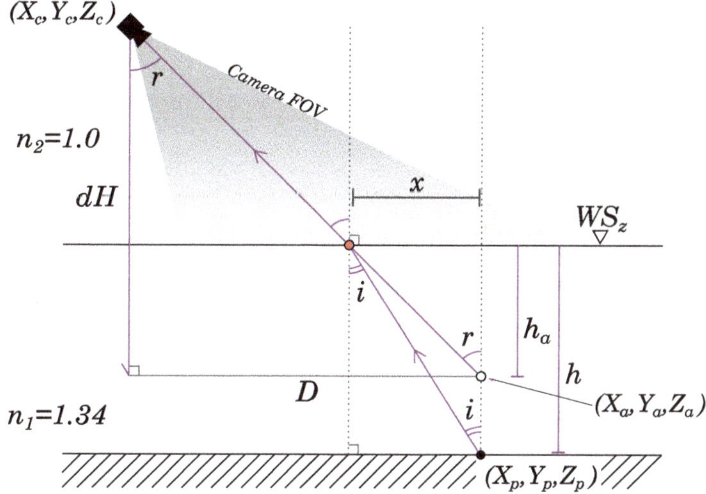

Figure 4. Diagram illustrating through water refraction geometry. Illustrating an off-nadir field of view (FOV) from the camera location (X_c, Y_c, Z_c) down to the water surface (WS_z). Point (X_a, Y_a, Z_a) is at the apparent depth (h_a) and represents the uncorrected SfM point cloud elevations. Point (X_p, Y_p, Z_p) is the "true" position as the "true" depth (h) that the refraction correction is solving. D is the point camera distance, dH is the elevation difference, r is the angle of refraction, i is the angle of incidence, x is the point/water interface distance, n_1 and n_2 are the refractive indicies of air and water. From [28], used with permission.

2.7. Small Angle Refraction Correction

For the small angle refraction correction process, we imported both the 'old' and the 'new' water surface elevation models into GIS as a raster, alongside the orthophoto and DEM outputs from the SfM process. Following the method of [27], the DEM elevations were subtracted from the water surface elevations to estimate the apparent water depth (h_a) for each epoch of data. The refraction correction is calculated using the "small angle substitution" by assuming the angles (r and i) are less than $10°$. From this assumption, the sines in Snell's Law can be substituted with the tangent ($\sin \theta \cong \tan \theta$) d Snell's Law simplifies to the refractive index of water (1.34) multiplied by the apparent depth:

$$h = 1.34 \times h_a \tag{1}$$

Thus, in GIS, the apparent water depth raster layer is multiplied by 1.34 to create a new, refraction corrected depth raster (h). The difference between these two depth rasters ($h - h_a$) is computed, and the result is subtracted from the original DEM to get corrected absolute elevations. This has the effect of lowering the channel elevations in submerged areas, to counter the underestimation of true bed elevations which occurs as a result of refraction. This process was carried out twice, using first the 'old' water surface and then the 'new' water surface, so that we could compare the two surfaces directly.

2.8. Multi-View Refraction Correction (BathySfM)

The multi-view refection correction is described in-depth in [28]. With this new research, the software was updated and renamed (now called 'pyBathySfM'), and is available for download on Github [43]. The new version has additional options and a new graphical user interface to make the tool more user-friendly. The minimum inputs for the bathymetric correction are designed to be software agnostic and include: (1) A point cloud dataset (in CSV format, with a water surface elevations attribute added), (2) the camera positions/orientations (X, Y, Z, Pitch, Roll, Yaw), (3) sensor description (focal length, sensor size [physical and pixel size]). The software calculates all of the points that each camera "sees" in its field of view (calculated via a pinhole camera model). For each point–camera combination, the horizontal distance (D) and difference in elevation (dH) are calculated and used to derive the angle of refraction (r). The point elevation (Z_a) is subtracted from the water surface elevation (WS_z) to get the apparent depth (h_a). From these variables, the other variables in Figure 4 can be calculated (via basic trigonometric functions) and incorporated into the equations derived from Snell's Law to calculate the true depth (h):

$$h = \frac{x}{\tan i} = \frac{h_a \tan r}{\tan\left[\sin^{-1}\left(\frac{n_2}{n_1} \sin r\right)\right]} \tag{2}$$

where i is the angle of incidence, n_1 is the refractive index of fresh water (1.34), n_2 is the refractive index of air (1.0). The average true depth for each point can then be subtracted from the water surface elevation to give the corrected bed elevation of the point ($Z_p = WS_z - \bar{h}$).

As discussed in [28], because each point has a fixed apparent depth with variable refraction angles to each camera, there is a different true depth calculated for each refraction angle (camera). The range of values is small at shallow depths and grows with depth. A range of descriptive statistics (e.g., mean, median, mode, percentile ranks) has been explored and we find that the mean value of all of the calculated h values for each point provides the best statistic. During the course of this research, we found decreased accuracy when the refraction angles (r) were higher (>$35°$ off-nadir). These high refraction angles correspond to greater point–camera distances, and also an increase in the amount of water the camera must 'see' through to estimate the bed surface. We decided to incorporate a filtering function into the software to add additional output that uses a limited set of the calculated h values, in our case those less than $35°$ off-nadir.

The output of the multi-view refraction correction is a new point cloud (CSV) that retains all the original attributes and new columns for h_a, h, and the corrected absolute elevations (Z_p).

2.9. Refraction Correction Validation

Using the validation points that we collected in the field (Figure 2), each of which has precise XYZ coordinates, we assessed the accuracy and precision of our models by subtracting the modelled elevations from the validation point elevations. A small number of outliers were identified and excluded from further analyses. The outliers occurred predominantly in areas of overhanging or dense vegetation over the submerged areas. We then subdivided our datasets into the following categories to quantify the distribution of errors resulting from each approach:

5. Exposed: Exposed areas only, where no refraction correction was necessary.
6. SmallAngle–Manual: Submerged areas only, using the small angle refraction correction (Equation (1)) and the manually digitised water surface elevations.
7. SmallAngle–Smooth: Submerged areas only, using the small angle refraction correction (Equation (1)) and the smoothed water surface elevations.
8. BathySfM–All: Submerged areas only, using the multi-view refraction correction and the smoothed water surface elevations.
9. BathySfM–Filtered: Submerged areas only, using the angle and distance filtered multi-view angle refraction correction and the smoothed water surface elevations.

We produced histograms and scatterplots to visualise the error distributions and calculated a variety of error statistics as measures of accuracy and precision. We compared the differences in mean errors and standard deviations between approaches to compare the different water surfaces and different refraction correction approaches. We compared the error statistics for all submerged datasets with those obtained for exposed areas, which to date have typically been much more accurate than those reported for submerged areas [27,28].

2.10. Spatially Variable Errors

We began our assessment of spatially variable elevation error by extracting several parameters at the location of each validation point, from both the 2016 and 2017 surveys. Table 1 details each of the 11 parameters and how they were extracted from the RPAS–SfM datasets. We chose these variables because they have been used previously for thresholding error when computing geomorphic change (e.g., [1]) or because we suspected that they have a bearing on error based on our prior experience. In addition, we computed the variability of some parameters (slope, roughness, image quality) within a 0.2 m circular window around each validation point, by extracting the maximum, mean, and standard deviation of values within this window. We performed this step using the Focal Statistics tool within ArcGIS (ESRI Inc.). The purpose of this step is to highlight how positioning errors in X and Y translate into elevation (Z) errors, particularly in areas where elevation changes rapidly (i.e., areas of high slope or roughness). For example, in theory, where there is a large range in slope angle and roughness values within the 0.2 m circular window, we expect greater topographic complexity and, therefore, also potentially greater elevation errors.

We tested our theories by assessing individual linear relationships between each variable and the associated error values obtained from our validation data. We also pioneered a new approach for predicting the spatial variability in elevation error, by combining variables together using multiple regression and through the use of machine learning classification algorithms.

We note that, where applicable, we used the DEMs and point clouds derived from the BathySfM–All refraction correction procedure and the smoothed water surfaces to generate the variables specified in Table 1. Given the lack of a consistently superior refraction correction approach (as reported in the results section below), this choice was subjective and one of the other approaches could have been chosen instead.

Table 1. Variables used for the prediction of elevation error within the RPAS–SfM surveys. Total number of validation samples for 2016 was 1522 and for 2017 was 2091. Sample size refers to the percentage of validation points which were assigned to each specific parameter (no data areas apply to some validation points). Where variables are binary, the number of validation points falling within each category is also given.

Parameter (Type)	Symbol	Method of Extraction	Type: Range/Units	Sample Size: Count (%) 2016	2017
Slope angle (topographic complexity)	S	RPAS–SfM DEMs imported to ArcGIS. 3D Analyst extension was used to compute slope angles. Focal (f) statistics for slope within a 0.2 m circular window of each validation point also computed.	Continuous: 0–90°	1522 (100%)	2091 (100%)
Point cloud roughness (topographic complexity)	R	RPAS–SfM point clouds imported to CloudCompare. 'Compute geometric features' tool used to compute roughness for each point in the cloud within spherical kernels with a 0.4 m radius. Roughness is defined as the distance between the point and the best fitting plane computing from all the surrounding points which fall within the kernel. Rasterised at 0.1 m pixel size and exported to ArcGIS. Smaller pixel sizes were found to produce too many holes in the resulting raster. Focal statistics for roughness within a 0.2 m circular window of each validation point also computed.	Continuous: 0–0.34 m	1515 (99.5%)	2090 (99.9%)
Point cloud density (survey quality/ landscape composition)	D	RPAS–SfM point clouds imported to CloudCompare and rasterised according to the number of points falling within each 0.05 m grid cell. Raster exported to ArcGIS.	Continuous: 0–64 count	1522 (100%)	2091 (100%)
Water depth (landscape composition)	h	Computed using the multi-view refraction correction process and smoothed water surface, as detailed within this paper (BathySfM–All).	Continuous: 0–1.97 m	355 (23.3%)	1170 (56%)
Image quality (survey quality)	CQ	Image quality estimates are provided by PhotoScan Pro on a scale of 0 to 1 and based on the level of image sharpness in the best focused area of the image. Images with quality values of less than 0.5 are generally not recommended for use in subsequent processing. Image quality data were exported from PhotoScan Pro and the average quality of cameras that 'see' each point was appended to that point. Point–camera connections were calculated in the same way as the BathySfM correction.	Continuous: 0–1	1522 (100%)	2086 (99.8%)
Tie point precision (survey quality/ landscape composition)	P	Computed using the tie point precision method presented by [49]. Tie points are rasterised at 1 m pixel size and exported from CloudCompare to ArcGIS. The tie points form the sparse point cloud from the PhotoScan Pro software, therefore exporting at a finer resolution would lead to large holes in the resulting raster.	Continuous: 0–4.8 m	1342 (88.2%)	1884 (90.1%)
Vegetation presence (landscape composition)	V	RPAS–SfM orthophotos imported to ArcGIS. Editor toolbar used to visually identify and map areas of particularly dense vegetation or tree coverage.	Binary: [0] = Not present [1] = Present	1522 (100%) [0] = 89.5% [1] = 10.5%	2091 (100%) [0] = 90.5% [1] = 9.5%
Presence of water surface reflection (survey conditions)	Rc	RPAS–SfM orthophotos imported to ArcGIS. Editor toolbar used to visually identify and map areas of notable water surface reflections.	Binary: [0] = Not present [1] = Present	1522 (100%) [0] = 98% [1] = 2%	2091 (100%) [0] = 94.3% [1] = 5.7%
Presence of shadows (survey conditions)	Sh	RPAS–SfM orthophotos imported to ArcGIS. Editor toolbar used to visually identify and map areas of dark shadowing.	Binary: [0] = Not present [1] = Present	1522 (100%) [0] = 99.5% [1] = 0.5%	2091 (100%) [0] = 98.3% [1] = 1.7%

2.11. Multiple Regression

The variables listed in Table 1 were considered for inclusion in a multiple linear regression against the elevation errors computed using the validation data. Variables were assessed for collinearity; where two variables demonstrated a strong correlation (defined as having an R^2 greater than 0.7), the variable with the weaker relationship with elevation error (as determined by single linear regression) was excluded from further analyses. This was necessary, given the strong correlations between variables such as roughness and focal roughness. The remaining continuous variables (max. slope focal, min. slope focal, water depth, point density, mean image quality focal, roughness focal, and tie point precision) were standardised by dividing by their standard deviation. Binary variables (vegetation, shadows, reflections) were not standardised. We used MS Excel to run a multiple regression using the chosen variables against (a) elevation errors, where both magnitude and direction of error are considered, and (b) absolute elevation errors, where only magnitude of error is considered. We performed multiple regressions on the 2016 and 2017 validation data separately and combined. We judged the performance of the multiple regressions by considering the multiple R, adjusted R square, standard error, and significance values.

2.12. Machine Learning Classification

For the first time in geomorphic change detection studies, we also investigated the potential of a machine learning classification approach for predicting elevation error for each validation point using the explanatory variables listed in Table 1. We had initially trialled an ML approach which aimed to predict absolute error values (rather than error categories), but we found that significantly larger training datasets would be required to produce meaningful outputs. Therefore, we binned the values into two sets of error classes: A set of 3 classes (<–0.2, –0.2 to 0.2, and >0.2, representing high magnitude negative error, low magnitude error, and high magnitude positive error) and a set of 10 more detailed classes (<–0.5, –0.5 to –0.2, –0.2 to –0.1, –0.1 to –0.05, –0.05 to 0.0, 0.0 to 0.05, 0.05 to 0.1, 0.1 to 0.2, 0.2 to 0.5, >0.5). This turns the problem into a classification problem, rather than a regression problem.

We started by performing an initial investigation of appropriate classification approaches using the Tree-based Pipeline Optimization Tool (TPOT; [64]), which uses a genetic algorithm to search across multiple common classification algorithms. Manual tweaking based on the TPOT results was used to obtain the best balance of accuracy and pipeline complexity. The resulting classifier was a Gaussian Naïve Bayes classifier using all components of a Principal Components Analysis transform of the explanatory variables as input, using the methods implemented in scikit-learn v.0.20.3 [65]. Separate models were created based on the 2016 validation data, the 2017 validation data, and the combined 2016 and 2017 data. During the ML phase, we observed that the class distributions in the training data were strongly imbalanced (e.g., for the three-class problem with the 2017 data, 90% of the validation points were in the central class). This meant that the classifiers were able to achieve a very high accuracy by simply always predicting outputs within this one class. To combat this, we oversampled the training data using the SMOTE algorithm [66], as implemented in the imblearn v.0.4.3 package [67]. Next, we used a stratified five-fold cross validation procedure for assessing the repeatability of the classification. We observed significant variation in results between runs due to very small numbers in some of the classes (particularly for the ten-class problem). As a result, we ran the five-fold cross-validation procedure five times and averaged the results. After this development and validation of the ML approach, we applied the models to the full raster datasets using the rasterio package [68] to handle raster processing. Finally, we assessed the performance of the ML classifier by comparing predicted outputs with those remaining in the validation dataset to compute overall accuracies and class-specific accuracies, the latter being displayed as confusion matrices.

All machine learning analysis was carried out using Python 3.7.1 using numpy v.1.16.2 [69], pandas v.0.24.1 [70], matplotlib v.3.0.3 [71], and Jupyter Notebook v.5.7.4 [72]. Our code is also available to download [73].

2.13. Error Propagation and Change Detection

To detect geomorphic change between the two epochs of data, we subtracted the earlier refraction-corrected elevation dataset (2016) from the later one (2017) to create a DEM of difference (DoD). For both epochs, we used elevation data which had been corrected for refraction in submerged areas using the BathySfM–All method. This choice was subjective, given that complex patterns of performance which exist between the different refraction correction methods (Table 5) and another method could have been chosen instead. To isolate real geomorphic change from noise within the derived DoD, we test the implementation of our multiple regression method for predicting spatially variable error. We had initially also intended to threshold the DoDs using the outputs from our machine learning classification approach. However, as detailed in the results, these outputs were not sufficiently robust for the detailed ten-class problem to be used for error propagation purposes. Still, we provide a map of error predictions from the three-class problems for comparison purposes in Figure 9.

First, we used the variables associated with each data epoch (Table 1) to predict the magnitude and direction of elevation error using the multiple linear regressions. We chose to predict both the magnitude and direction of error to maximise the reliability of error propagation during the subsequent quantification of elevation change. We took all input variables as raster layers in GIS and used Equation (3) to predict elevation error (SVE) for every pixel in the respective epochs of data, where $MaxSf$ and $MinSf$ are maximum and minimum slope angles within the 20 cm focal window, h is refraction corrected water depth (using BathySfM–All), D is point cloud density, $MeanCQf$ is mean image quality within the focal window, $MeanRf$ is the mean roughness within the focal window, P is tie point precision, V is vegetation presence, Sh is the presence of shadows, Rc is the presence of surface water reflections or water surface roughness, σ is the standard deviation of each variable (computed from the validation data for both epochs), a–g and i–k are the co-efficients associated with each variable, and θ is the intercept.

$$SVE = a\left(\frac{MaxSf}{\sigma_{MaxSf}}\right) + b\left(\frac{MinSf}{\sigma_{MinSf}}\right) + c\left(\frac{h}{\sigma_h}\right) + d\left(\frac{D}{\sigma_D}\right) + e\left(\frac{MeanCQf}{\sigma_{MeanCQf}}\right) + f\left(\frac{MeanRf}{\sigma_{MeanRf}}\right) + g\left(\frac{P}{\sigma_P}\right) \\ + (i \times V) + (j \times Sh) + (k \times Rc) + \theta \tag{3}$$

We performed this computation using the Raster Calculator tool in ArcGIS to output a new raster layer of spatially variable predicted elevation errors for every pixel within each epoch of data (SVE_{2016} and SVE_{2017}). We propagated these individual SVEs into the DoD using the 'Error Propagation' tool within the Geomorphic Change Detection v7 plugin for ArcGIS (http://gcd.riverscapes.xyz/). This is based on the approach detailed in Equation (4) [18], where ε_{DoD} is the propagated error in the DoD and SVE_{2016} and SVE_{2017} are the spatially variable elevation error predictions for each epoch respectively. Finally, we used the GCD toolbox to threshold the DoD on a cell-by-cell basis to quantify the direction and magnitude of change across the entire site.

$$\varepsilon_{DoD} = \sqrt{(SVE_{2016})^2 + (SVE_{2017})^2} \tag{4}$$

3. Results

3.1. SfM Modelling

The georeferencing accuracy and precision statistics associated with each RPAS–SfM survey are presented in Table 2. An evaluation of the validation points indicated that the errors were normally distributed and not affected by any systematic distortions.

Table 2. Georeferencing accuracy statistics (all values in metres).

		Mean	St Dev	95% Conf.	RMSE	MIN	MAX
	X	0.000	0.021	0.041	0.020	−0.047	0.030
2016	Y	0.000	0.027	0.052	0.026	−0.046	0.047
	Z	−0.001	0.017	0.034	0.017	−0.029	0.035
	X	0.000	0.035	0.068	0.034	−0.054	0.056
2017	Y	0.000	0.059	0.116	0.057	−0.096	0.076
	Z	0.000	0.006	0.013	0.006	−0.018	0.006

3.2. Water Surfaces

The statistically smoothed water surfaces, derived from the buffered water's edge points, produced smoothly varying surfaces that provided a better representation of the estimated water surface elevations when assessed visually (Figure 5). This is in contrast to the manually digitised ("point picking") methods used in past studies [27,28]. By picking individual points (from the raster or point cloud) to represent the water surface, there is an increased chance that small variations in the water's edge topography (e.g., topographic roughness of digitising blunders) can have a pronounced effect on the water surface elevations. In Figure 5, we see the effect of these digitising errors as cross-stream lines or also cross-stream mesh facets that translate directly into errors in the calculated apparent depth. These apparent depth errors then get transmitted through the refraction correction to the corrected depths.

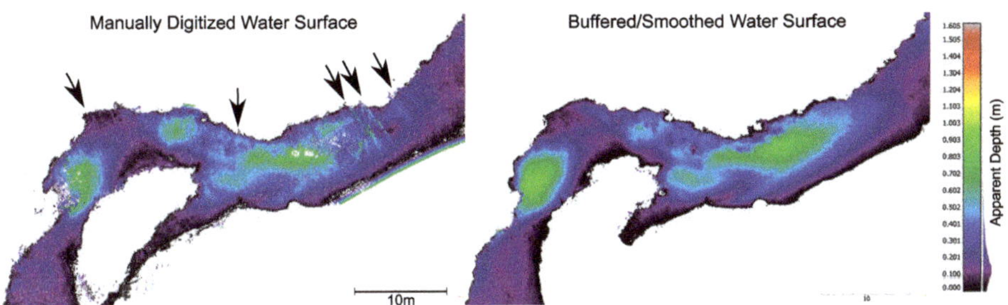

Figure 5. Comparison of two methods for water surface delineation, manually "picked" edge points (left, e.g., [27,28]) and the method used in this paper using statistically smoothed elevation profiles (right). Colours represent the apparent depth (h_a) calculated from each water surface method, the black arrows on the manually digitised surface highlight cross-stream errors/artefacts.

For each of the smoothing filters, the window size used for smoothing these data was 151 points (from the buffered point cloud points). Since the downstream densities of points were not uniform, the 151-point window size does not correspond to any particular length scale. For others using this method, we suggest testing several different window scales depending on the stream length, point cloud density, and topographic roughness.

The seven statistical filters varied in their representations of the water surface; the most pronounced differences were at transitions into and out of the steeper sections of the study reach. We calculated the range of water surface elevations that each smoothing filter produced along the interpolated centreline of the river. The maximum difference in 2016 was 0.038 m with a mean difference of 0.007 m. In 2017, the maximum difference was 0.18 m with a mean difference of 0.01 m. To illustrate the effect that the different water surface elevations have away from the centreline, we also calculated the range of values at each of the 2016 validation points. For the calculated apparent depth, the maximum difference between the different water surfaces was 0.074 m with an average of 0.006 m. After refraction correction, the maximum difference in the corrected elevations was 0.033 m with a mean difference of

0.003 m. Overall, a qualitative assessment found that the Savitzky–Golay filter gave the best visual match to the water's edges visible in both years' orthophotographs.

Table 3 presents the quantitative performance indicators for this smoothed water surface (column 3) compared to the manually digitised water surfaces (column 2) used within the small angle refraction correction calculations for both survey epochs. These results demonstrate an unmitigated improvement in accuracy and precision of submerged elevations with the implementation of the smoothed water surfaces. For the 2016 survey, mean error was reduced by ca. 73% and the standard deviation of error was reduced by ca. 69%. For the 2017 survey, mean error was reduced by 50% and the standard deviation was reduced by ca. 25%. Similar improvements can also be observed in the slope angle of the observed versus predicted relationships and the error ranges (as indicated by the min and max errors).

Table 3. Refraction correction accuracy and precision statistics (all values in metres, except slope).

Correction Method	2016					2017				
	(1) Exposed Only	(2) Small Angle– Manual	(3) Small Angle– Smooth	(4) Bathy SfM–All	(5) Bathy SfM–Filtered	(1) Exposed Only	(2) Small Angle– Manual	(3) Small Angle– Smooth	(4) Bathy SfM–All	(5) Bathy SfM–Filtered
Mean Error	0.024	0.150	0.041	0.016	0.027	0.065	0.034	0.017	−0.055	0.006
St Dev of Error	0.235	0.205	0.065	0.061	0.062	0.226	0.105	0.079	0.101	0.079
Regression Slope	0.942	0.915	0.997	0.989	0.992	0.940	0.903	1.068	1.057	1.060
Min Error	−1.710	−0.645	−0.130	−0.156	−0.152	−1.235	−0.293	−0.269	−0.439	−0.282
Max Error	1.097	1.403	0.501	0.445	0.482	1.566	2.261	0.520	0.485	0.513

3.3. Refraction Corrections

Table 3 and Figure 6 present the performance statistics associated with the comparison of different refraction correction approaches. The results in columns labelled 3–5 of Table 3 were all computed using the smoothed water surface elevations and Figure 6 presents only the smoothed water surface results. Here, we observe an inconsistent picture of accuracy and precision between the different refraction correction approaches and data epochs. For 2016, the small angle refraction correction produced the highest mean error (0.041 m) and standard deviations of error (0.065 m), indicating a poorer performance overall. However, for 2017, the poorest performance is observed from the BathySfM–All approach. When assessed by the strength of the observed-versus-predicted relationship (i.e., slope of the linear regression line), the small angle correction seems to perform better than the other correction methods in 2016, albeit with the caveat that the regression line has a slightly negative offset (Figure 6). However, this pattern is not evident in the 2017 data. Similar inconsistencies are observed in the minimum and maximum error values. As such, it is difficult to conclude that one particular refraction correction approach is superior for this particular site. It is important to note, however, that almost all of the approaches using the smoothed water surface provide significantly better representations of the submerged topography than the manually digitised water surface. Furthermore, depending on the exact choice of method, these representations are now typically in line with or better than those obtained in exposed areas.

Because the error statistics (Table 3) and distributions (Figure 6) are quite similar, we employed two statistical tests to evaluate if the errors between the different methods were statistically distinct (H_A) or came from the same parent distribution ($H_\varnothing$). The tests were a 2-sample Anderson–Darling (AD) and a 2-sample Kolmogorov–Smirnov (KS), both with a p-value threshold of 0.01. For the 2016 corrections, the small angle correction was statistically different than the multi-view corrections using both tests ($p = 0.001$). However, the two multi-view correction error distributions were not statistically distinct using the AD ($p = 0.017$) and KS ($p = 0.062$) tests, meaning that we cannot say with certainty that one method is better than the other. The 2017 corrections were all statistically distinct from one another ($p = 0.001$).

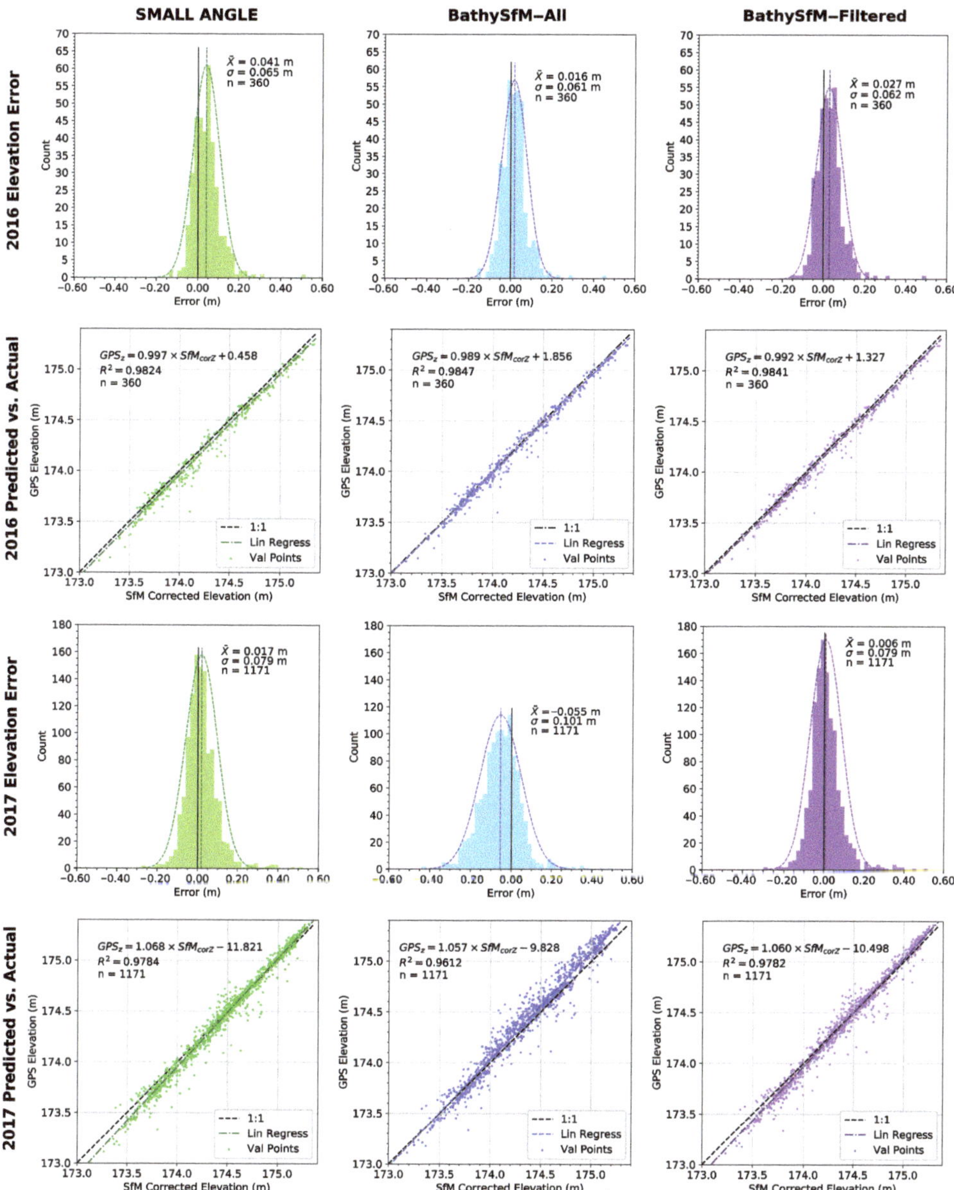

Figure 6. Error histograms and scatterplots for 2016 (top half) and 2017 (bottom half) comparing the three methods tested: Small angle correction (left column), unfiltered multi-view correction (BathySfM–All, centre column), filtered multi-view correction (BathySfM–Filt, right column).

3.4. Spatially Variable Error

3.4.1. Linear Regressions

Table 4 presents the multiple R values for the linear regressions between each variable and each of two measures of elevation error. The strongest correlations are between variables maximum slope within the focal window (0.52), slope (0.50), and standard deviation of slope within the focal window (0.44) with magnitude-only error. Other variables, including water depth, do not show a correlation with magnitude-only elevation error. When magnitude and direction of error are considered, there is a notable lack of correlation between error and any of the variables considered.

Table 4. Multiple R values for linear regressions of individual variables with elevation error.

Variable	(a) Elevation Error—Magnitude and Direction			(b) Elevation Error—Magnitude Only		
	Combined	2016	2017	Combined	2016	2017
Slope	−0.10	−0.20	0.00	0.50	0.47	0.56
Max. Slope Focal	−0.12	−0.24	0.00	0.52	0.53	0.53
Min. Slope Focal	−0.06	−0.09	−0.03	0.37	0.39	0.39
St. Dev. Slope Focal	−0.12	−0.25	0.04	0.44	0.44	0.45
Water Depth (BathySfM–All)	−0.05	0.01	−0.11	−0.15	−0.10	−0.19
Point Density	−0.12	−0.14	−0.10	0.28	0.23	0.43
Image Quality Mean	−0.04	−0.01	−0.07	0.05	0.07	0.02
Image Quality Mean Focal	−0.04	−0.01	−0.07	0.06	0.07	0.02
Roughness 40 cm	−0.04	−0.17	0.12	0.31	0.22	0.41
Roughness 40 cm Focal	−0.05	−0.19	0.14	0.38	0.30	0.49
Tie Point Precision	0.05	0.02	0.07	0.06	0.14	0.06

3.4.2. Multiple Regression

Table 5 and Figure 7 present the performance variables associated with the multiple regressions for each epoch of data and epochs combined. All regressions are statistically significant at the 99.9% level. The strongest relationship is between the 2016 variables with magnitude-only elevation error (0.65). Whilst the performance of these multiple regressions (Table 5) is notably better than for the linear regressions for individual variables (Table 4), none of the multiple regression permutations exhibit very strong relationships where the multiple R is greater than 0.7. Standard error is lowest for the 2017 variables and magnitude-only elevation error (0.11 m). Mean residual errors are low (<0.04 m) but the slopes of the observed versus predicted relationships (Figure 7) are far from 1:1. Furthermore, large minimum and maximum residual errors are evident, which push up the standard deviation of residual error (especially for the 2016 dataset). Overall, magnitude-only elevation error is better explained by the variables than is elevation error which has both a magnitude and direction. Combining the 2016 and 2017 datasets degrades the performance metrics in most scenarios.

Table 5. Performance of multiple regressions. NC denotes variables which were not computed.

Performance Variable	(a) Elevation Error—Magnitude and Direction			(b) Elevation Error—Magnitude Only		
	Combined	2016	2017	Combined	2016	2017
Multiple R	0.47	0.54	0.53	0.61	0.65	0.63
Adjusted R Square	0.22	0.29	0.28	0.37	0.42	0.40
Standard Error (m)	0.17	0.18	0.14	0.13	0.14	0.11
Significance (*p*-value)	<0.01	<0.01	<0.01	<0.01	<0.01	<0.01
Slope (obs vs. pred)	NC	0.1666	0.1297	NC	NC	NC
Mean Residual Error (m)	NC	−0.04	−0.01	NC	NC	NC
Max Residual Error (m)	NC	1.48	1.09	NC	NC	NC
Min Residual Error (m)	NC	−4.48	−1.23	NC	NC	NC
St Dev Residual Error (m)	NC	0.22	0.16	NC	NC	NC

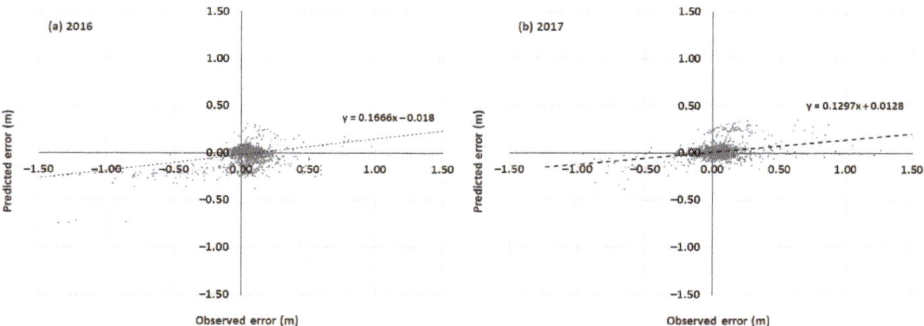

Figure 7. Observed versus predicted elevation errors for (**a**) 2016 and (**b**) 2017 datasets, using the multiple linear regression approach for both error magnitude and direction.

Both error magnitude and direction (i.e., positive or negative) are necessary for thresholding quantities of change. Therefore, it was these multiple regression relationships, for the 2016 and 2017 datasets separately, which we used to compute spatially variable error using Equation (3), despite their lower multiple R values. Of the variables included in this predictive equation, we found the presence of vegetation to be the most influential: Its exclusion from the multiple regression reduced the multiple R values from 0.54 and 0.53 to 0.42 and 0.35 for the 2016 and 2017 datasets, respectively. The exclusion of most other variables, one at a time, had minimal effect on the multiple R values for both data epochs, yet the inclusion of only the few most significant variables did not produce higher multiple R values than when all variables were included. As a result, we considered it important to include all variables within the multiple regression at this stage. However, future research would benefit from a more thorough exploration of the most significant combinations of variables.

Table 6 details the standard deviation values used to standardise the continuous numerical variables and the co-efficient and intercept outputs of the multiple regressions (Equation (3)). Figure 8 presents the SVE outputs for each epoch and the resultant propagated error.

Table 6. Standard deviations, co-efficients, and intercept values (multiple regression parameters).

Variable	Standard Deviation	Multiple Regression Co-Efficients/Intercept	
	Epochs Combined	2016	2017
Maximum Slope Focal (MaxSf)	20.3411	−0.0674	−0.0281
Minimum Slope Focal (MinSf)	5.4814	−0.0059	−0.0034
Refraction Corrected Water Depth (h)	0.2061	0.0140	−0.0228
Point Cloud Density (D)	4.1113	−0.0071	−0.0747
Mean Image Quality Focal (MeanCQf)	1.9848	−0.0010	−0.0266
Mean Roughness Focal (MeanRf)	0.0150	−0.0212	0.0333
Point Cloud Precision (P)	0.3863	0.0437	0.0019
Presence of Vegetation (V)	N/A	0.2604	0.2354
Presence of Shadows (Sh)	N/A	−0.4588	0.0616
Presence of Reflections (Rc)	N/A	0.0511	0.0084
Intercept (θ)	N/A	0.1533	1.1812

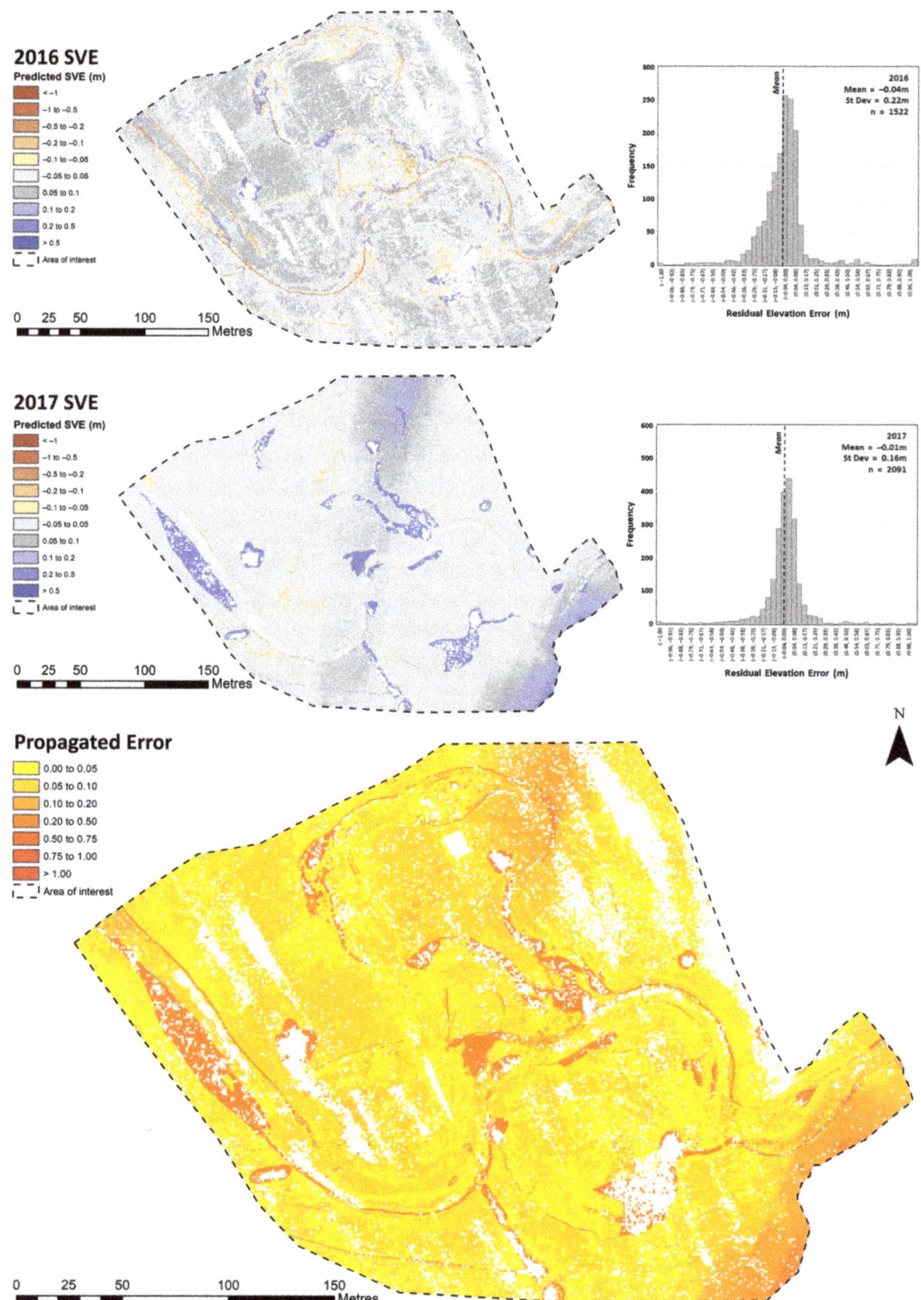

Figure 8. Predicted spatially variable elevation error maps for 2016 and 2017 datasets (with histograms of residual error associated with these predictions) and propagated error map (subsequently used to threshold elevation change)—all generated using data from the multiple regression approach.

3.5. Machine Learning Classification

Table 7 presents the accuracy results from the ML classification approach. Relatively high accuracies of ca. 75% and F1 scores are returned for the three-class problem, but significantly poorer values are observed for the ten-class problem. There are small differences in accuracy between the different epochs. More detailed information can be seen in the confusion matrices for the 2017 three-class problem (Figure 9a) and the 2017 ten-class problem (Figure 9b). For the three-class problem, these show good accuracy for the lower two classes (>80% accuracy), but a poorer accuracy for the higher class (with misclassifications split between the other two classes). Figure 10 shows the spatial distribution of error predicted using the three-class scenario. The ten-class problem confusion matrix gives highly mixed results, with reasonable accuracy for the lowest class (–2 to –0.5), middle class (0 to 0.05) and the highest class (0.5 to 6.5). The highest levels of misclassification occur for elevations between –0.1 and 0.1, which are often misclassified as being in the 0 to 0.05 class. The confusion matrices for the 2016 dataset and both years combined show similar patterns.

Table 7. Accuracy results for each of the classifications performed.

Classes	Epoch	Accuracy	F1 Score
	2016	78%	0.83
3 classes	2017	76%	0.81
	Both	74%	0.80
	2016	26%	0.22
10 classes	2017	31%	0.30
	Both	29%	0.27

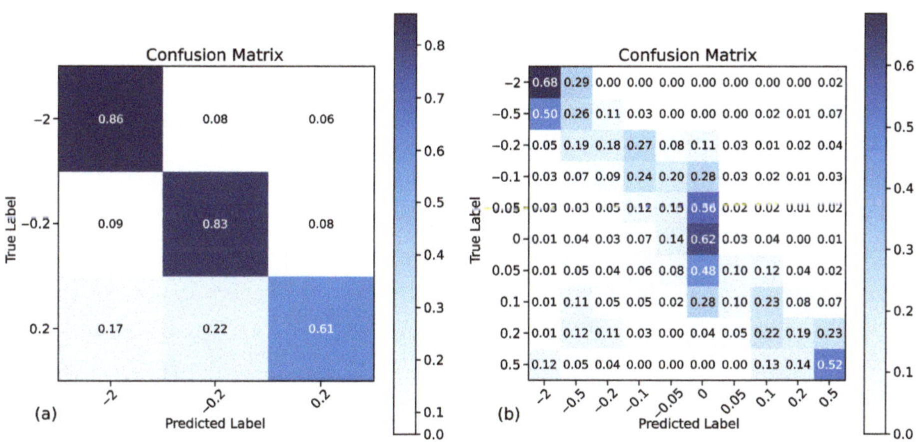

Figure 9. Confusion matrices for the 2017 data classification, for the (**a**) three-class and (**b**) ten-class problem. The matrices are normalised, summing to 1.0 along the rows. Classes are labelled with their lower bound (so in part (**a**), a label of –2 represents the class of –2 to –0.2).

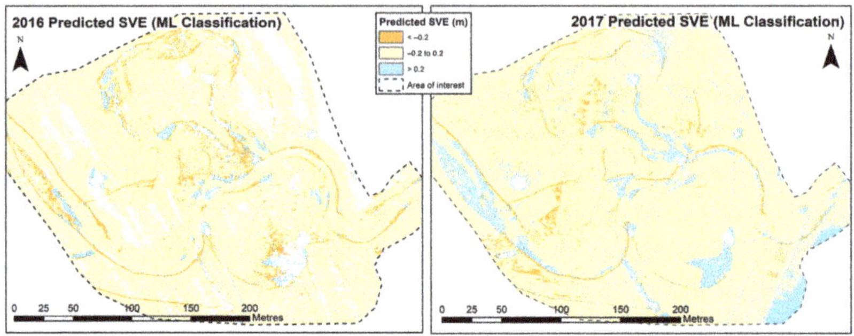

Figure 10. Predicted spatially variable elevation error maps for 2016 and 2017 datasets generated using data from the three-class machine learning classification approach.

3.6. Geomorphic Change

We used the multiple regression approach as a predictor of the direction and magnitude of elevation errors within our two epochs of data, given the low mean errors associated with this approach (Table 5). We chose not to use the outputs of the ML classification approach on this occasion, given the poor accuracies associated with the ten-class outputs. Whilst the classification results for three-classes are much better, our aim was to predict error with greater precision than is permitted by only three categories of error.

Figure 11 presents a map of geomorphic change at this section of the River Teme, thresholded by our error estimates using the multiple regression approach. Associated statistics which quantify areal and volume change are presented in Figure 12. Over the 1 year period between our surveys, this section of the River Teme experienced a net volume increase of 1155 m^3, as indicated by the teal-coloured areas within Figure 11. We note that a proportion of this increase will relate to growth of vegetation (as seen in the vicinity of trees when comparing Figure 11 with Figure 1) and does not solely represent sediment deposition.

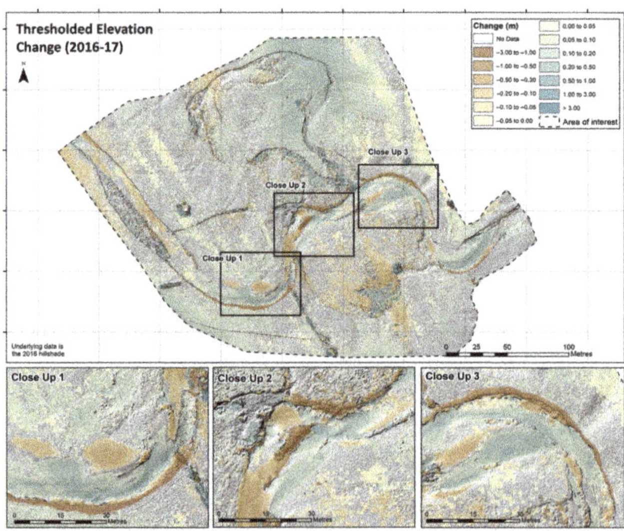

Figure 11. Elevation change map (2016–17), thresholded by elevation errors predicted using the multiple regression approach.

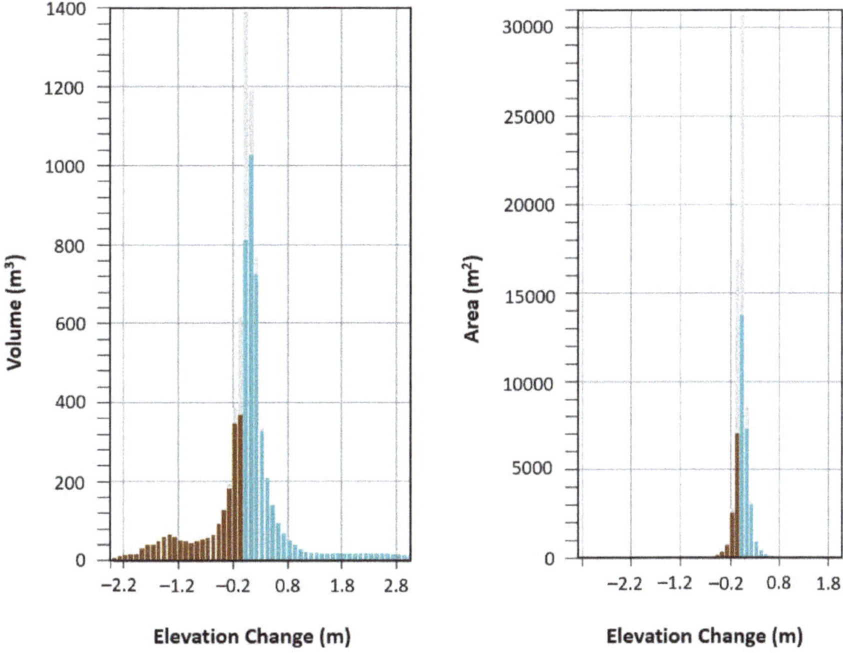

Figure 12. Quantities of elevation change (2016–17) in terms of volume (**left**) and area (**right**). Note that the x-axes have been truncated for clarity purposes. Total areal lowering was 8097 m^2, total areal raising was 17,570 m^2, and net volume difference was +1155 m^3.

4. Discussion

4.1. Refraction Correction

For the first time, our research has demonstrated that our RPAS–SfM refraction correction methods are capable of quantifying above- and below-water fluvial geomorphic change at hyperspatial resolutions, without the need for different methods between exposed and submerged areas or for acquisition of calibration data in submerged areas. This represents a significant time and cost saving compared to existing approaches (e.g., [37]) and has positive implications for surveyor safety, as there is no need to enter the water.

Our accuracy statistics (mean error) indicate that the below-water topographic reconstructions are now broadly as accurate as those obtained in exposed areas, and our precision values (standard deviation) demonstrate that they are significantly more reliable than for exposed areas (Table 3). This may result from a lack of vegetation in submerged areas which generates noise within the point cloud, thereby reducing precision of our topographic reconstructions in much of the exposed parts of the site.

Our results show no consistent pattern of compromise on the accuracy or precision of the topographic data compared with approaches which use different methods in submerged areas and/or those which require calibration data. For example, the accuracy and precision values we report here compare favourably with those reported by [37,42] (Table 8), where optical calibration approaches or site-specific refraction correction coefficients are used in submerged areas—both of which require extensive calibration data collected in-channel. Comparison of our results with those of [38] present a mixed picture, with our mean error and standard deviation values being higher but RMSE values being broadly comparable. Further exploration of this is restricted by the limited density of validation points and a lack of information on validation point distribution and water depths within the work of [38].

Table 8. Comparison of RPAS-based approaches to below-water topographic reconstruction and change detection.

	This Paper	Flener et al. [37]	Tamminga et al. [38]	Shintani and Fonstad [42]
Refraction correction approach	Co-efficient for clear water (1.34): Small angle & multiview	Optical calibration	Optical calibration	Site specific co-efficient
Calibration data needed?	No	Yes	Yes	Yes
Validation points	1522–2093	197 GPS points + extra ADCP points (not reported)	76–82	167
Survey size	600 m reach	ca. 150 m reach (20–30 m wide)	800 m reach	140 m reach
Validation survey type	Differential GNSS + total station	RTK GPS + ADCP	RTK GPS	RTK GPS
Validation point layout	Random distribution over a range of water depths	Zig-zag pattern along channel over range of water depths	Not reported	12 evenly spaced channel cross sections
Maximum water depth (m)	1.43	1.50	Not reported	1.25
Mean error (m)	0.006–0.041	0.117–0.1196	0.0001–0.0007	0.009
RMSE (m)	0.063–0.115	0.163–0.221	0.095–0.098	Not reported
Standard deviation (m)	0.061–0.101	Not reported	0.009–0.023	0.172

For the first time, our research has also demonstrated minimal differences in performance of the small angle and multi-view refraction correction methods applied at this site [27,28]. This suggests that either of these methods may be implemented by those wishing to produce similar outputs at their sites, assuming that the basic caveats for successful implementation are met (i.e., predominate camera angles [nadir vs. off-nadir], clear water, water depths < ca. 1.5 m, minimal water surface roughness). Those with a background in GIS may prefer to implement the small angle correction, whilst those familiar with running Python code may prefer the multi-view approach (the latest version has a user-friendly graphical interface). Here, we prefer to use the multi-angle approach as detailed in our GitHub repository [43] as it offers the flexibility to work with large, high-resolution 3D point clouds (rather than 2D rasters) and modify by parameters such as camera view angles, as we used within our BathySfM–Filtered scenario. We suggest that future research should focus on the comparison of these methods over a wider range of study conditions. In particular, establishing, quantitatively, the range of settings in which our refraction correction methods can be applied successfully is of importance. Dedicated laboratory and field experiments which aim to determine the limits of water depth, clarity, and surface roughness would be beneficial and would allow a quantification of the percentage of river environments in which our methods are applicable. Such data will be key for understanding the value of our methods for routine river monitoring and management purposes on a larger scale. Further exploration of the importance of camera view angles would also help inform routine RPAS surveys in such scenarios.

4.2. Water Surface Elevations

Our second key finding is that accurate water surface elevations are critical for accurate refraction correction in submerged areas and are capable of reducing mean elevation errors by 50–73%. Therefore, we recommend the use of our improved water surface estimation approach for all those wishing to reconstruct submerged fluvial topography using RPAS–SfM derived data.

Specifically, our investigations suggest the use of a Savitzky–Golay filter within a smoothing window produces the most accurate results; however, the low variability between the smoothed water surfaces would suggest that the local polynomial regression filters in general (Savitzky–Golay and normal/robust Loess and Lowess filters) are a good choice for characterising the water surfaces. The other filters that we tested (Moving Average and Gaussian) accounted for a large portion of the variability reported in the results, so these filters are not recommended.

One of the key variables in developing these filters is the smoothing/sampling window. In our dataset, the best results came from a 151-point smoothing window. We settled on this window via a trial and error process, along with a qualitative assessment of the "best-fit" of multiple smoothing window sizes. For other datasets, we would encourage researchers to test multiple filter windows to find the most appropriate size for their data. One of the challenges in choosing the correct window size was that smaller window sizes had a tendency to create "uphill" water surfaces in certain sections, influenced by small outliers in the edge point data. In our case, the larger window sizes helped moderate the influence of small-scale variations in the edge points.

Whilst this methodology represents a notable improvement over past studies (i.e., [27,28]), there remain some limitations. The method still relies on having clearly identifiable banks and bars to have sufficient data to perform the statistical smoothing. In river environments where the banks are heavily vegetated or the channel is incised and there are no visible banks, extracting enough information to utilise this method will be challenging. The assumption of a cross-stream planar water surface is a reasonable approximation for most streams, but there are situations where deviations from this assumption could impact the results. Super elevation of the water surface on the outside of meander bends would cause water surface elevations to be higher on the outside of the bend and potentially lower on the inside. In-stream features that cause local hydraulic disturbances (e.g., exposed and submerged sediment clasts, standing waves, or dune bedforms) will also cause locally variable water surface elevations that are not accounted for in our approximation.

Future work should focus on validating this methodology with detailed surveys of the water's edges, along with direct measurement of the water surface. The ultimate goal will be to develop methods and/or sensors that have application across a wide range of river types and that can be used to extract the water surface elevations, thereby enabling researchers to apply bathymetric corrections over longer river segments without the need for in situ water surface measurements. This will subsequently permit the collection of extensive measurements of river system variables.

4.3. Spatially Variable Error for Geomorphic Change Detection

Our third key finding is a lack of linear correlation between any individual variable and elevation error. This finding is not surprising, given numerous factors are known to contribute to such errors. It reinforces the existing observations that levels of detection used to threshold geomorphic change (in both magnitude and direction) in such settings should not be informed by single variables alone [1,48]. Furthermore, the lack of correlation between error and water depth evidences the success of our improved water surfaces, combined with our refraction correction approach. Previous studies had failed to break down this correlation so successfully (e.g., [27]).

Fourth, we find that multiple linear regressions (which regress elevation error against a combined total of 11 RPAS–SfM-derived variables) are capable of explaining up to 54% of the magnitude and direction of elevation error in our topographic surveys. This is in a scenario where we have controlled for sources of error within the SfM process in so far as is possible/known at the current time (including use of sufficient GCPs which are adequately distributed, a precise internal control network, and both oblique and nadir RPAS imagery to mitigate against systematic camera lens distortions). These multiple regressions allow us to model the spatial variability of error with mean accuracies of 0.01–0.04 m and standard deviations of 0.14–0.22 m. To our knowledge, this is the first such model of spatially variable elevation errors in RPAS–SfM surveys in a geomorphological setting, which allows us to quantify and threshold geomorphic change with a known level of accuracy and precision.

In the last few years, the RPAS–SfM approach has been described as a 'game-changer' for quantitative 3D analyses in geomorphology. Despite this, comprehensive, spatially-continuous, quantitative estimates of associated elevations errors at very high spatial resolutions have been lacking. Thresholding calculations of geomorphic change using spatially variable elevation errors is critical for the accuracy and reliability of subsequent quantitative analyses and decision making in a management context. Therefore, we anticipate that with further testing and development of our model over a range of different topographic settings, landscape compositions, and survey qualities/conditions, it has the potential for routine application for error estimation within the RPAS–SfM workflow.

However, further testing and development is required to improve on the current limitations of our approach. For instance, we note that 46% of elevation error remains unexplained by our model, and therefore unaccounted for within our maps and statistics of geomorphic change. Whilst mean residual errors are low, we also report notably large minimum and maximum residual elevation errors within our resulting predictions of spatially variable error (Table 5). Such large residual errors are rare (e.g., 85% (2016) to 92% (2017) of validation points have residual errors of ± 0.2 m) but deserve greater attention in future, as they tend to occur in areas of greater geomorphic complexity. Furthermore, we observe a significant lack of a 1:1 relationship between observed and predicted elevation error, as indicated by the slope values in Table 5 and Figure 7. We suspect that these poorer results relate to a number of key limitations to our approach:

Our method assumes a linear relationship between error and the included variables. Whilst this may be appropriate for some variables, such as slope or roughness, it makes less sense for the binary variables, such as vegetation and reflection/surface roughness. In future, rather than noting the presence or absence of vegetation, it may be worth exploring the inclusion of a continuous numerical variable like vegetation height above ground level. Whilst this is less easily computed using RPAS–SfM surveys than airborne LiDAR, for instance, the interpolation or extrapolation of terrain elevations across vegetated areas may go some of the way to ameliorating this current issue. This is likely to be easier in areas of reduced geomorphic complexity, such as for relatively flat floodplains rather than deep-seated landslides.

Our method does not account specifically for elevation error resulting from uncertainties in the co-alignment of the RPAS–SfM surveys with the calibration/validation data. This gives rise to the well-documented situation whereby the magnitude of elevation error is strongly influenced by slope (where relatively small offsets in X and Y result in large errors in Z). However, slope does not lead to a consistent positive or negative elevation error because the orientation of the slope varies in relation to the direction of the XY offset. This goes part of the way to explaining why our multiple regression, which predicted magnitude-only error, outperformed that which attempted to predict both magnitude and direction of error. Milan et al. [49] and Carbonneau et al. [74] suggest that the alignment of elevation data as point clouds using cloud-to-cloud registration or the M3C2 algorithm [21] within CloudCompare can ameliorate this issue and suggest that in future, the calibration/validation data used to develop and test models of spatially variable error should be available as point clouds. We anticipate that calibration/validation data which is larger and denser than our total station survey may be required here, such as that obtained using terrestrial laser scanning. This should be a priority for subsequent development of our model.

To give an indication of the impact of these limitations, we re-ran our model, this time excluding all validation points which fell within areas of dense vegetation or water surface reflections/roughness and/or on slopes of greater than or equal to 70 degrees. The output was a reduction in mean residual error of elevation error predictions from –0.04 to –0.03 m for the 2016 dataset and from –0.01 to –0.003 m for the 2017 dataset. Improved precision values were also observed: from 0.22 to 0.11 m (2016) and 0.16 to 0.09 m (2017). The slopes of the observed versus predicted elevation error relationships (Figure 7) were reduced from 0.1297 to 0.0327 (2016) and from 0.1666 to 0.0964 (2017). This suggests that, as we begin to remove the larger elevation errors from the model (resulting from dense vegetation, steep slopes (i.e., XY offsets) and reflections/roughness), we remove the systematic offsets which the model has 'learnt' and are left with what is likely to be random noise.

Whilst we can realistically aim to reconstruct systematic errors (or bias) within models, we will never be able to predict random errors (noise). For this reason, [1] argued that a deterministic model of elevation error can never be constructed unambiguously. As a result, they use a fuzzy inference system for error estimation based on membership of their GPS validation data to three functions: Slope, point density, and point quality. We suggest that a deterministic prediction of error is possible to an extent and, so long as we keep in mind the occurrence of random error in addition to the error we predict, that implementation of a model such as ours is appropriate. With the advent of artificial intelligence, we believe it is only a matter of time and effort, in collating sufficient and appropriate evidence, before machine/deep learning approaches are capable of predicting systematic errors within RPAS–SfM elevation datasets routinely.

Our initial findings suggest that variables associated with topographic complexity and landscape composition (e.g., slope, vegetation) have a greater bearing on elevation error than do variables associated with the SfM process itself (e.g., tie point precision). However, we also note that the relationships between these different variables are complex. For example, the SfM process relies on image texture for tie point identification, yet texture is itself determined by the composition and complexity of the landscape being surveyed. James et al. [49] argues that point cloud topographic datasets, thresholded by SfM tie point precision and aligned using the M3C2 approach produce "the most plausible distribution of topographic change" ([49], p. 1782). These results could be interpreted to mean that SfM tie point precision is the most reliable predictor of elevation error in RPAS–SfM surveys. However, whilst [49] accounts for horizontal offsets in this approach, they do not specifically explore the influence of other variables on elevation error. Furthermore, their judgement of the most 'plausible' approach is based on qualitative expectations of change at their site, rather than quantitative evidence collected contemporaneously. As such, we suggest that further investigation into the relative importance of a wide range of different variables on elevation error is required, with particular focus on how we might reduce or eliminate such errors. Within this paper, we have demonstrated such an approach for reducing topographic error which results from the presence and depth of water.

Our final key finding is that the machine learning classification approach implemented here is successful only in predicting elevation error within three broad classes: 'High negative error', 'high positive error', and 'low error', with overall accuracies of >75%. Whilst such a classification may be helpful in some broader scale analyses of change, our aim was to develop a method of quantifying spatially variable elevation error more precisely in order to threshold measures of geomorphic change at the high spatial resolutions provided by RPAS–SfM surveys. We anticipate our approach was significantly restricted by the size of the training dataset, which was particularly lacking in data points representing the larger elevation errors. Within the geomorphic-remote sensing community, the collection of validation/training data using traditional surveying approaches (e.g., total station/GPS surveys) is well established. However, as we move towards the exploration of more complex artificial intelligence (AI) tools for developing such predictive relationships, it is likely that we will need to acquire validation/training data on a completely different scale. That is, AI/ML approaches typically require millions of data points to 'learn' from, rather than the hundreds or thousands we might employ for standard regressions. Acquisition of such large datasets at scales comparable to high resolution RPAS surveys will require the use of terrestrial or airborne laser scanning. We anticipate that a more advanced ML approach, of which we are already seeing more widespread application within geomorphological fields (e.g., [75]), in combination with a significantly larger training dataset, will facilitate the development of a much improved AI model for predicting spatially variable elevation error for any given topographic dataset derived from the RPAS–SfM workflow.

5. Conclusions

In this research, we have shown that we are able to quantify submerged geomorphic change at levels of accuracy commensurate with the exposed areas of our river system with a single photogrammetric approach and without the need for calibration data. For the aerial imagery we collected (predominantly nadir views), there are minimal differences in results produced by different refraction correction procedures (small angle vs. multi-view), which may allow users a choice of software packages/processing complexity depending on their research objectives. Our improvements to estimations of water surface elevations with statistical smoothing in the downstream direction are critical for accurate topographic estimation in submerged areas and can reduce mean refraction corrected elevation errors by up to 73%. We have shown that machine learning, in the form of multiple linear regressions between error and 11 independent variables, can be used to generate high resolution, spatially continuous models of geomorphic change in submerged areas, constrained by spatially variable error estimates. Our machine learning model is capable of explaining up to 54% of magnitude and direction of topographic error, with accuracies of less than 0.04 m. With on-going testing and improvements, this machine learning approach has potential for routine application in spatially variable error estimation within the RPAS–SfM workflow.

Author Contributions: Conceptualization, A.S.W.; formal analysis, A.S.W., J.T.D. and R.T.W.; funding acquisition, A.S.W.; investigation, A.S.W. and J.T.D.; methodology, A.S.W. and J.T.D.; software, J.T.D. and R.T.W.; supervision, A.S.W.; validation, A.S.W. and J.T.D.; writing—original draft, A.S.W., J.T.D. and R.T.W.; writing—review and editing, A.S.W. and J.T.D.

Funding: This project was supported in part by funding from the University of Worcester (UK), where Amy Woodget was previously based.

Acknowledgments: We gratefully acknowledge the field assistance of Paul Attwood, Robbie Austrums, James Atkins, Fleur Visser, Mark, Sally and William Woodget, Carlos Puig-Mengual, Cat Fyffe, Steve Dugdale, Katie Long, and John Powell. We thank Mike James, of Lancaster University, for computing tie-point precision of our SfM surveys using his 2017 approach, and Olivia Wilson, of the Ordnance Survey, for expert guidance with the machine learning classification problem. Constructive comments from Patrice Carbonneau, Leonardo Camelo, and three anonymous reviewers helped to improve an earlier version of this manuscript. We also thank the landowners. The code produced in this publication is available online as follows. Machine learning: https://github.com/robintw/WoodgetDietrichWilson2019, Multi-view refraction correction: https://github.com/geojames/pyBathySfM. Please note that we do not provide the RPAS imagery associated with this article, due to landowner privacy concerns.

Conflicts of Interest: The authors declare no conflicts of interest.

References

1. Wheaton, J.M.; Brasington, J.; Darby, S.E.; Sear, D.A. Accounting for uncertainty in DEMs from repeat topographic surveys: Improved sediment budgets. *Earth Surf. Process. Landf.* **2010**, *35*, 136–156. [CrossRef]
2. Kociuba, W. Analysis of geomorphic changes and quantification of sediment budgets of a small Arctic valley with the application of repeat TLS surveys. *Z. Fur Geomorphol. Suppl. Issues* **2017**, *61*, 105–120. [CrossRef]
3. Rice, S.; Church, M. Grain size along two gravel-bed rivers: Statistical variation, spatial pattern and sedimentary links. *Earth Surf. Process. Landf.* **1998**, *23*, 345–363. [CrossRef]
4. Hodge, R.; Brasington, J.; Richards, K. In situ characterization of grain-scale fluvial morphology using Terrestrial Laser Scanning. *Earth Surf. Process. Landf.* **2009**, *34*, 954–968. [CrossRef]
5. Langhammer, J.; Lendzioch, T.; Miřijovský, J.; Hartvich, F. UAV-Based Optical Granulometry as Tool for Detecting Changes in Structure of Flood Depositions. *Remote Sens.* **2017**, *9*, 240. [CrossRef]
6. Woodget, A.S.; Austrums, R. Subaerial gravel size measurement using topographic data derived from a UAV-SfM approach. *Earth Surf. Process. Landf.* **2017**, *42*, 1434–1443. [CrossRef]
7. Fuller, I.C.; Large, A.R.G.; Milan, D.J. Quantifying channel development and sediment transfer following chute cutoff in a wandering gravel-bed river. *Geomorphology* **2003**, *54*, 307–323. [CrossRef]
8. Milan, D.J.; Heritage, G.L.; Hetherington, D. Application of a 3D laser scanner in the assessment of erosion and deposition volumes and channel change in a proglacial river. *Earth Surf. Process. Landf.* **2007**, *32*, 1657–1674. [CrossRef]
9. Verhaar, P.M.; Biron, P.M.; Ferguson, R.I.; Hoey, T.B. A modified morphodynamic model for investigating the response of rivers to short-term climate change. *Geomorphology* **2008**, *101*, 674–682. [CrossRef]
10. Slater, L.J.; Singer, M.B.; Kirchner, J.W. Hydrologic versus geomorphic drivers of trends in flood hazard. *Geophys. Res. Lett.* **2015**, *42*, 370–376. [CrossRef]
11. Coveney, S.; Roberts, K. Lightweight UAV digital elevation models and orthoimagery for environmental applications: Data accuracy evaluation and potential for river flood risk modelling. *Int. J. Remote Sens.* **2017**, *38*, 3159–3180. [CrossRef]
12. Newson, M.D.; Newson, C.L. Geomorphology, ecology and river channel habitat: mesoscale approaches to basin-scale challenges. *Prog. Phys. Geogr.* **2000**, *24*, 195–217. [CrossRef]
13. Woodget, A.S.; Austrums, R.; Maddock, I.P.; Habit, E. Drones and digital photogrammetry: From classifications to continuums for monitoring river habitat and hydromorphology. *Wiley Interdiscip. Rev. Water* **2017**, *4*, e1222. [CrossRef]

14. Lane, S.N.; Richards, K.S.; Chandler, J.H. Developments in monitoring and modelling small-scale river bed topography. *Earth Surf. Process. Landf.* **1994**, *19*, 349–368. [CrossRef]

15. Chandler, J. Effective application of automated digital photogrammetry for geomorphological research. *Earth Surf. Process. Landf.* **1999**, *24*, 51–63. [CrossRef]

16. Lane, S.N.; Westaway, R.M.; Hicks, D.M. Estimation of erosion and deposition volumes in a large, gravel-bed, braided river using synoptic remote sensing. *Earth Surf. Process. Landf.* **2003**, *28*, 249–271. [CrossRef]

17. Charlton, M.E.; Large, A.R.G.; Fuller, I.C. Application of airborne LiDAR in river environments: the River Coquet, Northumberland, UK. *Earth Surf. Process. Landf.* **2003**, *28*, 299–306. [CrossRef]

18. Brasington, J.; Langham, J.; Rumsby, B. Methodological sensitivity of morphometric estimates of coarse fluvial sediment transport. *Geomorphology* **2003**, *53*, 299–316. [CrossRef]

19. Fausch, K.D.; Torgersen, C.E.; Baxter, C.V.; Li, H.W. Landscapes to Riverscapes: Bridging the Gap between Research and Conservation of Stream Fishes. *BioScience* **2002**, *52*, 483–498. [CrossRef]

20. Harwin, S.; Lucieer, A. Assessing the Accuracy of Georeferenced Point Clouds Produced via Multi-View Stereopsis from Unmanned Aerial Vehicle (UAV) Imagery. *Remote Sens.* **2012**, *4*, 1573–1599. [CrossRef]

21. Lague, D.; Brodu, N.; Leroux, J. Accurate 3D comparison of complex topography with terrestrial laser scanner: Application to the Rangitikei canyon (N-Z). *ISPRS J. Photogramm. Remote Sens.* **2013**, *82*, 10–26. [CrossRef]

22. Lucieer, A.; Jong, S.M.d.; Turner, D. Mapping landslide displacements using Structure from Motion (SfM) and image correlation of multi-temporal UAV photography. *Prog. Phys. Geogr.* **2014**, *38*, 97–116. [CrossRef]

23. Clapuyt, F.; Vanacker, V.; Van Oost, K. Reproducibility of UAV-based earth topography reconstructions based on Structure-from-Motion algorithms. *Geomorphology* **2016**, *260*, 4–15. [CrossRef]

24. Cucchiaro, S.; Cavalli, M.; Vericat, D.; Crema, S.; Llena, M.; Beinat, A.; Marchi, L.; Cazorzi, F. Monitoring topographic changes through 4D-structure-from-motion photogrammetry: application to a debris-flow channel. *Environ. Earth Sci.* **2018**, *77*, 632. [CrossRef]

25. Micheletti, N.; Chandler, J.H.; Lane, S.N. Investigating the geomorphological potential of freely available and accessible structure-from-motion photogrammetry using a smartphone. *Earth Surf. Process. Landf.* **2015**, *40*, 473–486. [CrossRef]

26. Brasington, J.; Vericat, D.; Rychkov, I. Modeling river bed morphology, roughness, and surface sedimentology using high resolution terrestrial laser scanning. *Water Resour. Res.* **2012**, *48*, W11519. [CrossRef]

27. Woodget, A.S.; Carbonneau, P.E.; Visser, F.; Maddock, I.P. Quantifying submerged fluvial topography using hyperspatial resolution UAS imagery and structure from motion photogrammetry. *Earth Surf. Process. Landf.* **2015**, *40*, 47–64. [CrossRef]

28. Dietrich, J.T. Bathymetric Structure-from-Motion: extracting shallow stream bathymetry from multi-view stereo photogrammetry. *Earth Surf. Process. Landf.* **2017**, *42*, 355–364. [CrossRef]

29. Legleiter, C.J.; Harrison, L.R. Remote Sensing of River Bathymetry: Evaluating a Range of Sensors, Platforms, and Algorithms on the Upper Sacramento River, California, USA. *Water Resour. Res.* **2019**, *55*, 2142–2169. [CrossRef]

30. Vaaja, M.; Hyyppä, J.; Kukko, A.; Kaartinen, H.; Hyyppä, H.; Alho, P. Mapping Topography Changes and Elevation Accuracies Using a Mobile Laser Scanner. *Remote Sens.* **2011**, *3*, 587–600. [CrossRef]

31. Schaffrath, K.R.; Belmont, P.; Wheaton, J.M. Landscape-scale geomorphic change detection: Quantifying spatially variable uncertainty and circumventing legacy data issues. *Geomorphology* **2015**, *250*, 334–348. [CrossRef]

32. Miřijovský, J.; Langhammer, J. Multitemporal Monitoring of the Morphodynamics of a Mid-Mountain Stream Using UAS Photogrammetry. *Remote Sens.* **2015**, *7*, 8586–8609. [CrossRef]

33. Cook, K.L. An evaluation of the effectiveness of low-cost UAVs and structure from motion for geomorphic change detection. *Geomorphology* **2017**, *278*, 195–208. [CrossRef]

34. Hamshaw, S.D.; Bryce, T.; Rizzo, D.M.; O'Neil-Dunne, J.; Frolik, J.; Dewoolkar, M.M. Quantifying streambank movement and topography using unmanned aircraft system photogrammetry with comparison to terrestrial laser scanning. *River Res. Appl.* **2017**, *33*, 1354–1367. [CrossRef]

35. Javemick, L.; Brasington, J.; Caruso, B. Modeling the topography of shallow braided rivers using Structure-from-Motion photogrammetry. *Geomorphology* **2014**, *213*, 166–182. [CrossRef]

36. Starek, M.J.; Giessel, J. Fusion of uas-based structure-from-motion and optical inversion for seamless topo-bathymetric mapping. In Proceedings of the 2017 IEEE International Geoscience and Remote Sensing Symposium (IGARSS), Fort Worth, TX, USA, 23–28 July 2017; pp. 2999–3002.

37. Flener, C.; Vaaja, M.; Jaakkola, A.; Krooks, A.; Kaartinen, H.; Kukko, A.; Kasvi, E.; Hyyppä, H.; Hyyppä, J.; Alho, P. Seamless Mapping of River Channels at High Resolution Using Mobile LiDAR and UAV-Photography. *Remote Sens.* **2013**, *5*, 6382–6407. [CrossRef]

38. Tamminga, A.D.; Eaton, B.C.; Hugenholtz, C.H. UAS-based remote sensing of fluvial change following an extreme flood event. *Earth Surf. Process. Landf.* **2015**, *40*, 1464–1476. [CrossRef]

39. Fonstad, M.A.; Dietrich, J.T.; Courville, B.C.; Jensen, J.L.; Carbonneau, P.E. Topographic structure from motion: a new development in photogrammetric measurement. *Earth Surf. Process. Landf.* **2013**, *38*, 421–430. [CrossRef]

40. Carrivick, J.L.; Smith, M.W. Fluvial and aquatic applications of Structure from Motion photogrammetry and unmanned aerial vehicle/drone technology. *Wiley Interdiscip. Rev. Water* **2019**, *6*, e1328. [CrossRef]

41. Bagheri, O.; Ghodsian, M.; Saadatseresht, M. Reach scale application of UAV+SfM methods in shallow rivers hyperspatial bathymetry. In Proceedings of the ISPRS—International Archives of the Photogrammetry, Remote Sensing and Spatial Information Sciences, Kish Island, Iran, 23–25 November 2015; Copernicus GmbH: Göttingen, Germany, 2015; Volume XL-1-W5, pp. 77–81.

42. Shintani, C.; Fonstad, M.A. Comparing remote-sensing techniques collecting bathymetric data from a gravel-bed river. *Int. J. Remote Sens.* **2017**, *38*, 2883–2902. [CrossRef]

43. Dietrich, J.T. *pyBathySfM v4.0*; GitHub: San Francisco, CA, USA, 2019. [CrossRef]

44. Fisher, P.F.; Tate, N.J. Causes and consequences of error in digital elevation models. *Prog. Phys. Geogr.* **2006**, *30*, 467–489. [CrossRef]

45. Sear, D.A.; Milne, J.A. Surface modelling of upland river channel topography and sedimentology using GIS. *Phys. Chem. Earthpart B Hydrol. Ocean. Atmos.* **2000**, *25*, 399–406. [CrossRef]

46. Brasington, J.; Rumsby, B.T.; McVey, R.A. Monitoring and modelling morphological change in a braided gravel-bed river using high resolution GPS-based survey. *Earth Surf. Process. Landf.* **2000**, *25*, 973–990. [CrossRef]

47. Jaud, M.; Grasso, F.; Le Dantec, N.; Verney, R.; Delacourt, C.; Ammann, J.; Deloffre, J.; Grandjean, P. Potential of UAVs for Monitoring Mudflat Morphodynamics (Application to the Seine Estuary, France). *ISPRS Int. J. Geo-Inf.* **2016**, *5*. [CrossRef]

48. Milan, D.J.; Heritage, G.L.; Large, A.R.G.; Fuller, I.C. Filtering spatial error from DEMs: Implications for morphological change estimation. *Geomorphology* **2011**, *125*, 160–171. [CrossRef]

49. James, M.R.; Robson, S.; Smith, M.W. 3-D uncertainty-based topographic change detection with structure-from-motion photogrammetry: precision maps for ground control and directly georeferenced surveys. *Earth Surf. Process. Landf.* **2017**, *42*, 1769–1788. [CrossRef]

50. Seier, G.; Stangl, J.; Schöttl, S.; Sulzer, W.; Sass, O. UAV and TLS for monitoring a creek in an alpine environment, Styria, Austria. *Int. J. Remote Sens.* **2017**, *38*, 2903–2920. [CrossRef]

51. Heritage, G.L.; Milan, D.J.; Large, A.R.G.; Fuller, I.C. Influence of survey strategy and interpolation model on DEM quality. *Geomorphology* **2009**, *112*, 334–344. [CrossRef]

52. Chollet, F. *Deep Learning with Python*, 1st ed.; Manning Publications: Shelter Island, NY, USA, 2017; ISBN 978-1-61729-443-3.

53. Rivas-Casado, M.; González, R.B.; Ortega, J.F.; Leinster, P.; Wright, R. Towards a Transferable UAV-Based Framework for River Hydromorphological Characterization. *Sensors* **2017**, *17*. [CrossRef]

54. Buscombe, D.; Ritchie, A.C. Landscape Classification with Deep Neural Networks. *Geosciences* **2018**, *8*, 244. [CrossRef]

55. Liu, T.; Abd-Elrahman, A.; Morton, J.; Wilhelm, V.L. Comparing fully convolutional networks, random forest, support vector machine, and patch-based deep convolutional neural networks for object-based wetland mapping using images from small unmanned aircraft system. *GIScience Remote Sens.* **2018**, *55*, 243–264. [CrossRef]

56. Milani, G.; Volpi, M.; Tonolla, D.; Doering, M.; Robinson, C.; Kneubühler, M.; Schaepman, M. Robust quantification of riverine land cover dynamics by high-resolution remote sensing. *Remote Sens. Environ.* **2018**, *217*, 491–505. [CrossRef]

57. Boonpook, W.; Tan, Y.; Ye, Y.; Torteeka, P.; Torsri, K.; Dong, S. A Deep Learning Approach on Building Detection from Unmanned Aerial Vehicle-Based Images in Riverbank Monitoring. *Sensors* **2018**, *18*. [CrossRef] [PubMed]

58. Baron, J.; Hill, D.J.; Elmiligi, H. Combining image processing and machine learning to identify invasive plants in high-resolution images. *Int. J. Remote Sens.* **2018**, *39*, 5099–5118. [CrossRef]

59. Heritage, G.L.; Hemsworth, M.; Hicks, L. *Restoring the River Teme SSSI: A River Restoration Plan—Technical Report Draft (v4.2)*; JBA for Natural England: Skipton, UK, 2013.

60. James, M.R.; Robson, S. Mitigating systematic error in topographic models derived from UAV and ground-based image networks. *Earth Surf. Process. Landf.* **2014**, *39*, 1413–1420. [CrossRef]

61. Wackrow, R.; Chandler, J.H. Minimising systematic error surfaces in digital elevation models using oblique convergent imagery. *Photogramm. Rec.* **2011**, *26*, 16–31. [CrossRef]

62. Chandler, J.H.; Fryer, J.G.; Jack, A. Metric capabilities of low-cost digital cameras for close range surface measurement. *Photogramm. Rec.* **2005**, *20*, 12–26. [CrossRef]

63. Legleiter, C.J.; Kyriakidis, P.C. Forward and Inverse Transformations between Cartesian and Channel-fitted Coordinate Systems for Meandering Rivers. *Math. Geol.* **2006**, *38*, 927–958. [CrossRef]

64. Olson, R.S.; Urbanowicz, R.J.; Andrews, P.C.; Lavender, N.A.; Kidd, L.C.; Moore, J.H. Automating Biomedical Data Science through Tree-Based Pipeline Optimization. In *Applications of Evolutionary Computation, Proceedings of EvoApplications 2016*; Springer: Berlin, Germany, 2016.

65. Pedregosa, F.; Varoquaux, G.; Gramfort, A.; Michel, V.; Thirion, B.; Grisel, O.; Blondel, M.; Prettenhofer, P.; Weiss, R.; Dubourg, V.; et al. Scikit-learn: Machine Learning in Python. *J. Mach. Learn. Res.* **2011**, *12*, 2825–2830.

66. Chawla, N.V.; Bowyer, K.W.; Hall, L.O.; Kegelmeyer, W.P. SMOTE: Synthetic Minority Over-sampling Technique. *J. Artif. Intell. Res.* **2002**, *16*, 321–357. [CrossRef]

67. Lemaître, G.; Nogueira, F.; Aridas, C.K. Imbalanced-learn: A Python Toolbox to Tackle the Curse of Imbalanced Datasets in Machine Learning. *J. Mach. Learn. Res.* **2017**, *18*, 559–563.

68. Mapbox. Rasterio v1.0. 2018. Available online: https://github.com/mapbox/rasterio (accessed on 12 December 2018).

69. van der Walt, S.; Colbert, S.C.; Varoquaux, G. The NumPy Array: A Structure for Efficient Numerical Computation. *Comput. Sci. Eng.* **2011**, *13*, 22–30. [CrossRef]

70. McKinney, W. Data Structures for Statistical Computing in Python. In Proceedings of the 9th Python in Science Conference, Austin, TX, USA, 28 June–3 July 2010; pp. 51–56. [CrossRef]

71. Hunter, J.D. Matplotlib: A 2D Graphics Environment. *Comput. Sci. Eng.* **2007**, *9*, 90–95. [CrossRef]

72. Kluyver, T.; Ragan-Kelley, B.; Pérez, F.; Granger, B.E.; Bussonnier, M.; Frederic, J.; Kelley, K.; Hamrick, J.B.; Grout, J.; Corlay, S.; et al. Jupyter Notebooks-a publishing format for reproducible computational workflows. In Proceedings of the 20th International Conference on Electronic Publishing, Göttingen, Germany, 9 June 2016; pp. 87–90.

73. Wilson, R.T.; Woodget, A.S. *Code for Woodget, Dietrich and Wilson*; GitHub: San Francisco, CA, USA, 2019. [CrossRef]

74. Carbonneau, P.E.; Dietrich, J.T. Cost-effective non-metric photogrammetry from consumer-grade sUAS: Implications for direct georeferencing of structure from motion photogrammetry. *Earth Surf. Process. Landf.* **2017**, *42*, 473–486. [CrossRef]

75. Buscombe, D. SediNet: A configurable deep learning model for mixed qualitative and quantitative optical granulometry. *EarthArXiv* **2019**. [CrossRef]

Review

Remote Sensing of River Discharge: A Review and a Framing for the Discipline

Colin J Gleason [1,*] and Michael T Durand [2]

[1] Department of Civil and Environmental Engineering, University of Massachusetts Amherst, Amherst, MA 01002, USA

[2] School of the Environment, The Ohio State University; Columbus, OH 43210, USA; durand.8@osu.edu

*   Correspondence: cjgleason@umass.edu

Received: 13 February 2020; Accepted: 25 March 2020; Published: 31 March 2020

Abstract: Remote sensing of river discharge (RSQ) is a burgeoning field rife with innovation. This innovation has resulted in a highly non-cohesive subfield of hydrology advancing at a rapid pace, and as a result misconceptions, mis-citations, and confusion are apparent among authors, readers, editors, and reviewers. While the intellectually diverse subfield of RSQ practitioners can parse this confusion, the broader hydrology community views RSQ as a monolith and such confusion can be damaging. RSQ has not been comprehensively summarized over the past decade, and we believe that a summary of the recent literature has a potential to provide clarity to practitioners and general hydrologists alike. Therefore, we here summarize a broad swath of the literature, and find after our reading that the most appropriate way to summarize this literature is first by application area (into methods appropriate for gauged, semi-gauged, regionally gauged, politically ungauged, and totally ungauged basins) and next by methodology. We do not find categorizing by sensor useful, and everything from un-crewed aerial vehicles (UAVs) to satellites are considered here. Perhaps the most cogent theme to emerge from our reading is the need for context. All RSQ is employed in the service of furthering hydrologic understanding, and we argue that nearly all RSQ is useful in this pursuit provided it is properly contextualized. We argue that if authors place each new work into the correct application context, much confusion can be avoided, and we suggest a framework for such context here. Specifically, we define which RSQ techniques are and are not appropriate for ungauged basins, and further define what it means to be 'ungauged' in the context of RSQ. We also include political and economic realities of RSQ, as the objective of the field is sometimes to provide data purposefully cloistered by specific political decisions. This framing can enable RSQ to respond to hydrology at large with confidence and cohesion even in the face of methodological and application diversity evident within the literature. Finally, we embrace the intellectual diversity of RSQ and suggest the field is best served by a continuation of methodological proliferation rather than by a move toward orthodoxy and standardization.

Keywords: remote sensing; rivers; discharge; hydrology; modelling; geomorphology; ungauged basins

---

1. Introduction

Remote sensing (RS) provides value to the earth science community as a source of primary data (electromagnetic radiation recorded directly by the satellite/aircraft) obtained from a unique frame of reference. Although raw electromagnetic radiation received by sensors needs to be carefully calibrated to transform these observations into useable science signals, remote sensing platforms provide high quality primary data at a variety of spatial, temporal, and spectral scales. Hydrology in particular has been traditionally open to the use of RS (e.g., ([1–4]); Calmant et al., 2008; Lettenmaier et al. 2015; Doell et al. 2016; Lettenmaier et al., 2006), in part because this physically integrative discipline is quite often subject to political data restrictions or difficulties in field data collection.

RS can contribute to both basic and applied hydrology equally from its unique vantage point. Take for example recent work by Allen et al. ([5]); 2018), which provides a new and interesting insight into stream dynamics and organization at the global scale only possible via combining detailed field observations with global remote sensing of rivers. Likewise, Smith et al. ([6]; 2017) used fixed-wing un-crewed aerial vehicles (UAVs) to reveal previously unknown perennial stream networks atop the Greenland Ice Sheet, which suggested a complete revamp of understanding in how ice sheet meltwater is routed to the global ocean. At broader spatial scales, Lehner and Doll ([7]; 2004) offered an early global mapping of lakes and reservoirs, Prigent et al. ([8]; 2001) similarly mapped wetland dynamics, Allen and Pavelsky ([9]; 2015) and Yamazaki et al. ([10]; 2019) provided new maps of global rivers, and Pekel et al. ([11]; 2016) mapped all global hydromorphic features as seen by the Landsat family of satellites over the past 32 years. These important interventions in our understanding of the observed, rather than modelled, positions of the planet's water features represent insight into the earth system only possible with remote sensing. We argue that remote sensing hydrology is therefore poised to respond to the charge laid by McDonnell et al. ([12]; 2007) to use primary data and hypothesis driven science to push the discipline forward.

Remote sensing is also well positioned to respond to the charge of Wood et al. ([13]; 2011) to focus on hydrologic model improvement in pursuit of advancing knowledge of water resources. Lin et al. ([14]; 2019) recently published new simulations of global hydrology that pull from tens of thousands of river gauges and numerous remote sensing products to bring remote sensing to bear in the fullest expression of global hydrologic modelling to date. Such efforts will continue to play important roles in both climate reanalysis and prediction. Remote sensing has also been noted for its ability to observe human impacts on hydrology (e.g., [4,15,16]; Lettenmaier and Famligetti, 2006; Martin et al., 2016; Yoon and Beighley, 2015), and the impact of hydrology on humans in the form of floods and flood forecasting (e.g., [17–21]; Barton and Bathols, 1989; Biancamaria et al., 2011; Grimaldi et al., 2016, Li et al., 2016; Schumann et al., 2016). These approaches reflect the role hydrology plays in society: water is fundamental to all civilizations and any advance in our predictive capability is welcomed. Ultimately, the unique vantage point of remotely sensed platforms (especially spaceborne sensors) feed hydrology's need for primary data to drive both fundamental discovery and practical application.

River discharge is one of the most important and frequent targets for remote sensing in hydrology. Discharge is the product of river flow area and velocity, or the volume of water passing a specified point at any instant in time. The only method truly capable of *measuring* discharge is a bucket and a stopwatch: literally quantifying a volume of collected water for some given time period. This is obviously impractical for all but the smallest streams, and as such the most respected discharge *estimates* come from an Acoustic doppler current profiler (ADCP) or weir equations. These techniques are commonly referred to as measurements, but they are in fact approximations of discharge (albeit accurate approximations), and are themselves subject to error, especially at very high and very low flows. Stream gauges are discharge estimates as well that transform automated measurements of river stage to discharge via an empirically calibrated rating curve. The data points to calibrate such curves are normally driven by ADCP estimates of discharge. These gauge discharge estimates form the backbone of human water management decisions and hydrologic science.

Remote sensing of river discharge (RSQ) is thus not surprisingly a vibrant field of study in the scholarly literature, but this vibrancy has led to some confusion. There are varying degrees of processing involved in turning primary remotely sensed information into discharge. Remote sensors record electromagnetic radiation, which is then converted to a signal of interest to the hydrologist. These raw signals include recorded reflectances, range and interferometric phase observations, and emittances. Usually, these signals are radiometrically calibrated to conditions at the top of the atmosphere and then georeferenced to a regular grid on the earth's surface to provide the most basic form of primary data in RSQ. At the other end of the spectrum, gravity recovery and climate experiment (GRACE) observations of relative satellite positions are first turned into a model of the geoid and then processed into mass anomalies via the application of a hydrologic model: the remote sensing signal has indeed driven the data, but many other sources of information and model physics have been invoked to produce the desired output. Thus, many users are confused about remotely sensed data products, and it can be difficult for the non-specialist to determine which products rely on ancillary data and which do not. In the case of RSQ, all methodologies must transform observations to discharge, as we cannot instantaneously and simultaneously measure river depth, cross sectional area, and depth-averaged velocity from space. Accordingly, hundreds of manuscripts have been written on this subject, all fundamentally seeking to use remotely sensed data to estimate river discharge.

We believe that a review of the RSQ literature is timely and necessary. A recent explosion of the literature (and journals willing to publish that literature, including this one) requires careful summarization to understand advances of the past decade. Along with this recent proliferation, many common misconceptions about RSQ have emerged, with authors sometimes incorrectly citing other work and sometimes missing wide swaths of the relevant RSQ literature when introducing new studies. We argue that a fruitful examination of the literature is not based on differentiation via sensor or sensor class as in past sketches of the field (e.g., active vs passive, microwave vs. optical), but based on application area and methodological approach, what is the purpose for the manuscript, and how does a remotely sensed signal turn into discharge? We will thus be sensor agnostic in this regard, and everything from UAVs to satellites are considered here. Such a lens affords us a broad swath of the literature to review, and we are able to bring numerous and seemingly disparate subfields of RSQ together under unifying themes. Ultimately, we hope this paper can serve as a guide to hydrologists in choosing what methods and data might work for them.

The goals of this paper are as follows. (1) We review the literature (mostly since 2010) without downloading and reading every single paper available: we were not uncritically exhaustive. It is inevitable that we have not canvassed some of the relevant literature or very recent literature, but we do believe we have captured a snapshot of important work of the past decade and beyond. Our goal is a broad survey rather than deep discussion in any specific area of the RSQ literature. (2) We introduce terminology and a context for RSQ to differentiate different meanings of the term 'ungauged,' and argue this is necessary for a comprehensive understanding of the RSQ literature. (3) Further, we categorize the literature on two axes: a "gauged" axis and a methodological axis, and find that these distinctions place the literature into the most useful context. (4) Finally, we attempt to debunk common misconceptions about RSQ and discuss its ethics and politics. We believe that what emerges is a holistic picture of the literature as it stands now and where it might fruitfully go next.

Overview and Organization

Sections 2.1 and 2.2 divide the literature into those techniques that are appropriate for politically and totally ungauged basins (Section 2.2) and those that are not (Section 2.1). These terms are introduced in Figure 1. Section 2.1 contains almost twice the number of papers as Section 2.2, reflecting the priorities of many authors. We believe that this high level separation is an important axis of differentiation that should be used to clearly categorize RSQ work in the past and future, and the discussion in Section 3 interrogates our use of this classifying scheme. Following this broad division, we have lumped the almost 170 manuscripts reviewed here into further broad categories within the gauged/ungauged divide as appropriate from our reading (Figure 2). For gauged, semi-gauged, and regionally gauged basins, these categories include calibration/assimilation into hydrologic models that use ancillary in situ data (Section 2.1.1), calibration of hydraulic models (Section 2.1.2), and the largest subsection in the entire RSQ literature: calibration of local channel hydraulics (Section 2.1.3). For politically and totally ungauged basins our subsections include calibration/assimilation into hydrologic (Section 2.2.1) and hydraulic (Section 2.2.2) models as well as geomorphic inverse problems (Section 2.2.3). We have made these categorizations after careful scrutiny of the literature, and the length of each subsection loosely reflects the volume of the literature in that subfield.

Basin Type	Description	Role of RSQ
Gauged	A watershed with gauges at the outlet and numerous other channels. Well understood hydrologically.	Extend gauge record in space and time
Semi-Gauged	A watershed with a few gauges relative to the entire network. Some reaches well understood.	Extend gauge record in space and time, and provide calibrated input at ungauged reaches
Regionally Gauged	An ungauged watershed nearby other well gauged watersheds. Hydrologically similar to gauged watersheds.	Provide discharge informed by nearby basins
Politically Ungauged	A watershed that may be well gauged, but for which gauge data are not shared publicly.	Improve on existing knowledge
Totally Ungauged	A watershed with no gauges, either due to remoteness, harshness of the environment, or resource deficiency.	Improve on existing knowledge

Figure 1. RSQ adds value to hydrology in different ways based on what we already know about the basin, from improving on that knowledge in the most basic sense (ungauged) to finely augmenting existing data in space and time for specific purposes (gauged). We note rivers in the same basin may change category depending on the specific study reach or sub-watershed of interest: a sub-watershed may be gauged while the basin as a whole is semi-gauged.

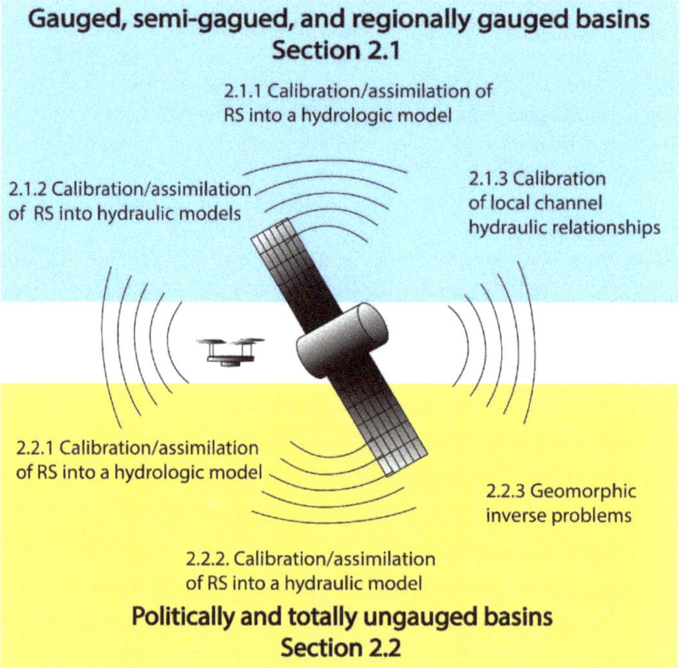

Figure 2. Our classification of the remote sensing of the discharge literature after our reading of the manuscripts cited here. We have purposely brought hydrologic modelling and channel remote sensing together, as we believe they are united by the common goal of discharge prediction.

We note that our reading is broader than previous reviews, and is especially different from an early and influential view of RSQ offered by Alsdorf et al. ([22]; 2007). They noted the rise of then-nascent satellite hydrology and theorized that measurements of water surface elevation and surface slope made from an active sensor would be the best and perhaps only way to provide needed data to address data gaps in hydrology. This perspective on the river channel and its hydraulics as the object of sensing has had a strong influence on the field, and even today our inclusion of Sections 2.1.1 and 2.2.1 bridging between remote sensing and modeling will be surprising to some readers. As recently as 2016, for example, Sichangi et al. [23] begin their introduction to a new RSQ method with their view on the field, which offers a succinct synthesis of many methods boiled down to four bullet points. These authors summarized RSQ into four basic approaches: use of altimetry data to calibrate rating curves, inference of discharge from corrolatory indices based on inundation area, remote sensing of hydraulic components of discharge directly, and the at-many-stations hydraulic geometry (AMHG) approach of Gleason and Smith ([24]; 2014), later to become known as an example of a mass conserved flow law inversion (McFLI, [25]; Gleason et al., 2017). Sichangi et al. thus offer a classification scheme that is effective and focused but excludes much of the relevant literature, particularly on hydrologic modelling. Our reading has indicated that the global hydrologic modelling community relies heavily on remote sensing and has many similar outcomes and goals to the traditionally channel-focused RSQ community, and thus we wish to use this manuscript as an opportunity to view the problem of discharge estimation in as holistic a paradigm as possible.

2. Summary of Remote Sensing of River Discharge

Our organization of the literature relies first on the application area of RSQ: in either a gauged or an ungauged basin. Rather than view an ungauged basin as a binary condition, we believe ungauged status to be a continuum after our reading: ranging from a true lack of data at any relevant spatial or temporal proximity (totally ungauged) to rivers well-monitored by continuous gauge data (gauged, Figure 1). Between these extremes, we consider other examples of ungauged status: some watersheds have sparse gauges through a watershed but not in a channel of interest (semi-gauged), others have gauges in nearby climatologically and geologically similar watersheds (regionally ungauged), while still others are well monitored by gauge data that are not publicly shared (politically ungauged). These tags represent distinct hydrologic realities that all manifest as 'ungauged' or 'gauged' in much of the literature. We argue that the RSQ goals and appropriate methodologies for these cases are distinct from one another, and that this organizational context is essential for understanding RSQ as a whole.

2.1. Gauged, Semi-Gauged, and Regionally Ungauged Basins

In all of the approaches in Section 2.1, the goal of RSQ is to provide as-accurate-as-possible discharge estimates. These approaches leverage whatever data they can find in order to make the best estimates of discharge possible, and in doing so gain accuracy and precision but become more beholden to the ancillary data/assumptions needed to make each approach successful. Thus, RSQ in this section could be driven almost entirely by a hydrologic model or an in situ gauging station, with RS data used to extend this knowledge in space and time. For example, RSQ could be used to extend gauges to ungauged headwater streams or rivers in a neighboring basin, particularly useful in important but sparsely gauged locations (e.g., the Tibetan Plateau, Siberia). This is in contrast to Section 2.2, where RSQ methods assume very little is known about the watershed according to the context given in Section 3.1.

2.1.1. Calibration/Assimilation of RS into a Hydrologic Model

Hydrologic models seek to parse the components of the hydrologic cycle (precipitation (P), evapotranspiration (ET), terrestrial storage) to calculate water excess (i.e., runoff), which is then routed through a channel network to become river discharge. Often, these models rely on a land-surface module that explicitly parameterizes water and energy fluxes across schema of varying complexity, and the hydrologic literature has shown itself to be open to bold conclusions made at global scales by models integrating remotely sensed data (e.g., [4,26]; Lettenmaier and Famligetti, 2006; Rodell et al., 2009;). Models frequently require calibration to function well, and remote sensing was seen as an early source of independent calibration/validation data for such models ([27,28]; Gupta et al., 1998; Sivapalan et al., 2003). In fact, Gelati et al. ([29]; 2018) refer to RS data as "ideal benchmarks for spatially distributed evaluations of land surface models." As a consequence, we believe that no review of RSQ is complete without inclusion of the literature coupling hydrologic modelling and remote sensing to produce discharge. There are hundreds if not thousands of modelling studies using RS information as targets for calibration and assimilation with a goal of producing discharge. We have here distilled the literature into a non-exhaustive survey that we believe captures the spirit of that scholarship. This paradigm for RSQ is not channel based: remote sensing is not asked to provide data on river channels themselves, but rather on the interconnected whole of land surface hydrology in a quest to provide accurate local, regional, or global discharge estimates. Previous characterizations of RSQ (e.g., [22,23,30]; Alsdorf et al., 2007; Sichangi et al., 2016; Tarpanelli et al., 2019) do not include this paradigm in their sketches of the field.

We are aware of the differences between calibration and assimilation. Chief among these is a distinction between tuning parameters that produce a model state variable (calibration), updating this model state directly (assimilation), or updating states and parameters simultaneously (also assimilation: [31]; Reichle, 2008). However, for the purposes of RSQ, we argue that these non-trivial differences are in service of the same goal—to use RS to improve a model of discharge. We thus lump them together here. Also, as Jodar et al. ([32]; 2018) note, a classic approach in hydrologic modelling is to turn to regionalization, where in situ information in one basin is mapped onto another ungauged basin. This regionalization is, for our purposes, another form of calibration. Finally, both Reichle et al., ([33]; 2014) and Maxwell et al., ([34]; 2018) urge caution in assimilation work, as the likelihood function 'under the hood' of data assimilation is complex and requires care and consideration in its formulation rather than off the shelf deployment.

Chen et al. ([35]; 1998) were among the first to show how models may be combined with RS to produce discharge, using Topex/Poseidon data to track changes in sea levels and understand anomalies in sea surface heights. They argued that if the ocean is the ultimate reservoir for all terrestrial water, then tracking ocean anomalies is a way to understand necessary changes in the global hydrologic cycle that produced those anomalies. At smaller scales, Dziubanksi and Franz [36] 2016) assimilated RS into a snow model to improve estimates of snow water equivalent, but in turn improve estimates of discharge, and Fortin et al. [37] explored the concept of this coupling as early as 2001. Similarly, watershed storage change can be addressed through RS using GRACE satellite observations. GRACE geoid observations can resolve mass fluxes at relevant hydrologic timescales ([38]; Rowlands et al., 2005), and these large-area fluxes of water are used for assimilation and calibration in many hydrology models with a goal of producing river discharge ([39–46]; Syed et al., 2005; 2007; 2009; 2010; Schmidt et al., 2008; Werth et al., 2009; Frappart et al., 2011; Eom et al., 2017). We note that GRACE mass anomalies are themselves often a product of hydrologic modelling ([47]; Wiese et al., 2016), and thus by using GRACE as an RS signal, a hydrologist has perhaps already invoked a calibrated model whether they had intended to or not.

Ultimately, many successful attempts at hydrologic calibration/assimilation invoke more than one RS signal (e.g., [48–51]; Siquera et al., 2018; Chandanpurker et al., 2017; Zhang et al., 2016; Silvestro et al., 2015). The most recent of these efforts ([14]; Lin et al., 2019) delivers on the promise of a previous decade of work. In their study, Lin et al. use large quantities of both ground and RS data (mainly for precipitation and ET) in conjunction with the latest in RS hydrography to produce a coherent global reanalysis of daily river discharge at almost three million river reaches. This level of temporal and spatial precision had not previously been demonstrated, and illustrates the power of RS for global hydrologic modelling. Lin et al. used a calibration approach that considers uncertainty: calibrating with gauges where available, calibrating with RS products where gauges are unavailable, and calibrating with reanalysis data where RS and gauges are available. This allows optimal use of both in situ and RS data to produce discharge that would be impossible without the use of remote sensing. We note that the approach taken by Lin et al. ([14]; 2019) need not be global; in fact, previous work has taken the same tack toward addressing water resources challenges in the Congo, Himalayas, and Tibetan Plateau ([52–54]; Lee et al., 2011; Wang et al., 2015; Wulf et al., 2016). All of these studies used RS data where gauges were not available to calibrate their model, but all rely on in situ data to provide the bulk of model skill.

We have barely scratched the surface of this literature, but have included these brief examples to make our assertion that the regional and global hydrologic modelling community are as much a part of RSQ as traditional channel-based RSQ. In fact, hydrologic modelling can overcome the spatial and temporal limitations of RS data (which themselves augment the spatial limitations of gauges), and clever assimilation and calibration can account for uncertainties in both RS and in situ parameters and thus lead to a final product greater than the sum of its parts. Rivers are but one part of a hydrologic whole, and authors in this section move forward with this in mind.

2.1.2. Calibration/Assimilation RS into Hydraulic Models

More traditionally, RSQ is grounded in the fields of fluid mechanics, hydraulics, and fluvial geomorphology. These large fields of study have long been interested in how river channels respond to channel form, sediment transport, and landscape evolution and have yielded numerous empirical and first principles equations that govern precisely how water in a river channel responds to different environmental conditions. The famous Manning's equation is one such example that seeks to balance friction losses with gravity-driven flow in a river channel, despite its simplified insistence on a fixed velocity-depth exponent and a stage-constant roughness. Similarly, hydraulic geometry predicts responses in width, depth, and velocity given changes in discharge (e.g [55,56]; Ferguson, 1986; Gleason, 2015), and many additional empirical fluvial geomorphic phenomena relating satellite-observable quantities with discharge have been observed beyond these traditional ground-based parameters. These hydraulic phenomena thus form hydraulic models of river behavior, which can be as simple as the Manning's equation or as complex as a full 3D conservation of mass and momentum in a finite element model. These hydraulics can be coupled with information from the hydrologic cycle, as the shape of hydrographs themselves has information about watershed processes that may inform RSQ ([57]; Fleischmann et al., 2016). The studies in Sections 2.1.2 and 2.1.3 invoke these hydraulic relations as the basis for RSQ in conjunction with data and assumptions only available for gauged, semi-gauged, and regionally gauged basins. An earlier review germane to this topic and specific to the Amazon is given by Hall et al. ([58]; 2011).

Among the most interesting and innovative hydraulically based approaches was pioneered by hydrologist Dave Bjerklie. Bjerklie and his collaborators were writing principally in the early 2000s, and if their approach had been devised today, we would likely label it as a 'big data' approach. Bjerklie et al. ([59]; 2003) noted that discharge can be reduced to a function of width, depth, slope, and Manning's n assuming the restrictive and limiting assumption of a parabolic channel shape. Using thousands of field observations, Bjerklie et al. built statistical relationships between these parameters, and argued on the strength of their training data that mean values of these parameters are reliably estimated globally by this purely empirical approach and RS observations. Bjerklie et al. ([60]; 2005) built on this to fit a calibration between maximum channel width and slope as a power function using multiple regression on a ~1000 river dataset of hydraulic observations, and Bjerklie ([61]; 2007) derived an equation for bankfull velocity from channel slope and lengths of meander bends. Thus, Bjerklie et al. have simplified the underconstrained RSQ hydraulic problem by using big-data empirical geomorphic relationships that theoretically represent all global rivers, creating new generalized hydraulic models ripe for remote sensing in the process. Further work has explicitly compared this Bjerklie approach to space-based rating curves ([62] Kebede et al., 2020). This raises an interesting question about whether or not this approach is best placed in Section 2.1 or Section 2.2. This approach would not be possible without mining thousands of in situ observations (hence Section 2.1), but once these empirical relations are defined, they are globally applicable without further calibration (Section 2.2). However, any statistical learning is subject to the representativeness of the training data, which in the case of

Bjerklie et al.'s work comes entirely from the US. Frasson et al. ([63]; 2019) have recently updated this framework with additional observations, this time built at the truly global scale, which places their work better in the context of Section 2.2. We therefore argue that the works of Bjerklie are only truly applicable within their training samples, and this work is more continental than global, while the work of Frasson et al. is a better example of how this paradigm might apply in ungauged basins. Further, we argue that Bjerklie's work is closer to regionalization than to global modelling as in Section 2.1, and this distinction will become important as we seek context for the field of RSQ in Section 3.

Neal et al. ([64]; 2009) offer a different tack for calibrating hydraulic models. In their and the rest of the papers in this section, the goal is to remotely sense the components of discharge separately (e.g., depth, velocity, width, flow resistance) before passing these to a model. In these cases, the hydraulic model in question is a finite element or finite difference model capable of prediction in space and time, rather than the generalized models of Bjerklie. Neal et al. ([64]; 2009) used SAR to estimate discharge using a Kalman Filter and assumptions of a given initial flow, the river as linear reservoir, stress and flow thresholds, a lidar DEM, ground surveys, an assumed Manning's n, and a soil moisture deficit. Given all of these assumptions, their method worked well. However, the authors acknowledge the amount of 'expert judgement' required for a good result, and this approach is useful only in situations in which a good deal of information is already known about the channel. Similarly, Temimi et al. ([65]; 2011) assimilated different RS signals into a model coupled to assumptions of hydraulic geometry to assess flood discharge. They too were successful, but like Neal et al. required in situ data to correctly parameterize the channel. King et al. ([66]; 2018) improved on this approach to eliminate many of these assumptions using UAV stereo photogrammetry to drive the US Army Corps' Hydraulic Engineering Center's River Analysis System (HEC RAS) hydraulic model. Given an input flow from an upstream gauge and a rating curve, they adjusted the water levels in the hydraulic model until they matched area changes observed by the UAV, relying on the fine-scale DEM created photogrammetrically. This labor intensive RSQ algorithm is effective, but only at the scales for which it can be replicated. Harada and Li ([67]; 2018) combined numerous field inputs with multispectral imagery to calibrate sediment grain size distributions and shear stresses along with discharge, and Try et al. ([68]; 2018) used an empirical width-depth formula in their model to estimate discharge from river width. In all of these cases, authors start from a few known and a few unknown hydraulic parameters of a particular hydraulic model, and use RS for the unknowns. The success of such an approach is, not surprisingly, a function of how well the problem can be posed, with the caveat that the better posed the problem, the more data are required.

2.1.3. Calibration of Local Channel Hydraulic Relationships

Remotely Sensed Rating Curves

Perhaps the most logical transition from traditional hydrology to RSQ is the realization that empirical relationships might be built from RS signals and calibrated directly to observed discharge for a specific channel. In its simplest form, this paradigm takes shape as a space-based rating curve. Among the earliest examples of the space rating curve uses altimetry estimates of river stage directly in a rating curve together with in situ measured flows ([69]; Koblinsky et al., 1993). This is an analog to the gauging station, and forms a powerful discharge monitoring approach provided sufficient in situ data are available to populate the rating curve. Other early work included SAR studies of floodplains, rivers, and lakes interested in deriving levels and/or widths in service of these rating goals ([70–75]; Smith et al., 1996; Alsdorf, 2003; Alsdorf et al., 2001; 2001; Frappart et al., 2005; LeFavour and Alsdorf, 2005), and Smith ([76]; 1997) reviewed work to date at the time of writing. Kouraev et al. ([77]; 2004) and similar studies continued to expand and refine the capabilities of traditional radar altimetry to compute discharge at locations where in situ data are available, and this work continues through today (e.g., [78–84]; Pavelsky, 2014; Pavelsky and Smith, 2009; Schneider et al., 2017; Young et al., 2015; Paris et al., 2016; Nathanson et al., 2012; Feng et al., 2019). The field has matured to the point where detailed considerations of RS signal quality are becoming important (e.g., [85]; Normandin et al., 2018), rather than repeated proof of concept of the viability of space-based rating. We note also that altimetry signals often used in space based rating curves are generally only applicable over large rivers; most altimeters have been designed for ocean applications and reprocessing these signals over smaller rivers is often challenging. Width measurements can also drive rating curves. These are straightforward to derive for RSQ purposes on rivers as narrow as 12 m with the advent of CubeSats, but these signals are often of less radiometric quality than those available from conventional satellites ([84]; Feng et al., 2019). Research in this area also goes beyond the use of satellites. Ashmore and Sauks ([86]; 2006), Gleason et al. ([87]; 2015), and Young et al. ([81]; 2016) all used time lapse cameras to provide an RS signal for river effective width extraction for remote Arctic rivers coupled with in situ discharge measurements. Huang et al. ([88]; 2018) used UAVs in conjunction with satellite data and a gauge on the information-sparse Tibetan plateau, similar to King et al. ([66]; 2018). These small-area applications highlight how not all RSQ work is driven by either global or ungauged basin interests, and provide innovative solutions to practical fieldwork issues via RSQ. In sum, this section represents the most straightforward and non-controversial application of RSQ reviewed here, as RS has simply replaced a ground measurement in a traditional discharge estimation paradigm.

Correlations between RS Observations and Discharge

Whatever the RS signal, the idea that RS observations correlate with discharge is powerful. This can move the idea of a rating curve beyond measuring channel stage and/or width toward measuring other targets that can reliably drive RSQ. The work pioneered by Brakenridge et al. ([89]; 2007) illustrates this concept excellently. In this 2007 paper, building off their previous work, Brakenridge et al. identify a simple and powerful phenomenon. They observed that the response of passive microwave satellite observations of river channels relative to a nearby 'dry' land area (what they call the ratio between a calibration area C and measurement area M) is strongly correlated with river discharge. The logic behind the Brakenridge ratio is that rivers get wider as discharge increases (the same as the width-based rating curves noted above), but it is not necessary to precisely quantify this top width change (see Figure 3). Rivers do indeed get wider with discharge, but soil moisture near the river also increases and there are numerous spatial changes to the river surface not captured by cross sectional top width changes. A 'wet' pixel thus reflects quantities changing with discharge, even if it does not track changes in width directly, while the nearby 'dry' pixel is relatively insensitive to these changes. Using a ratio between wet and dry allows a finer response to hydrologic forcing than tracking uncalibrated changes to the 'wet' area alone. Further, this approach has the advantage of spatial resolution—coarser ground areas may be considered rather than precise measurements of river width, thus overcoming errors in discharge that propagate from width measurement errors. With this ratio in hand, Brakenridge et al. made rating curves from in situ discharges using higher order polynomial regression.

Since this 2007 publication, this ratio approach has grown into a successful subfield of RSQ that was born from the unique hydrologic vantage point of RS, unlike space based rating curves, which are extensions of ground hydrology. The McFLI approaches covered in Section 2.2.2 are similarly fundamentally tied to a RS vantage point, and both of these 'schools' of RSQ are conceived with RS data in mind from the onset. In 2012, Brakenridge et al. [90] pushed their approach to thousands of global stations, using a previously in-situ calibrated hydrology model to provide the training data for discharge predictions. In this case the model provides 'truth' to guide the remote sensing, so applicability of this method is dependent on faith in the model output, and in this we see a strong parallel with the studies of Section 2.1.1. Tarpanelli et al. [91]; 2013) proved the same ratio concept is viable from visible and near-infrared observations, and Van Dijk et al. ([92]; 2016) investigated the global potential for the ratio approach while explicitly considering tradeoffs between passive and active sensors. Tarpanelli et al. have led efforts to refine this approach, with papers in 2017 [93] and 2018 [94] that consider fusion between optical imagery and altimetry while also investigating machine learning (rather than regression) for training algorithms. This work also branches toward the spatial extension of training data through hydrologic modelling, which moves this application squarely beyond well- and semi-gauged basins and into regionally gauged basins.

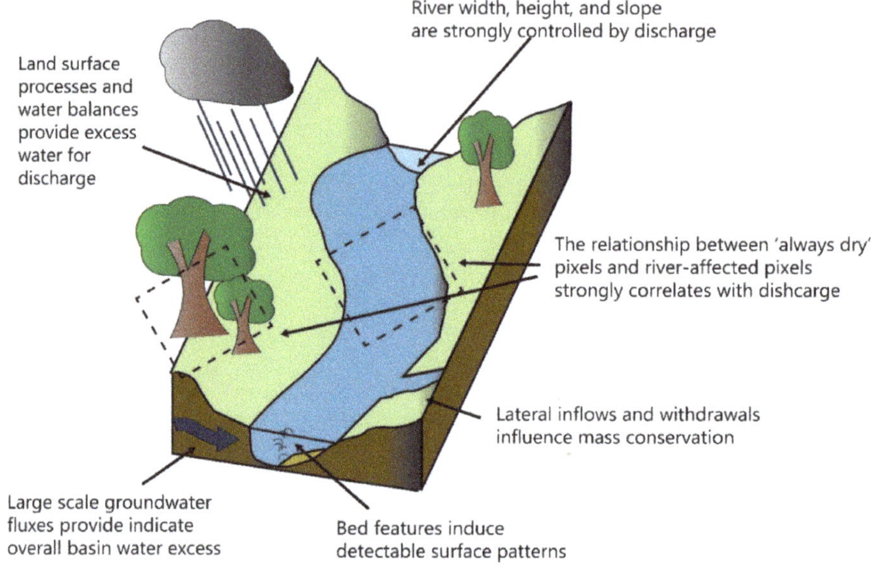

Figure 3. This graphical representation of a river channel encompasses nearly all of the different RS signals considered by authors reviewed here. By presenting them in one image, we hope to highlight the similarities across different techniques.

In Pursuit of Local Channel Hydraulic Parameters

A final channel hydraulic calibration was recognized by scientists as early as the 1970s, when Lynzenga ([95]; 1978) calibrated the reflectance of the bottom of a water column to field measurements to yield a remotely sensed estimate of depth. The principle behind remote sensing of depth is driven by attenuation of light in a water column (Beer's Law). Lee et al. ([96,97]; 1998; 1999) describe this process for vertically homogenous, optically shallow waters (i.e., water columns in which light is able to reflect from the channel bottom and back through the water surface). They separated observed upwelling radiance into three parts—bottom reflectance, surface reflectance, and air column radiance, while controlling for water column attenuation. Lee et al. noted that slight changes in turbidity affect the performance of depth retrieval, so the model must be parameterized according to the properties of the water in both space and time, making this approach impractical in a global sense. Contemporary work continued in this vein (e.g., [98]; Gould et al., 1999), until Fonstad and Marcus ([99]; 2005) and Marcus and Fonstad ([100]; 2008) were among the first to translate this approach to rivers. Remote sensing scientist Carl Legleiter has built on this heritage with a calibrated approach termed optical band ratio analysis (OBRA), first published in 2009 [101]. This approach is built on the same principles of reflection and attenuation as earlier work, and requires calibration or assumption of optical properties to back out river depth. This OBRA approach has been successfully demonstrated in numerous contexts ([102–104]; Legleiter, 2015; Legleiter et al., 2017; Smith et al., 2015). Alternatively, Johnson and Cowen ([105]; 2016) used highly controlled flume experiments to provide estimates of bathymetry based on observed turbulence structures. For example, riffles or hydraulic jumps on a water surface are indications of changes in channel structure. In all cases, these observations need be calibrated with some in situ information to yield discharge.

If calibrated optical attenuation and other approaches can be used to provide depth estimates, then only channel velocity is needed for a discharge estimate. Costa et al. ([106]; 2000) provide an early remote sensing of velocity method, but this approach requires sophisticated equipment, highly trained personnel, and a bridge: it is not a practical approach for numerous rivers. Further, surface velocity is not the same as the depth-averaged velocity needed to estimate discharge, and more reductive assumptions are needed to transform one to the other. However, Costa et al. provide a proof of concept for remote sensing of channel velocity. More recent work has focused on two basic techniques: Particle Image Velocimetry (tracking particles on water surfaces using image processing techniques) and leveraging turbulence structures. Legleiter et al. ([103]; 2017), for instance, used the OBRA technique for river depth and bridge-based thermal imagery to track surface velocity. A number of recent papers flesh this out, in the context of applying entropy theory to the cross-sectional velocity profile ([107–110]; Chiu, 1991; Fulton and Ostrowski, 2008; Moramarco et al., 2017; 2019). All of these investigations are aided by better geomorphic knowledge of channels, and RS can be useful here too, often through lidar mapping of topography and sometimes bathymetry (e.g., [111]; Hilldale and Raff, 2008). These and the other papers in Section 2.1 show that highly accurate discharge can be estimated from RS, and there seems to be a tradeoff in discharge accuracy vs. the amount of prior data we have on a river, making the most accurate RSQ estimates in the best-understood basins.

2.2. Politcally and Totally Ungauged Basins

Fundamentally, RSQ in ungauged basins is an ill posed problem. We cannot measure enough hydraulics from RS to solve for discharge directly, as bathymetry, friction, and discharge will always be unknown in a channel without calibration. The approaches in Section 2.1 might use reasonable values for friction based on knowledge of the basin/channels, or pull bathymetry from a regional regression, or use a hydrologic model, or bypass the hydraulics and calibrate discharge directly to a RS signal to circumvent this problem. The approaches in Section 2.2, however, seek to overcome the problem solely from outer space, limiting themselves to assumptions about rivers that may be made only from RS data or other global data that are independent of specific calibration. The accuracy of these approaches tends to be much lower than the approaches in Section 2.1, yet as we note in Figure 1 the goal of these authors is to improve on what we know in ungauged basins. Given that in many ungauged basins we know very little, even approaches with high errors in discharge can be useful. This literature is much smaller than the scholarly output for Section 2.1, and as such we cover this literature in slightly more depth.

2.2.1. Calibration/Assimilation of an RS Signal into a Hydrologic Model

We first consider RSQ as seen again from a hydrologic perspective, yet this time we focus on RS as the *sole* source of calibration data, as opposed to use of RS data in conjunction with an in situ calibrated model as in Section 2.1. The balance of these fluxes is water excess, and so a valid RSQ approach is to emulate a hydrologic model solely from outer space. Various authors have investigated RS of each of these specific components in an effort to better constrain each one. Parr et al. ([112]; 2015) used RS ET and leaf area index products in conjunction with the VIC model, while Lopez-Lopez et al. ([113]; 2018) explored downscaling and in 2017 calibrated the PCR GLOBWB model for a basin in Morocco with RS ET and soil moisture, and both concluded that their approach is viable and improves discharge accuracy [114]. However, Mendiguren et al. ([115]; 2017) and Bowman et al. ([116]; 2016) explicitly compared RS ET energy balance models against traditionally calibrated hydrological models and found low correlation between the two products. These authors further argued that the spatial component of errors in ET are important and well-suited to remote sensing. Remote sensing of precipitation is a huge field given the importance of precipitation to flood forecasting, and thus this literature is not covered here despite providing important grounding and uncertainty analysis to RSQ; see Lettenmaier et al. ([2]; 2015) for a recent review. While this precipitation literature is certainly germane, any satisfactory review would render this manuscript overlong. The GRACE signals covered in Section 2.1.1 can provide the final storage term, and thus a simple water balance may be made at very large spatial footprints for totally and politically ungauged basins from RS estimates of precipitation, ET, and storage.

Moving beyond a simple water balance, Sun et al. ([117]; 2015) provide a basin-scale approach that seeks to use no in situ data to calibrate their hydrologic model, instead using river width observations as a calibration target. This decision required a reconfiguration of traditional model physics to represent width as the state variable (i.e., rearranging the discharge equation so width is on the left hand side). Sun et al. then attempt to force their hydrology model with only globally available data, using the dynamic river width signal to calibrate. Emery et al. ([118]; 2018) follow Brakenridge et al. ([90]; 2012) in using a hydrologic model 'off the shelf' to provide calibration data, where their calibrating model was itself previously calibrated using altimetry to track water levels ([82]; Paris et al. 2016). Emery et al. then use this RS-driven model and new, independent altimetry estimates of river stage as inputs to a second hydrologic model, tuning the model based on the Paris et al. ([82]; 2016) discharges. In this approach we see the coupling of the work in Sections 2.1.1 and 2.1.3: using RS signals for both hydraulic and hydrologic components. We have placed this work in the 'ungauged' category. On the one hand, Sun et al. and Emery et al. needed no in situ data in their specific approaches and can rightly claim their techniques work in ungauged basins. However, their forcing data, while uniformly available, relied on in situ calibration data from either precipitation gauges or stream gauges in the US and Europe (Sun et al.), or from a specific region (Emery et al.). This idea can be taken further in the context of the Lin et al. ([14]; 2019) paper discussed earlier: that study used over 10,000 in situ gauges to produce the best possible global discharge estimates at almost 3 million river reaches. These Lin et al. data are globally available and consistent, and offer a starting point for any model assimilation/calibration study, be it hydrologic, hydraulic, or both. However, using these outputs necessarily invokes the gauges used to produce them: even discharge information in gauge-sparse locations like Siberia implicitly relies on gauge knowledge via coupling with well-gauged basins. Thus, we expect future work to struggle with what it means for a basin to be ungauged when all global basins have been modelled in a fully coupled land surface framework. We leave further discussion for Section 3.

2.2.2. Assimilation/Calibration of an RS Signal into a Hydraulic Model

RS signals can also be incorporated into models that explicitly represent some aspect of channel hydraulics without in situ data, just as in Section 2.1.2. These representations of hydraulics could range from a full computational fluid dynamics finite element model solving for full four dimensional conservation of energy, mass, and momentum in time, to simple box-channel routing models capable of reproducing channel stage but not width, to Manning's equation. This channel-based paradigm is perhaps more familiar to readers, as two of the river variables with the longest remote sensing heritage, width and stage, are often employed in these approaches. Bates et al. ([119]; 1997) summarized early work on RS and flood hydraulic modelling, noting the sharp change in fluvial behavior once a river enters its floodplain. Bates et al. noted that computationally efficient one-dimensional flow modelling is acceptable within bank, but not in overbank situations, but also argued that RS signals are ideally poised to capture actual flood events (also noted by [120]; Brakenridge et al., 1998). Thus, RS is well suited to locations where traditional one-dimensional hydraulic models poorly represent hydraulics. Bates et al. conclude by noting that floodplains in particular are excellently suited to RS, as when they are dry, they may be mapped and their elevations recorded for future flood models (furthered in nuance by [121–123]; Horritt and Bates, 2001, Poole et al., 2002, and Horritt, 2006). Then, RS can capture actual floodwater occupation of the floodplain area and transform this into floodplain water depth in order to calibrate the model. Following these early works, many examples of calibration of hydraulic models with RS information have since focused on a flooding context (e.g., [63,124–131], Gumley and King, 1995; Mason et al., 2007; Schumann et al., 2007; 2008; 2009; Coe et al., 2008, Di Baldassarre et al., 2009; Khan et al., 2011; Frasson et al., 2019). Some of these studies assume upstream flow is known and then use RS to improve downstream flow, but it is possible to iteratively force a model with different flows and an explicit hydraulic model until model output matches RS observations in the floodplain, as Sun et al. did with width in Section 2.2.1. Thus, here we see calibration of a model to RS data, rather than calibrating RS data to a gauge or a model as in Section 2.1.

More recently, advances in global mapping, computational power, and the lengthening of the satellite data record has made a quantum leap in the amount of RS signals available for calibration of hydraulic models. Consider the Pekel et al. ([11]; 2016) product, which classified every single Landsat image in the archive into areas of water and non-water, allowing researchers to track global changes in inundation dynamics and river surface areas (i.e., width) directly for a 32 year period. Using the logic of Bates et al. ([119,132]; 1997; 2003), these changes in surface area can reveal flood stage extents and even heights when mapped onto newly available DEMs built for hydrology (e.g., MERIT DEM, [10] Yamazaki 2019). These inundation signals can also be obtained from non-optical satellites (e.g., [133]; Du et al., 2018), or even from non-satellites (e.g., UAVs, [134]; Niedzielski et al., 2016). As the satellite archive grows, and especially if new commercial CubeSats continue to improve their radiometric calibration and geolocation errors, we predict that approaches built on assimilating/calibrating these signals will proliferate.

Some of this proliferation is underway and is potentially overcoming reliance on gauges we see in the work of the late 2000s. Sichangi et al. ([23]; 2016) offer an interesting approach coupling altimetry, MODIS, and a DEM for eight very large rivers globally. They assume a wide channel with a constant width (greatly simplifying hydraulics), which allows a set of hydraulic equations to be driven by altimetry, which they calibrated with the nearest available altimetry gauge station. This rightly places this paper in Section 2.1.3, but this example shows that the field is moving toward consciously focusing on attempting to eliminate the need for in situ data, presumably to better understand ungauged basins in future.

Andreadis et al. ([135]; 2007) and Durand et al. ([136]; 2008) laid groundwork for RSQ from hydraulic models without any in situ data or global model inputs via detailed experiments to understand how well data assimilation was able to reproduce unknown channel parameters (e.g., friction, bathymetry, flow) from synthetic RS measurements of channel width, slope, and height within a simple hydraulic model. Biancamaria et al. [137]; 2011) and Yoon et al. ([138]; 2012) performed the next generation of this work, which lead to '4D' variational data assimilation (VDA) within an uncalibrated hydraulic model ([139–142]; Gejadze and Malaterre, 2017; Oubanas et al. 2018a; b; Larnier et al., 2020), and these recent efforts offer the most sophisticated take on this problem. This computationally expensive approach is able to consider model states both forward and backward in time and consider uncertainties of both initial model expectations and RS data. 4D VDA considers that initial flow is a static hydrograph, thus removing a dependency on external hydrologic modelling. These authors have shown that accurate discharges can be estimated using this approach, and that these approaches are viable indeed in politically and totally ungauged basins, but sensitivity to the initial hydrograph and computational burden are challenges for large-area application.

2.2.3. Geomorphic Inverse Problems

Finally, we consider approaches to RSQ that again independently consider hydraulic components of discharge (as in Section 2.1.3), but here explicitly rely on geomorphology, RS, and global databases. These approaches are driven by many of the same geomorphic hydraulic models of Section 2.1.2. A special case of geomorphically driven inverse problems is the mass conserved flow law inversion (McFLI) approach. Gleason et al. ([25]; 2017) coined the term as a response to a growing body of the RSQ literature that made the same basic assumptions about how to approach RSQ. Many of these approaches were motivated by the upcoming surface water and ocean topography (SWOT) mission as authors thought about how to best to use SWOT's novel and simultaneous measurements of river width, height, and slope, but McFLI approaches are fundamentally independent from SWOT. Durand et al. ([143]; 2010) laid the groundwork for McFLI with a mathematical approach that required known friction coefficients a priori, but the first true examples of McFLI capable of working in totally ungauged basins without this restrictive assumption were all published in 2014 or 2015. In a McFLI approach, authors first assume that a set of cross sections in a reach, or alternatively a set of short connected reaches, are mass conserved. What remains is to solve for unknown parameters in a flow law (e.g., Manning's, hydraulic geometry) given RS observations, meaning the problem remains ill-posed but is now more constrained and solutions may be approximated *exclusively* from remotely sensed data. Like the ratio approach pioneered by Brakenridge et al. ([89]; 2007), McFLI approaches are fundamentally a product of a remotely sensed paradigm. To a McFLI, a remotely sensed snapshot of a river is just that—a moment frozen in time capturing data over a large spatial area. Traditional geomorphology tends to consider evolution of rivers at a single point over long time periods or considers geologically synoptic spatial patterns. McFLI considers the temporal co-evolution of interconnected fluvial entities, which matches RS observations perfectly.

Both Garambois and Monnier ([144]; 2015) and Durand et al. ([145]; 2014) published McFLI algorithms based on Manning's equation assuming known (vis RS) river surface elevation, width, and slope. Unknown parameters are thus Manning's n, some unobserved area below the lowest water surface elevation observation, and discharge. We must note that these data are currently not available from RS without painstaking fusion of optical and altimetry datasets confounded by cloud issues and orbit geometry. Bjerklie et al. ([146]; 2018) were able to produce such signals for one river in Alaska, but most Manning's based McFLI approaches use model data to simulate RS measurements. Durand et al. used these simulated data to solve this problem in a Bayesian context by using likelihood functions in a Markov chain to allow estimates of these unobserved parameters to converge toward a likely posterior distribution. Tuozzolo et al. ([147]; 2019) have explored the impact of reach vs cross section based formulations of Manning's equation for this purpose. Garambois and Monnier used an optimization algorithm to define the pareto parameters that best minimized errors given a set of constraints (including mass conservation) for the same simulated data. Moving toward real data, Altenau et al. ([148]; 2017) showed that SWOT's Ka-Band observations should function as intended, using an airborne Ka-band radar to produce the same observations as SWOT, and Tuozzolo et al. ([149]; 2019) were able to use these airborne observations to successfully give first demonstration of a Manning based McFLI from wide swath altimetry.

Manning's equation is not the only Flow Law invoked in McFLI. Gleason et al. ([24,150]; 2014a;b) proposed a McFLI based on hydraulic geometry (both at a station and at many stations), which requires only river width as an observable input. Unknown parameters in this McFLI are the width exponent of the hydraulic geometry power law and two at many stations' hydraulic geometry constants specific to each river. This width-based McFLI has been successfully demonstrated for Landsat, Sentinel, and Planet optical imagery ([24,84,151]; Gleason and Smith, 2014; Gleason et al., 2014; Gleason and Hamdan, 2015; Feng et al., 2019), first using a genetic algorithm to solve for discharge and later using a Bayesian formulation capable of width-only or width, height, and slope discharge inversion ([152]; Hagemann et al., 2016). Currently, this width-based McFLI is the only known technique capable of leveraging existing satellite data to run a McFLI at all totally ungauged basins at the global scale given our definitions here.

Both Bonnema et al. ([153]; 2016) and Durand et al. ([154]; 2016) have intercompared McFLI approaches using a variety of hydraulic model data to stand in for remotely sensed observations. Durand et al. in particular provided the first ever exhaustive comparison of multiple McFLI approaches on model output representing almost 20 rivers. This controlled experiment allowed Durand et al. to more deeply understand controls on error and accuracy in McFLI approaches. They concluded that approaches that use multiple inputs (i.e., width, height, and slope) generally outperform approaches that use only a single input (i.e., width alone), as expected. They also concluded that all McFLI approaches provide good estimates of discharge dynamics, but all frequently have large biases. These biases are difficult to overcome without better prior estimates of unknown parameters because of the equifinal nature of McFLI. Finally, Durand et al. found that methods are sensitive to prior expectations of unknown parameters (either the 'first guess' in an optimization problem or the Bayesian 'prior distribution'), but that all methods improve upon these priors. That is, McFLI methods estimate discharge on rivers more accurately than our initial estimate of discharge without in situ or calibration data of any kind, but the more we know about a river to start with, the better the final estimate of discharge will be. This point becomes a critical determinant in choosing an RSQ method in the context of Figures 1 and 4.

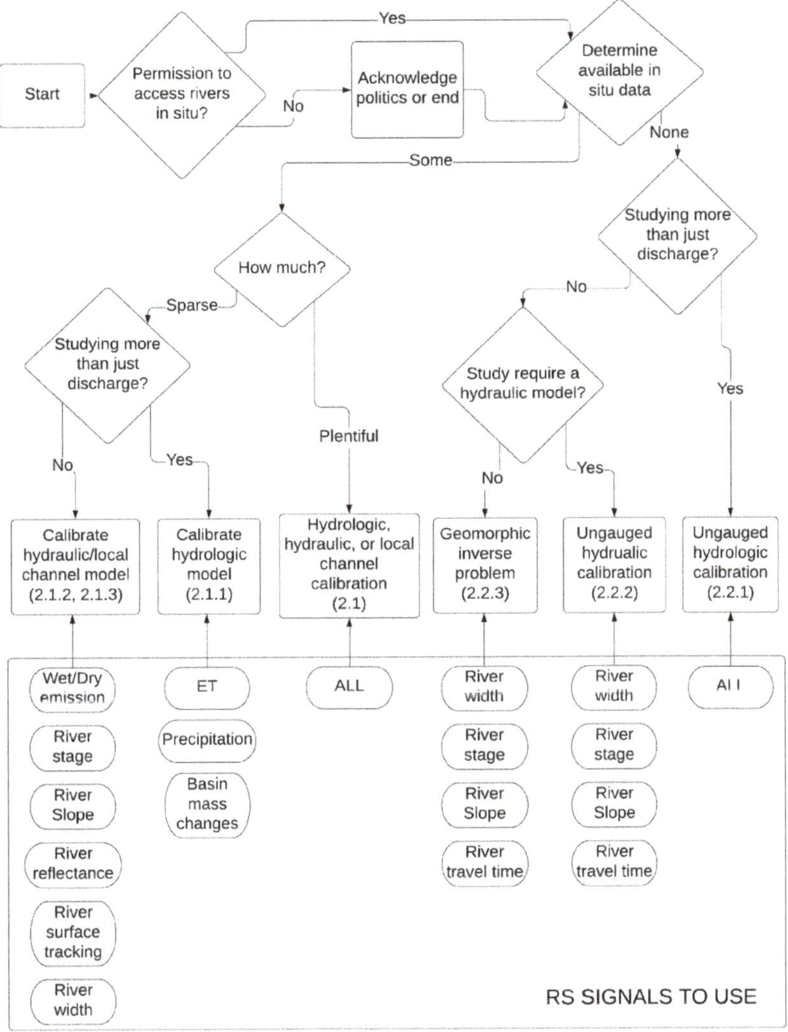

Figure 4. We suggest that this decision chart could be useful in organizing this paper and as a guide for future studies in helping new practitioners decide which approach to use. This organization also reflects our conclusion that the most important determinant between RSQ methods is their need for in situ data. At the bottom of the chart, the RS signals from Figure 3 label each method based on our reading here.

Sichangi et al., ([155]; 2018) offer a different take on underconstrained geomorphic inversion. They used MODIS to back out the travel time of a flood wave by looking at the peak of a width signal at multiple stations in a MODIS river width-time plot. That is, they assumed that maximum widths at any cross section correspond to maximum discharge (a safe assumption per hydraulic geometry) and that the peaks in both stations correspond to the same hydrologic event. Thus, two stations 500 km apart yielded a bulk channel velocity (travel time) by tracking the time delay between width peaks at the two stations. This is an innovative and clever approach to sensing channel velocity from space, but did require Sichangi et al. to assume a Manning's n from look up tables of values ([156]; Chow, 1964) to back out channel depth, and relies on a conversion from celerity to velocity. This rather exciting idea is only useful on large rivers where bulk velocity is stable over long distances and where the assumption of a travelling maximum width is valid. This again also highlights how not all ungauged approaches are applicable globally, and how some assumptions are considered restrictive and others are not. If, for instance, an author can study imagery of a river and a global DEM to determine probable Rosgen ([157]; 1994) classes, then assumptions about friction and width/depth ratios, for example, are well founded and did not invoke in situ data.

3. Discussion

Our broad sketch of the literature has hopefully revealed the numerous ways hydrologists have approached RSQ. We have deliberately included as wide a swath of the literature as possible to fully contextualize the field for practitioners and other interested parties alike. We have raised several discussion points above without bringing them to conclusion, saving them instead for this present section. Specifically, we wish to further probe several ideas that render even our nuanced classification of what it means to be 'gauged' or 'ungauged' problematic vis a vis RSQ; we wish to address common misconceptions about RSQ; and finally we wish to consider human political and economic realities of RSQ.

3.1. A Framework for Understanding RSQ, Ungauged Basins, and Hydrologic Knowledge

Although we did not keep a tally, a majority of the literature surveyed mentioned ungauged basins in their introduction as a justification for the work. We have thus used our gauge continuum to explicitly parse the literature at this highest level, placing each study into our understanding of the appropriate context. However, we have repeatedly raised the question of what it means to be ungauged in an era of global hydrologic modelling. Traditionally, RSQ authors have argued only those studies invoking absolutely no in situ data whatsoever qualify as appropriate for ungauged. RSQ via regionalization (parameter transfer) thus fails this test, and so too does parameterization from a global or regional model as there are almost no global hydrology datasets available that do not invoke in situ calibration data. Even the most basic RS water balance relies on precipitation gauges to calibrate the precipitation signal—without it, these estimates would likely be so wildly wrong as to be not useable. While we maintain an explicit difference between regionalization/ regional calibration vs. coupled modelling, we recognize that global hydrology has advanced tremendously in the RSQ era and thus believe that any RSQ investigation in an ungauged basin will (and should) likely start from an off the shelf global hydrology model to provide a first guess or Bayesian prior. Therefore, following our survey of the literature, we propose the following four conditions for when an RSQ approach is truly capable of tackling totally or politically ungauged basins.

In order to be applicable in totally and politically ungauged basins:

1. The study uses only globally available hydrologic (e.g., P, ET, runoff) forcing
2. If a global model is used, the study must not calibrate an RS signal explicitly to it
3. The study does not transfer parameters or calibrate to regionally available gauges
4. Specific assumptions about channel hydraulics must be obtained exclusively from remotely sensed platforms

These tests should be applied to each new ungauged study to ensure that the primary RS data is indeed driving the discharge result, rather than in situ data available elsewhere. We argue that this independence is essential for politically and totally ungauged basins, as otherwise reviewers and the public will rightly question whether or not discharges in these basins are simply a regurgitation of what we knew before. Hence, we believe that if basins are truly ungauged, models can only guide or bound RSQ, not drive it. Applying these restrictions could limit the accuracy of the resulting discharges ([144,154]; Garambois and Monnier, 2014; Durand et al., 2016), but we argue that these less accurate but independent discharges represent the largest furthering of hydrologic knowledge in these cases and allow hydrologists to make declarative statements about the water cycle.

This is not to say that studies that do not meet these criteria are not RSQ: on the contrary, the vast majority of the literature surveyed here is uninterested entirely with meeting these criteria. Rather, we introduce these tests as a useful filter for future reviewers and authors to help decide what technique is best for a particular application. Thus, RSQ methods need not change if these tests are failed, rather, their introduction should highlight the context of regionally or locally available information and place the study into the appropriate category. Our emphasis on avoiding transferability is grounded in field hydrology, as we, like McDonnel et al. ([12]; 2007), are wary of assuming hydrologic process in unknown and unstudied watersheds. For example, using gauges from one watershed to calibrate a fully coupled earth system model at the continental scale is, to us, fundamentally different than assuming the parameters calibrated for that one watershed should apply to all basins across the continent. The lines between 'regionally gauged' and 'totally ungauged' are becoming blurred as coupled earth system models produce fluxes with greater accuracy and precision. Even so, we maintain a fundamental difference between using a global hydrology product as a starting point for an RSQ method vs. calibrating an RS signal to that model to produce discharge. In the latter, the resulting discharges are fundamentally a product of the original model: at best, RSQ would give you the model discharge back again. In this case, RS would be better used in conjunction with other data to produce the global product in a hydrologic model as in Section 2.1.1, rather than exist as a derived product. Finally, we assert that while in situ data are important to improve the accuracy of RSQ, they are not indispensable. The breadth of the literature covered here makes this abundantly clear, and we see value in pursuing RSQ research across the entire gauge spectrum.

We believe that review of the myriad of techniques for RSQ herein is secondary to definition of the paradigm in which they are employed. While methodological innovation is essential, the literature has revealed that the more data an author has in hand to begin an RSQ study, the more accurate their final discharges. While we have very purposefully avoided the tedium of reporting individual discharge retrieval accuracies across the literature given the impossibility of direct comparison following specific sets of assumptions, RS data, and ancillary data used in each case, we can assert in general that the literature in Section 2.1 is more accurate than the literature in Section 2.2. This is of course intuitive—studies in Section 2.1 all used local or regional gauge data or tailored assumptions of hydraulics specifically to study channels given prior knowledge. However, the methods of Section 2.2 can perform as accurately as those of Section 2.1 given the same calibration data ((141,144); Garambois and Monnier, 2015; Oubanas et al., 2018). Thus, we have chosen to organize the literature by application area first and methodology second, as differences in accuracy are driven largely by prior data and not methodology. Nowhere have we split the literature via sensor, as we find papers arguing for or against specific sensing platforms at the cost of others unhelpful. We do recognize the advantages of passive/active sensors at varying resolutions for specific tasks, but argue that ultimately what matters is whether or not the study has achieved its goal. Figure 3 exists in service of this goal, highlighting all the usable RS signals we have reviewed here in a single figure.

At its core, RSQ seeks to advance hydrologic understanding. In basins that are gauged, semi-gauged, or regionally gauged, hydrologists start with some of this understanding in hand, and RSQ should improve our ability to parse the hydrologic cycle in space and time. For water resource managers, even a well-gauged basin can have insufficient understanding when attempting to balance hydropower, environmental flows, drought/flood protection, and recreational use ([158]; Brown et al., 2015). RSQ can thus provide real advances in hydrologic understanding that improve human decision making, even in gauged basins. In politically and totally ungauged basins, we suggest that methods that produce hydrographs in error by even a factor of two or more (although we must rely on their performance in gauged basins to characterize this error) might still be an improvement on our existing knowledge. Therefore, we assert that nearly all RSQ is useful to the hydrology community provided authors place their study in the proper context. This context, given here and in Figure 1, is essential for practitioners, authors, reviewers, and editors. Without it, non RS hydrologists can be confused as to whether or not a given technique will work for their application, and editors and reviewers may accept otherwise excellent RSQ work that makes claims of being "ungauged" that it cannot support.

3.2. "The Goal of Remote Sensing of Discharge is to Entirely Replace Gauges": A Common Misconception

RSQ is designed to produce river discharge and the advantages of air and spaceborne platforms over grueling fieldwork are clear and much touted by RSQ authors. This might lead some to believe that a goal of RSQ is to replace expensive and politically sensitive gauge data. Thus, this section title represents a pernicious misconception about RSQ. The majority of the work cited above requires gauges to produce successful discharge estimates, and even those methods in Section 2.2 that do not require gauge data improve dramatically when informed by gauge data. There will never be any hydrologic substitute for gauge data—these data are carefully monitored by dedicated staff of numerous agencies worldwide, and many agencies go to great lengths to ensure that only the highest quality data are presented as the gauge record. This can have political consequences, as many gauge data receive the 'seal of approval' of officially sanctioned government agencies, and a hydrologic world without sanctioned discharges would place much of climate science and water treaties in jeopardy. With that stated, there has been a precipitous decline in available gauge data globally ([159]; Hannah, 2011), and public gauge records are practically nonexistent in places where water politics are thorny ([151]; Gleason and Hamdan, 2015). Gauges that are here today may be gone tomorrow at the whims of economics and politics.

Thus, we believe that RSQ is a natural *complement* to river gauges and other in situ discharge monitoring. In this we agree with Fekete et al. ([160]; 2012), even if we have drawn this conclusion via very different means. The benefits of in situ data to RSQ are clear and have been discussed extensively above. Here, we discuss the benefits of RS signals to in situ monitoring. Consider a well-gauged river network, with a gauge at the outlet and across a large proportion of ungauged upstream reaches. RSQ may actually be at its most beneficial in this case, as high quality gauge calibration data allow RS signals to extend point gauge measurements in space and time. As networks move along the continuum from completely gauged to ungauged, RSQ provides more novel primary information, but RSQ retrievals reduce in accuracy. Alsdorf et al. ([22]; 2007) also highlight that RSQ can document water resources in diverse channel forms and in flood contexts: situations where gauges struggle. Discontinued gauges are another area where RSQ can shine, as gauges that overlap the RS record in the past can be calibrated and then used to parameterize models as they move into the future (e.g., [161,162]; Birkinshaw et al., 2010; Gleason et al., 2018).

In sum, we argue that no RSQ practitioner can or should cogently argue that RSQ is a complete replacement for gauges. Both forms of discharge monitoring have strengths and it is only by merging the two that hydrology can move forward toward more complete understanding of the global hydrologic cycle. The Arctic and much of Asia are almost complete hydrologic black boxes, despite the almost certain existence of un-shared high quality gauge data in these places. Lin et al. ([14]; 2019) highlight the benefits of even sparse global gauging by considering a coupled land-atmosphere system, primary data with low uncertainty might be used to reduce uncertainty in poorly monitored areas, and in this case RS data provide a welcome and irreplaceable source of primary data. We believe that hydrology, like many geophysical disciplines, should rest on primary data, and loss of either gauge or remotely sensed data is a detriment to the field.

3.3. Ethics and Politics

Finally, we wish to make a brief note on the ethics and politics of RSQ. Gleason and Hamdan ([151]; 2016) and Alvarez-Leon and Gleason ([163]; 2017) have recently explored this topic and have concluded that RSQ is an inherently political act. They reached this conclusion following two different logical tacks. First, since states (the preferred term for 'country' in the geographical literature) own their water resources, they are allowed to express their national sovereignty by withholding hydrometric data (i.e., politically ungauged basins). There is no international legal mechanism to force states to release water data, and even bi- or multi-lateral treaties have proven ineffective at forcing water data transparency, despite assertations that such transparency is essential for good governance ([164–168]; Sneddon and Fox, 2006; 2011; 2012; Dore and Lebel, 2010; Ho, 2014). Circumventing these data restrictions via the use of satellites is thus in fact a direct violation of national sovereignty and an inherently political act, regardless of the intent of the scientist ([151] Gleason and Hamdan, 2016). This is not to say that practitioners should not engage in RSQ—on the contrary, there are humanitarian and climate reasons why scientists should perform this very political act. However, scientists have no recourse to state that they are 'simply observing the earth' are thus somehow immune to politics. This is a naïve view that has long been debunked in social sciences and the humanities, but is slow to catch on in the non-social sciences, especially in remote sensing. Second, access to satellite data themselves is a result of political decisions ([163] Alvarez-Leon and Gleason, 2017). The United States government has launched a fleet of earth observing satellites, and made it their policy that all of these data are available to anyone with access to the internet. Despite this, the United States government could change its mind at any point and restrict access to these data, thus blanketing access to these crucial primary data in an instant. We do not suggest this is a likely scenario, but it is possible. Consider the numerous other commercial and state-satellites that exist but for whom data are not free or simply not available outside of the home state. Public access to high quality earth system data is ultimately subject to the whims of the owners of the satellites themselves, and this access has profound impacts on hydrologists' ability to perform RSQ. Taken together, these two arguments suggest that practitioners should take care and make thoughtful choices when applying RSQ in contexts where it would be politically impossible to gather the same data in the field. This forms a good test, and 'Would I have permission to do this RS work as fieldwork?' should be a question practitioners ask themselves when beginning a study. If the answer is 'no,' practitioners should proceed if believe it is appropriate, but thcy should be aware of the implications of the study.

4. Conclusions

The existing RSQ literature has frequently justified itself based on whether it is highly accurate or whether it is applicable in ungauged basins. Each new manuscript generally begins with an explicit statement on the utility of RSQ and why this new work belongs in the pantheon of that particular application, yet much work claiming to function in ungauged basins often cannot, by our definitions, truly work in ungauged basins. This leads to evident division in citations within RSQ studies—authors typically only cite work within the categories (i.e., sections and subsections) presented here, and papers frequently do not incorporate information from different 'families' of RSQ. What we have shown, however, is that many of these seemingly disparate methods are in fact slight variations on the same paradigm, and that the entirety of the literature has more in common than individual papers would suggest. Our inclusion of 'non channel' RSQ here is perhaps controversial to some along this vein, as many RSQ practitioners might suggest that using RS signals within a hydrologic model is somehow not RSQ. It is our hope that this review serves as a reminder that in all cases reviewed here, remotely sensed data were used to estimate discharge in service of furthering discharge understanding. There are nuances in how this was achieved, but all ~170 papers here have the same goal. Finally, we suggest that our newly defined framing for application of RSQ (e.g., Figure 1 and Section 3.1, Figure 4) might offer much clarity to readers, authors, editors, reviewers, and importantly users of RSQ.

Our review has also implicitly outlined some future directions for the field, which we make explicit here. Data mining/big data approaches seem particularly understudied despite the success of Bjerklie in the early 2000s, and the availability of powerful computer science techniques available to hydrologists today makes this area ripe for potential powerful interventions in RSQ. Satellites dedicated to surface water and sponsored by large state entities (like SWOT), and commercial fleets of CubeSats alike also stand to transform the field. While SWOT has received much attention in the literature from many schools of RSQ, CubeSats are relatively underused. Diverse signals can only improve what we know about the hydrologic system. Finally, we repeat our call for a broadening of perspective of RSQ practitioners. We have shown here that the field is robust, rapidly growing, and rich with an extraordinary diversity of ideas. The field would suffer, we believe, from a descent into orthodoxy, which might stifle creativity. Instead, we hope that new works can use this review as a basis to situate their work in broader context.

Author Contributions: C.J.G. drafted the manuscript and figures. C.J.G. and M.T.D. shared in every aspect of producing this manuscript. All authors have read and agreed to the published version of the manuscript.

Funding: Research was funded by NASA New Investigator Grant 80NSSC18K0741 and NSF CAREER Grant 1748653 to CJ Gleason, and NASA SWOT Science Team grant number NNX16AH82G to MT Durand and CJ Gleason.

Acknowledgments: We apologize to authors of those relevant works we have not cited. Our survey is necessarily broad, and while we have cited more than 160 published works, we have of course missed some relevant literature. The authors disagree strongly on whether or not the letter 'u' should appear in the word 'gauge.' The first author's preference has been adopted. We thank also two anonymous reviewers for their prompt comments. This review deviates from journal style to include both numbered citation format and author-date citation information for ease of cross reference by readers.

Conflicts of Interest: The authors declare no conflict of interest.

References

1.	Calmant, S.; Seyler, F.; Cretaux, J.F. Monitoring Continental Surface Waters by Satellite Altimetry. *Surv. Geophys.* **2008**, *29*, 247–269. [CrossRef]
2.	Lettenmaier, D.P.; Alsdorf, D.; Dozier, J.; Huffman, G.J.; Pan, M.; Wood, E.F. Inroads of remote sensing into hydrologic science during the WRR era. *Water Resour. Res.* **2015**, *51*, 7309–7342. [CrossRef]
3.	Doell, P.; Douville, H.; Guentner, A.; Mueller Schmied, H.; Wada, Y. Modelling Freshwater Resources at the Global Scale: Challenges and Prospects. *Surv. Geophys.* **2016**, *37*, 195–221. [CrossRef]
4.	Lettenmaier, D.P.; Famiglietti, J.S. Hydrology—Water from on high. *Nature* **2006**, *444*, 562–563. [CrossRef]
5.	Allen, G.H.; Pavelsky, T.M.; Barefoot, E.A.; Lamb, M.P.; Butman, D.; Tashie, A.; Gleason, C.J. Similarity of stream width distributions across headwater systems. *Nat. Commun.* **2018**, *9*, 610. [CrossRef]
6.	Smith, L.C.; Yang, K.; Pitcher, L.H.; Overstreet, B.T.; Chu, V.W.; Rennermalm, Å.K.; Ryan, J.C.; Cooper, M.G.; Gleason, C.J.; Tedesco, M.; et al. Direct measurements of meltwater runoff on the Greenland ice sheet surface. *Proc. Natl. Acad. Sci. USA* **2017**, *114*, E10622–E10631. [CrossRef]
7.	Lehner, B.; Döll, P. Development and validation of a global database of lakes, reservoirs and wetlands. *J. Hydrol.* **2004**, *296*, 1–22. [CrossRef]
8.	Prigent, C.; Matthews, E.; Aires, F.; Rossow, W.B. Remote sensing of global wetland dynamics with multiple satellite data sets. *Geophys. Res. Lett.* **2001**, *28*, 4631–4634. [CrossRef]
9.	Allen, G.H.; Pavelsky, T.M. Patterns of river width and surface area revealed by the satellite-derived North American River Width data set. *Geophys. Res. Lett.* **2015**, *42*, 395–402. [CrossRef]
10.	Yamazaki, D.; Ikeshima, D.; Sosa, J.; Bates, P.D.; Allen, G.H.; Pavelsky, T.M. MERIT Hydro: A High-Resolution Global Hydrography Map Based on Latest Topography Dataset. *Water Resour. Res.* **2019**, *55*, 5053–5073. [CrossRef]
11.	Pekel, J.-F.; Cottam, A.; Gorelick, N.; Belward, A.S. High-resolution mapping of global surface water and its long-term changes. *Nature* **2016**, *540*, 418–422. [CrossRef] [PubMed]
12.	McDonnell, J.J.; Sivapalan, M.; Vaché, K.; Dunn, S.; Grant, G.; Haggerty, R.; Hinz, C.; Hooper, R.; Kirchner, J.; Roderick, M.L.; et al. Moving beyond heterogeneity and process complexity: A new vision for watershed hydrology. *Water Resour. Res.* **2007**, *43*. [CrossRef]

13. Wood, E.F.; Roundy, J.K.; Troy, T.J.; van Beek, L.P.H.; Bierkens, M.F.P.; Blyth, E.; de Roo, A.; Döll, P.; Ek, M.; Famiglietti, J.; et al. Hyperresolution global land surface modeling: Meeting a grand challenge for monitoring Earth's terrestrial water. *Water Resour. Res.* **2011**, *47*. [CrossRef]

14. Lin, P.; Pan, M.; Beck, H.E.; Yang, Y.; Yamazaki, D.; Frasson, R.; David, C.H.; Durand, M.; Pavelsky, T.M.; Allen, G.H.; et al. Global Reconstruction of Naturalized River Flows at 2.94 Million Reaches. *Water Resour. Res.* **2019**, *55*, 6499–6516. [CrossRef] [PubMed]

15. Martin, E.; Gascoin, S.; Grusson, Y.; Murgue, C.; Bardeau, M.; Anctil, F.; Ferrant, S.; Lardy, R.; Le Moigne, P.; Leenhardt, D.; et al. On the Use of Hydrological Models and Satellite Data to Study the Water Budget of River Basins Affected by Human Activities: Examples from the Garonne Basin of France. *Surv. Geophys.* **2016**, *37*, 223–247. [CrossRef]

16. Yoon, Y.; Beighley, E. Simulating streamflow on regulated rivers using characteristic reservoir storage patterns derived from synthetic remote sensing data. *Hydrol. Process.* **2015**, *29*, 2014–2026. [CrossRef]

17. Barton, I.J.; Bathols, J.M. Monitoring floods with AVHRR. *Remote Sens. Environ.* **1989**, *30*, 89–94. [CrossRef]

18. Biancamaria, S.; Hossain, F.; Lettenmaier, D.P. Forecasting transboundary river water elevations from space. *Geophys. Res. Lett.* **2011**, *38*, L11401. [CrossRef]

19. Grimaldi, S.; Li, Y.; Pauwels, V.R.N.; Walker, J.P. Remote Sensing-Derived Water Extent and Level to Constrain Hydraulic Flood Forecasting Models: Opportunities and Challenges. *Surv. Geophys.* **2016**, *37*, 977–1034. [CrossRef]

20. Li, Y.; Grimaldi, S.; Walker, J.P.; Pauwels, V.R.N. Application of Remote Sensing Data to Constrain Operational Rainfall-Driven Flood Forecasting: A Review. *Remote Sens.* **2016**, *8*, 456. [CrossRef]

21. Schumann, G.J.-P.; Stampoulis, D.; Smith, A.M.; Sampson, C.C.; Andreadis, K.M.; Neal, J.C.; Bates, P.D. Rethinking flood hazard at the global scale. *Geophys. Res. Lett.* **2016**, *43*, 10249–10256. [CrossRef]

22. Alsdorf, D.E.; Rodriguez, E.; Lettenmaier, D.P. Measuring surface water from space. *Rev. Geophys.* **2007**, *45*. [CrossRef]

23. Sichangi, A.W.; Wang, L.; Yang, K.; Chen, D.; Wang, Z.; Li, X.; Zhou, J.; Liu, W.; Kuria, D. Estimating continental river basin discharges using multiple remote sensing data sets. *Remote Sens. Environ.* **2016**, *179*, 36–53. [CrossRef]

24. Gleason, C.J.; Smith, L.C. Toward global mapping of river discharge using satellite images and at-many-stations hydraulic geometry. *Proc. Natl. Acad. Sci. USA* **2014**, *111*, 4788–4791. [CrossRef] [PubMed]

25. Gleason, C.J.; Garambois, P.A.; Durand, M. Tracking River Flows from Space. Available online: https://eos.org/science-updates/tracking-river-flows-from-space (accessed on 25 March 2020).

26. Rodell, M.; Velicogna, I.; Famiglietti, J.S. Satellite-based estimates of groundwater depletion in India. *Nature* **2009**, *460*, 999–1002. [CrossRef] [PubMed]

27. Gupta, H.V.; Sorooshian, S.; Yapo, P.O. Toward improved calibration of hydrologic models: Multiple and noncommensurable measures of information. *Water Resour. Res.* **1998**, *34*, 751–763. [CrossRef]

28. Sivapalan, M.; Takeuchi, K.; Franks, S.W.; Gupta, V.K.; Karambiri, H.; Lakshmi, V.; Liang, X.; McDonnell, J.J.; Mendiondo, E.M.; O'Connell, P.E.; et al. IAHS decade on Predictions in Ungauged Basins (PUB), 2003-2012: Shaping an exciting future for the hydrological sciences. *Hydrol. Sci. J. Des.* **2003**, *48*, 857–880. [CrossRef]

29. Gelati, E.; Decharme, B.; Calvet, J.-C.; Minvielle, M.; Polcher, J.; Fairbairn, D.; Weedon, G.P. Hydrological assessment of atmospheric forcing uncertainty in the Euro-Mediterranean area using a land surface model. *Hydrol. Earth Syst. Sci.* **2018**, *22*, 2091–2115. [CrossRef]

30. Tarpanelli, A.; Camici, S.; Nielsen, K.; Brocca, L.; Moramarco, T.; Benveniste, J. Potentials and limitations of Sentinel-3 for river discharge assessment. *Adv. Space Res.* **2019**, in press. [CrossRef]

31. Reichle, R.H. Data assimilation methods in the Earth sciences. *Adv. Water Resour.* **2008**, *31*, 1411–1418. [CrossRef]

32. Jodar, J.; Carpintero, E.; Martos-Rosillo, S.; Ruiz-Constan, A.; Marin-Lechado, C.; Cabrera-Arrabal, J.A.; Navarrete-Mazariegos, E.; Gonzalez-Ramon, A.; Lamban, L.J.; Herrera, C.; et al. Combination of lumped hydrological and remote-sensing models to evaluate water resources in a semi-arid high altitude ungauged watershed of Sierra Nevada (Southern Spain). *Sci. Total Environ.* **2018**, *625*, 285–300. [CrossRef] [PubMed]

33. Reichle, R.H.; De Lannoy, G.J.M.; Forman, B.A.; Draper, C.S.; Liu, Q. Connecting Satellite Observations with Water Cycle Variables Through Land Data Assimilation: Examples Using the NASA GEOS-5 LDAS. *Surv. Geophys.* **2014**, *35*, 577–606. [CrossRef]

34. Maxwell, D.H.; Jackson, B.M.; McGregor, J. Constraining the ensemble Kalman filter for improved streamflow forecasting. *J. Hydrol.* **2018**, *560*, 127–140. [CrossRef]
35. Chen, J.L.; Wilson, C.R.; Chambers, D.P.; Nerem, R.S.; Tapley, B.D. Seasonal global water mass budget and mean sea level variations. *Geophys. Res. Lett.* **1998**, *25*, 3555–3558. [CrossRef]
36. Dziubanski, D.J.; Franz, K.J. Assimilation of AMSR-E snow water equivalent data in a spatially-lumped snow model. *J. Hydrol.* **2016**, *540*, 26–39. [CrossRef]
37. Fortin, J.; Turcotte, R.; Massicotte, S.; Moussa, R.; Fitzback, J.; Villeneuve, J. Distributed Watershed Model Compatible with Remote Sensing and GIS Data. I: Description of Model. *J. Hydrol. Eng.* **2001**, *6*, 91–99. [CrossRef]
38. Rowlands, D.D.; Luthcke, S.B.; Klosko, S.M.; Lemoine, F.G.R.; Chinn, D.S.; McCarthy, J.J.; Cox, C.M.; Anderson, O.B. Resolving mass flux at high spatial and temporal resolution using GRACE intersatellite measurements. *Geophys. Res. Lett.* **2005**, *32*. [CrossRef]
39. Syed, T.H.; Famiglietti, J.S.; Zlotnicki, V.; Rodell, M. Contemporary estimates of Pan-Arctic freshwater discharge from GRACE and reanalysis. *Geophys. Res. Lett.* **2007**, *34*. [CrossRef]
40. Syed, T.H.; Famiglietti, J.S.; Chen, J.; Rodell, M.; Seneviratne, S.I.; Viterbo, P.; Wilson, C.R. Total basin discharge for the Amazon and Mississippi River basins from GRACE and a land-atmosphere water balance. *Geophys. Res. Lett.* **2005**, *32*. [CrossRef]
41. Syed, T.H.; Famiglietti, J.S.; Chambers, D.P. GRACE-Based Estimates of Terrestrial Freshwater Discharge from Basin to Continental Scales. *J. Hydrometeorol.* **2009**, *10*, 22–40. [CrossRef]
42. Syed, T.H.; Famiglietti, J.S.; Chambers, D.P.; Willis, J.K.; Hilburn, K. Satellite-based global-ocean mass balance estimates of interannual variability and emerging trends in continental freshwater discharge. *Proc. Natl. Acad. Sci. USA* **2010**, *107*, 17916–17921. [CrossRef] [PubMed]
43. Schmidt, R.; Petrovic, S.; Guentner, A.; Barthelmes, F.; Wuensch, J.; Kusche, J. Periodic components of water storage changes from GRACE and global hydrology models. *J. Geophys. Res. Solid Earth* **2008**, *113*. [CrossRef]
44. Werth, S.; Güntner, A.; Petrovic, S.; Schmidt, R. Integration of GRACE mass variations into a global hydrological model. *Earth Planet. Sci. Lett.* **2009**, *277*, 166–173. [CrossRef]
45. Frappart, F.; Papa, F.; Güntner, A.; Werth, S.; Santos da Silva, J.; Tomasella, J.; Seyler, F.; Prigent, C.; Rossow, W.B.; Calmant, S.; et al. Satellite-based estimates of groundwater storage variations in large drainage basins with extensive floodplains. *Remote Sens. Environ.* **2011**, *115*, 1588–1594. [CrossRef]
46. Eom, J.; Seo, K.-W.; Ryu, D. Estimation of Amazon River discharge based on EOF analysis of GRACE gravity data. *Remote Sens. Environ.* **2017**, *191*, 55–66. [CrossRef]
47. Wiese, D.N.; Landerer, F.W.; Watkins, M.M. Quantifying and reducing leakage errors in the JPL RL05M GRACE mascon solution. *Water Resour. Res.* **2016**, *52*, 7490–7502. [CrossRef]
48. Siqueira, V.A.; Paiva, R.C.D.; Fleischmann, A.S.; Fan, F.M.; Ruhoff, A.L.; Pontes, P.R.M.; Paris, A.; Calmant, S.; Collischonn, W. Toward continental hydrologic-hydrodynamic modeling in South America. *Hydrol. Earth Syst. Sci.* **2018**, *22*, 4815–4842. [CrossRef]
49. Chandanpurkar, H.A.; Reager, J.T.; Famiglietti, J.S.; Syed, T.H. Satellite- and Reanalysis-Based Mass Balance Estimates of Global Continental Discharge (1993–2015). *J. Clim.* **2017**, *30*, 8481–8495. [CrossRef]
50. Zhang, Y.; Pan, M.; Wood, E.F. On Creating Global Gridded Terrestrial Water Budget Estimates from Satellite Remote Sensing. *Surv. Geophys.* **2016**, *37*, 249–268. [CrossRef]
51. Silvestro, F.; Gabellani, S.; Rudari, R.; Delogu, F.; Laiolo, P.; Boni, G. Uncertainty reduction and parameter estimation of a distributed hydrological model with ground and remote-sensing data. *Hydrol. Earth Syst. Sci.* **2015**, *19*, 1727–1751. [CrossRef]
52. Lee, H.; Beighley, R.E.; Alsdorf, D.; Jung, H.C.; Shum, C.K.; Duan, J.; Guo, J.; Yamazaki, D.; Andreadis, K. Characterization of terrestrial water dynamics in the Congo Basin using GRACE and satellite radar altimetry. *Remote Sens. Environ.* **2011**, *115*, 3530–3538. [CrossRef]
53. Wang, S.; Liu, S.; Mo, X.; Peng, B.; Qiu, J.; Li, M.; Liu, C.; Wang, Z.; Bauer-Gottwein, P. Evaluation of Remotely Sensed Precipitation and Its Performance for Streamflow Simulations in Basins of the Southeast Tibetan Plateau. *J. Hydrometeorol.* **2015**, *16*, 2577–2594. [CrossRef]
54. Wulf, H.; Bookhagen, B.; Scherler, D. Differentiating between rain, snow, and glacier contributions to river discharge in the western Himalaya using remote-sensing data and distributed hydrological modeling. *Adv. Water Resour.* **2016**, *88*, 152–169. [CrossRef]
55. Ferguson, R.I. Hydraulics and hydraulic geometry. *Prog. Phys. Geogr.* **1986**, *10*, 1–31. [CrossRef]

56. Gleason, C.J. Hydraulic geometry of natural rivers a review and future directions. *Prog. Phys. Geogr.* **2015**, *39*, 337–360. [CrossRef]

57. Fleischmann, A.S.; Paiva, R.C.D.; Collischonn, W.; Sorribas, M.V.; Pontes, P.R.M. On river-floodplain interaction and hydrograph skewness. *Water Resour. Res.* **2016**, *52*, 7615–7630. [CrossRef]

58. Hall, A.C.; Schumann, G.J.-P.; Bamber, J.L.; Bates, P.D. Tracking water level changes of the Amazon Basin with space-borne remote sensing and integration with large scale hydrodynamic modelling: A review. *Phys. Chem. Earth Parts A/B/C* **2011**, *36*, 223–231. [CrossRef]

59. Bjerklie, D.M.; Lawrence Dingman, S.; Vorosmarty, C.J.; Bolster, C.H.; Congalton, R.G. Evaluating the potential for measuring river discharge from space. *J. Hydrol.* **2003**, *278*, 17–38. [CrossRef]

60. Bjerklie, D.M.; Moller, D.; Smith, L.C.; Dingman, S.L. Estimating discharge in rivers using remotely sensed hydraulic information. *J. Hydrol.* **2005**, *309*, 191–209. [CrossRef]

61. Bjerklie, D.M. Estimating the bankfull velocity and discharge for rivers using remotely sensed river morphology information. *J. Hydrol.* **2007**, *341*, 144–155. [CrossRef]

62. Kebede, M.G.; Wang, L.; Li, X.; Hu, Z. Remote sensing-based river discharge estimation for a small river flowing over the high mountain regions of the Tibetan Plateau. *Int. J. Remote Sens.* **2020**, *41*, 3322–3345. [CrossRef]

63. De Frasson, R.P.M.; Pavelsky, T.M.; Fonstad, M.A.; Durand, M.T.; Allen, G.H.; Schumann, G.; Lion, C.; Beighley, R.E.; Yang, X. Global Relationships Between River Width, Slope, Catchment Area, Meander Wavelength, Sinuosity, and Discharge. *Geophys. Res. Lett.* **2019**, *46*, 3252–3262. [CrossRef]

64. Neal, J.; Schumann, G.; Bates, P.; Buytaert, W.; Matgen, P.; Pappenberger, F. A data assimilation approach to discharge estimation from space. *Hydrol. Process.* **2009**, *23*, 3641–3649. [CrossRef]

65. Temimi, M.; Lacava, T.; Lakhankar, T.; Tramutoli, V.; Ghedira, H.; Ata, R.; Khanbilvardi, R. A multi-temporal analysis of AMSR-E data for flood and discharge monitoring during the 2008 flood in Iowa. *Hydrol. Process.* **2011**, *25*, 2623–2634. [CrossRef]

66. King, T.V.; Neilson, B.T.; Rasmussen, M.T. Estimating Discharge in Low-Order Rivers with High-Resolution Aerial Imagery. *Water Resour. Res.* **2018**, *54*, 863–878. [CrossRef]

67. Harada, S.; Li, S.S. Combining remote sensing with physical flow laws to estimate river channel geometry. *River Res. Appl.* **2018**, *34*, 697–708. [CrossRef]

68. Try, S.; Lee, G.; Yu, W.; Oeurng, C.; Jang, C. Large-Scale Flood-Inundation Modeling in the Mekong River Basin. *J. Hydrol. Eng.* **2018**, *23*, 05018011. [CrossRef]

69. Koblinsky, C.J.; Clarke, R.T.; Brenner, A.C.; Frey, H. Measurement of river level variations with satellite altimetry. *Water Resour. Res.* **1993**, *29*, 1839–1848. [CrossRef]

70. Smith, L.C.; Isacks, B.L.; Bloom, A.L.; Murray, A.B. Estimation of discharge from three braided rivers using synthetic aperture radar satellite imagery: Potential application to ungaged basins. *Water Resour. Res.* **1996**, *32*, 2021–2034. [CrossRef]

71. Alsdorf, D.E. Water Storage of the Central Amazon Floodplain Measured with GIS and Remote Sensing Imagery. *Ann. Assoc. Am. Geogr.* **2003**, *93*, 55–66. [CrossRef]

72. Alsdorf, D.E.; Smith, L.C.; Melack, J.M. Amazon floodplain water level changes measured with interferometric SIR-C radar. *IEEE Trans. Geosci. Remote Sens.* **2001**, *39*, 423–431. [CrossRef]

73. Alsdorf, D.; Birkett, C.; Dunne, T.; Melack, J.; Hess, L. Water level changes in a large Amazon lake measured with spaceborne radar interferometry and altimetry. *Geophys. Res. Lett.* **2001**, *28*, 2671–2674. [CrossRef]

74. Frappart, F.; Seyler, F.; Martinez, J.-M.; León, J.G.; Cazenave, A. Floodplain water storage in the Negro River basin estimated from microwave remote sensing of inundation area and water levels. *Remote Sens. Environ.* **2005**, *99*, 387–399. [CrossRef]

75. LeFavour, G.; Alsdorf, D. Water slope and discharge in the Amazon River estimated using the shuttle radar topography mission digital elevation model. *Geophys. Res. Lett.* **2005**, *32*. [CrossRef]

76. Smith, L.C. Satellite remote sensing of river inundation area, stage, and discharge: A review. *Hydrol. Process.* **1997**, *11*, 1427–1439. [CrossRef]

77. Kouraev, A.V.; Zakharova, E.A.; Samain, O.; Mognard, N.M.; Cazenave, A. Ob' river discharge from TOPEX/Poseidon satellite altimetry (1992–2002). *Remote Sens. Environ.* **2004**, *93*, 238–245. [CrossRef]

78. Pavelsky, T.M.; Durand, M.T.; Andreadis, K.M.; Beighley, R.E.; Paiva, R.C.D.; Allen, G.H.; Miller, Z.F. Assessing the potential global extent of SWOT river discharge observations. *J. Hydrol.* **2014**, *519*, 1516–1525. [CrossRef]

79. Pavelsky, T.M.; Smith, L.C. Remote sensing of suspended sediment concentration, flow velocity, and lake recharge in the Peace-Athabasca Delta, Canada. *Water Resour. Res.* **2009**, *45*. [CrossRef]

80. Schneider, R.; Godiksen, P.N.; Villadsen, H.; Madsen, H.; Bauer-Gottwein, P. Application of CryoSat-2 altimetry data for river analysis and modelling. *Hydrol. Earth Syst. Sci.* **2017**, *21*, 651–664. [CrossRef]

81. Young, D.S.; Hart, J.K.; Martinez, K. Image analysis techniques to estimate river discharge using time-lapse cameras in remote locations. *Comput. Geosci.* **2015**, *76*, 1–10. [CrossRef]

82. Paris, A.; de Paiva, R.D.; da Silva, J.S.; Moreira, D.M.; Calmant, S.; Garambois, P.-A.; Collischonn, W.; Bonnet, M.-P.; Seyler, F. Stage-discharge rating curves based on satellite altimetry and modeled discharge in the Amazon basin. *Water Resour. Res.* **2016**, *52*, 3787–3814. [CrossRef]

83. Nathanson, M.; Kean, J.W.; Grabs, T.J.; Seibert, J.; Laudon, H.; Lyon, S.W. Modelling rating curves using remotely sensed LiDAR data. *Hydrol. Process.* **2012**, *26*, 1427–1434. [CrossRef]

84. Feng, D.; Gleason, C.J.; Yang, X.; Pavelsky, T.M. Comparing Discharge Estimates Made via the BAM Algorithm in High-Order Arctic Rivers Derived Solely From Optical CubeSat, Landsat, and Sentinel-2 Data. *Water Resour. Res.* **2019**, *55*, 7753–7771. [CrossRef]

85. Normandin, C.; Frappart, F.; Diepkile, A.T.; Marieu, V.; Mougin, E.; Blarel, F.; Lubac, B.; Braquet, N.; Ba, A. Evolution of the Performances of Radar Altimetry Missions from ERS-2 to Sentinel-3A over the Inner Niger Delta. *Remote Sens.* **2018**, *10*, 833. [CrossRef]

86. Ashmore, P.; Sauks, E. Prediction of discharge from water surface width in a braided river with implications for at-a-station hydraulic geometry. *Water Resour. Res.* **2006**, *42*. [CrossRef]

87. Gleason, C.J.; Smith, L.C.; Finnegan, D.C.; LeWinter, A.L.; Pitcher, L.H.; Chu, V.W. Technical Note: Semi-automated effective width extraction from time-lapse RGB imagery of a remote, braided Greenlandic river. *Hydrol. Earth Syst. Sci.* **2015**, *19*, 2963–2969. [CrossRef]

88. Huang, Q.; Long, D.; Du, M.; Zeng, C.; Qiao, G.; Li, X.; Hou, A.; Hong, Y. Discharge estimation in high-mountain regions with improved methods using multisource remote sensing: A case study of the Upper Brahmaputra River. *Remote Sens. Environ.* **2018**, *219*, 115–134. [CrossRef]

89. Brakenridge, G.R.; Nghiem, S.V.; Anderson, E.; Mic, R. Orbital microwave measurement of river discharge and ice status. *Water Resour. Res.* **2007**, *43*. [CrossRef]

90. Brakenridge, G.; Cohen, S.; Kettner, A.J.; De Groeve, T.; Nghiem, S.V.; Syvitski, J.P.M.; Fekete, B.M. Calibration of satellite measurements of river discharge using a global hydrology model. *J. Hydrol.* **2012**, *475*, 123–136. [CrossRef]

91. Tarpanelli, A.; Brocca, L.; Lacava, T.; Melone, F.; Moramarco, T.; Faruolo, M.; Pergola, N.; Tramutoli, V. Toward the estimation of river discharge variations using MODIS data in ungauged basins. *Remote Sens. Environ.* **2013**, *136*, 47–55. [CrossRef]

92. Van Dijk, A.I.J.M.; Brakenridge, G.R.; Kettner, A.J.; Beck, H.E.; De Groeve, T.; Schellekens, J. River gauging at global scale using optical and passive microwave remote sensing. *Water Resour. Res.* **2016**, *52*, 6404–6418. [CrossRef]

93. Tarpanelli, A.; Amarnath, G.; Brocca, L.; Massari, C.; Moramarco, T. Discharge estimation and forecasting by MODIS and altimetry data in Niger-Benue River. *Remote Sens. Environ.* **2017**, *195*, 96–106. [CrossRef]

94. Tarpanelli, A.; Santi, E.; Tourian, M.J.; Filippucci, P.; Amarnath, G.; Brocca, L. Daily River Discharge Estimates by Merging Satellite Optical Sensors and Radar Altimetry Through Artificial Neural Network. *IEEE Trans. Geosci. Remote Sens.* **2018**, *57*, 329–341. [CrossRef]

95. Lyzenga, D.R. Passive remote sensing techniques for mapping water depth and bottom features. *Appl. Opt.* **1978**, *17*, 379. [CrossRef]

96. Lee, Z.; Carder, K.L.; Mobley, C.D.; Steward, R.G.; Patch, J.S. Hyperspectral Remote Sensing for Shallow Waters. I. A Semianalytical Model. *Appl. Opt.* **1998**, *37*, 6329. [CrossRef]

97. Lee, Z.; Carder, K.L.; Mobley, C.D.; Steward, R.G.; Patch, J.S. Hyperspectral Remote Sensing for Shallow Waters. 2. Deriving Bottom Depths and Water Properties by Optimization. *Appl. Opt.* **1999**, *38*, 3831. [CrossRef]

98. Gould, R.W., Jr.; Arnone, R.A.; Martinolich, P.M. Spectral Dependence of the Scattering Coefficient in Case 1 and Case 2 Waters. *Appl. Opt.* **1999**, *38*, 2377. [CrossRef]

99. Fonstad, M.A.; Marcus, W.A. Remote sensing of stream depths with hydraulically assisted bathymetry (HAB) models. *Geomorphology* **2005**, *72*, 320–339. [CrossRef]

100. Marcus, W.A.; Fonstad, M.A. Optical remote mapping of rivers at sub-meter resolutions and watershed extents. *Earth Surf. Process. Landf.* **2008**, *33*, 4–24. [CrossRef]
101. Legleiter, C.J.; Roberts, D.A.; Lawrence, R.L. Spectrally based remote sensing of river bathymetry. *Earth Surf. Process. Landf.* **2009**, *34*, 1039–1059. [CrossRef]
102. Legleiter, C.J. Calibrating remotely sensed river bathymetry in the absence of field measurements: Flow REsistance Equation-Based Imaging of River Depths (FREEBIRD). *Water Resour. Res.* **2015**, *51*, 2865–2884. [CrossRef]
103. Legleiter, C.J.; Kinzel, P.J.; Nelson, J.M. Remote measurement of river discharge using thermal particle image velocimetry (PIV) and various sources of bathymetric information. *J. Hydrol.* **2017**, *554*, 490–506. [CrossRef]
104. Smith, L.C.; Chu, V.W.; Yang, K.; Gleason, C.J.; Pitcher, L.H.; Rennermalm, A.K.; Legleiter, C.J.; Behar, A.E.; Overstreet, B.T.; Moustafa, S.E.; et al. Efficient meltwater drainage through supraglacial streams and rivers on the southwest Greenland ice sheet. *Proc. Natl. Acad. Sci. USA* **2015**, *112*, 1001–1006. [CrossRef] [PubMed]
105. Johnson, E.D.; Cowen, E.A. Remote monitoring of volumetric discharge employing bathymetry determined from surface turbulence metrics. *Water Resour. Res.* **2016**, *52*, 2178–2193. [CrossRef]
106. Costa, J.E.; Spicer, K.R.; Cheng, R.T.; Haeni, F.P.; Melcher, N.B.; Thurman, E.M.; Plant, W.J.; Keller, W.C. measuring stream discharge by non-contact methods: A Proof-of-Concept Experiment. *Geophys. Res. Lett.* **2000**, *27*, 553–556. [CrossRef]
107. Chiu, C.-L. Application of Entropy Concept in Open-Channel Flow Study. *J. Hydraul. Eng.* **1991**, *117*, 615–628. [CrossRef]
108. Fulton, J.; Ostrowski, J. Measuring real-time streamflow using emerging technologies: Radar, hydroacoustics, and the probability concept. *J. Hydrol.* **2008**, *357*, 1–10. [CrossRef]
109. Moramarco, T.; Barbetta, S.; Tarpanelli, A. From Surface Flow Velocity Measurements to Discharge Assessment by the Entropy Theory. *Water* **2017**, *9*, 120. [CrossRef]
110. Moramarco, T.; Barbetta, S.; Bjerklie, D.M.; Fulton, J.W.; Tarpanelli, A. River Bathymetry Estimate and Discharge Assessment from Remote Sensing. *Water Resour. Res.* **2019**, *55*, 6692–6711. [CrossRef]
111. Hilldale, R.C.; Raff, D. Assessing the ability of airborne LiDAR to map river bathymetry. *Earth Surf. Process. Landf.* **2008**, *33*, 773–783. [CrossRef]
112. Parr, D.; Wang, G.; Bjerklie, D. Integrating Remote Sensing Data on Evapotranspiration and Leaf Area Index with Hydrological Modeling: Impacts on Model Performance and Future Predictions. *J. Hydrometeorol.* **2015**, *16*, 2086–2100. [CrossRef]
113. Lopez, P.L.; Immerzeel, W.W.; Sandoval, E.A.R.; Sterk, G.; Schellekens, J. Spatial Downscaling of Satellite Based Precipitation and Its Impact on Discharge Simulations in the Magdalena River Basin in Colombia. *Front. Earth Sci.* **2018**, *6*, 68. [CrossRef]
114. Lopez, P.L.; Sutanudjaja, E.H.; Schellekens, J.; Sterk, G.; Bierkens, M.F.P. Calibration of a large-scale hydrological model using satellite-based soil moisture and evapotranspiration products. *Hydrol. Earth Syst. Sci.* **2017**, *21*, 3125–3144. [CrossRef]
115. Mendiguren, G.; Koch, J.; Stisen, S. Spatial pattern evaluation of a calibrated national hydrological model—A remote-sensing-based diagnostic approach. *Hydrol. Earth Syst. Sci.* **2017**, *21*, 5987–6005. [CrossRef]
116. Bowman, A.L.; Franz, K.J.; Hogue, T.S.; Kinoshita, A.M. MODIS-Based Potential Evapotranspiration Demand Curves for the Sacramento Soil Moisture Accounting Model. *J. Hydrol. Eng.* **2016**, *21*, 04015055. [CrossRef]
117. Sun, W.; Ishidaira, H.; Bastola, S.; Yu, J. Estimating daily time series of streamflow using hydrological model calibrated based on satellite observations of river water surface width: Toward real world applications. *Environ. Res.* **2015**, *139*, 36–45. [CrossRef]
118. Emery, C.M.; Paris, A.; Biancamaria, S.; Boone, A.; Calmant, S.; Garambois, P.-A.; Santos da Silva, J. Large-scale hydrological model river storage and discharge correction using a satellite altimetry-based discharge product. *Hydrol. Earth Syst. Sci.* **2018**, *22*, 2135–2162. [CrossRef]
119. Bates, P.D.; Horritt, M.S.; Smith, C.N.; Mason, D. Integrating remote sensing observations of flood hydrology and hydraulic modelling. *Hydrol. Process.* **1997**, *11*, 1777–1795. [CrossRef]
120. Brakenridge, G.R.; Tracy, B.T.; Knox, J.C. Orbital SAR remote sensing of a river flood wave. *Int. J. Remote Sens.* **1998**, *19*, 1439–1445. [CrossRef]
121. Horritt, M.S.; Bates, P.D. Predicting floodplain inundation: Raster-based modelling versus the finite-element approach. *Hydrol. Process.* **2001**, *15*, 825–842. [CrossRef]

122. Poole, G.C.; Stanford, J.A.; Frissell, C.A.; Running, S.W. Three-dimensional mapping of geomorphic controls on flood-plain hydrology and connectivity from aerial photos. *Geomorphology* **2002**, *48*, 329–347. [CrossRef]
123. Horritt, M.S. A methodology for the validation of uncertain flood inundation models. *J. Hydrol.* **2006**, *326*, 153–165. [CrossRef]
124. Gumley, L.E.; King, M.D. Remote Sensing of Flooding in the U.S. Upper Midwest during the Summer of 1993. *Bull. Am. Meteorol. Soc.* **1995**, *76*, 933–943. [CrossRef]
125. Mason, D.C.; Horritt, M.S.; Dall'Amico, J.T.; Scott, T.R.; Bates, P.D. Improving River Flood Extent Delineation From Synthetic Aperture Radar Using Airborne Laser Altimetry. *IEEE Trans. Geosci. Remote Sens.* **2007**, *45*, 3932–3943. [CrossRef]
126. Schumann, G.; Matgen, P.; Hoffmann, L.; Hostache, R.; Pappenberger, F.; Pfister, L. Deriving distributed roughness values from satellite radar data for flood inundation modelling. *J. Hydrol.* **2007**, *344*, 96–111. [CrossRef]
127. Schumann, G.; Matgen, P.; Cutler, M.E.J.; Black, A.; Hoffmann, L.; Pfister, L. Comparison of remotely sensed water stages from LiDAR, topographic contours and SRTM. *ISPRS J. Photogramm. Remote Sens.* **2008**, *63*, 283–296. [CrossRef]
128. Schumann, G.; Baldassarre, G.D.; Bates, P.D. The Utility of Spaceborne Radar to Render Flood Inundation Maps Based on Multialgorithm Ensembles. *IEEE Trans. Geosci. Remote Sens.* **2009**, *47*, 2801–2807. [CrossRef]
129. Coe, M.T.; Costa, M.H.; Howard, E.A. Simulating the surface waters of the Amazon River basin: Impacts of new river geomorphic and flow parameterizations. *Hydrol. Process.* **2008**, *22*, 2542–2553. [CrossRef]
130. Di Baldassarre, G.; Schumann, G.; Bates, P.D. A technique for the calibration of hydraulic models using uncertain satellite observations of flood extent. *J. Hydrol.* **2009**, *367*, 276–282. [CrossRef]
131. Khan, S.I.; Hong, Y.; Wang, J.H.; Yilmaz, K.K.; Gourley, J.J.; Adler, R.F.; Brakenridge, G.R.; Policelli, F.; Habib, S.; Irwin, D. Satellite Remote Sensing and Hydrologic Modeling for Flood Inundation Mapping in Lake Victoria Basin: Implications for Hydrologic Prediction in Ungauged Basins. *IEEE Trans. Geosci. Remote Sens.* **2011**, *49*, 85–95. [CrossRef]
132. Bates, P.D.; Marks, K.J.; Horritt, M.S. Optimal use of high-resolution topographic data in flood inundation models. *Hydrol. Process.* **2003**, *17*, 537–557. [CrossRef]
133. Du, J.; Kimball, J.S.; Galantowicz, J.; Kim, S.-B.; Chan, S.K.; Reichle, R.; Jones, L.A.; Watts, J.D. Assessing global surface water inundation dynamics using combined satellite information from SMAP, AMSR2 and Landsat. *Remote Sens. Environ.* **2018**, *213*, 1–17. [CrossRef] [PubMed]
134. Niedzielski, T.; Witek, M.; Spallek, W. Observing river stages using unmanned aerial vehicles. *Hydrol. Earth Syst. Sci.* **2016**, *20*, 3193–3205. [CrossRef]
135. Andreadis, K.M.; Clark, E.A.; Lettenmaier, D.P.; Alsdorf, D.E. Prospects for river discharge and depth estimation through assimilation of swath-altimetry into a raster-based hydrodynamics model. *Geophys. Res. Lett.* **2007**, *34*. [CrossRef]
136. Durand, M.; Andreadis, K.M.; Alsdorf, D.E.; Lettenmaier, D.P.; Moller, D.; Wilson, M. Estimation of bathymetric depth and slope from data assimilation of swath altimetry into a hydrodynamic model. *Geophys. Res. Lett.* **2008**, *35*. [CrossRef]
137. Biancamaria, S.; Durand, M.; Andreadis, K.M.; Bates, P.D.; Boone, A.; Mognard, N.M.; Rodríguez, E.; Alsdorf, D.E.; Lettenmaier, D.P.; Clark, E.A. Assimilation of virtual wide swath altimetry to improve Arctic river modeling. *Remote Sens. Environ.* **2011**, *115*, 373–381. [CrossRef]
138. Yoon, Y.; Durand, M.; Merry, C.J.; Clark, E.A.; Andreadis, K.M.; Alsdorf, D.E. Estimating river bathymetry from data assimilation of synthetic SWOT measurements. *J. Hydrol.* **2012**, *464–465*, 363–375. [CrossRef]
139. Gejadze, I.; Malaterre, P.-O. Discharge estimation under uncertainty using variational methods with application to the full Saint-Venant hydraulic network model. *Int. J. Numer. Methods Fluids* **2017**, *83*, 405–430. [CrossRef]
140. Oubanas, H.; Gejadze, I.; Malaterre, P.-O.; Mercier, F. River discharge estimation from synthetic SWOT-type observations using variational data assimilation and the full Saint-Venant hydraulic model. *J. Hydrol.* **2018**, *559*, 638–647. [CrossRef]
141. Oubanas, H.; Gejadze, I.; Malaterre, P.-O.; Durand, M.; Wei, R.; Frasson, R.P.M.; Domeneghetti, A. Discharge Estimation in Ungauged Basins Through Variational Data Assimilation: The Potential of the SWOT Mission. *Water Resour. Res.* **2018**, *54*, 2405–2423. [CrossRef]

142. Larnier, K.; Monnier, J.; Garambois, P.-A.; Verley, J. River Discharge and Bathymetry Estimations from SWOT Altimetry Measurements. 2019. Available online: https://hal.archives-ouvertes.fr/hal-01811683v2 (accessed on 1 February 2020).

143. Durand, M.; Fu, L.-L.; Lettenmaier, D.P.; Als, D.E.; Rodriguez, E.; Esteban-Fernandez, D. The Surface Water and Ocean Topography Mission: Observing Terrestrial Surface Water and Oceanic Submesoscale Eddies. *Proc. IEEE* **2010**, *98*, 766–779. [CrossRef]

144. Garambois, P.-A.; Monnier, J. Inference of effective river properties from remotely sensed observations of water surface. *Adv. Water Resour.* **2015**, *79*, 103–120. [CrossRef]

145. Durand, M.; Neal, J.; Rodríguez, E.; Andreadis, K.M.; Smith, L.C.; Yoon, Y. Estimating reach-averaged discharge for the River Severn from measurements of river water surface elevation and slope. *J. Hydrol.* **2014**, *511*, 92–104. [CrossRef]

146. Bjerklie, D.M.; Birkett, C.M.; Jones, J.W.; Carabajal, C.; Rover, J.A.; Fulton, J.W.; Garambois, P.-A. Satellite remote sensing estimation of river discharge: Application to the Yukon River Alaska. *J. Hydrol.* **2018**, *561*, 1000–1018. [CrossRef]

147. Tuozzolo, S.; Langhorst, T.; de Moraes Frasson, R.P.; Pavelsky, T.; Durand, M. The impact of reach averaging Manning's equation for an in-situ dataset of water surface elevation, width, and slope. *J. Hydrol.* **2019**, *578*, 123866. [CrossRef]

148. Altenau, E.H.; Pavelsky, T.M.; Moller, D.; Lion, C.; Pitcher, L.H.; Allen, G.H.; Bates, P.D.; Calmant, S.; Durand, M.; Smith, L.C. AirSWOT measurements of river water surface elevation and slope: Tanana River, AK. *Geophys. Res. Lett.* **2017**, *44*, 181–189. [CrossRef]

149. Tuozzolo, S.; Lind, G.; Overstreet, B.; Mangano, J.; Fonstad, M.; Hagemann, M.; Frasson, R.P.M.; Larnier, K.; Garambois, P.-A.; Monnier, J.; et al. Estimating River Discharge With Swath Altimetry: A Proof of Concept Using AirSWOT Observations. *Geophys. Res. Lett.* **2019**, *46*, 1459–1466. [CrossRef]

150. Gleason, C.J.; Smith, L.C.; Lee, J. Retrieval of river discharge solely from satellite imagery and at-many-stations hydraulic geometry: Sensitivity to river form and optimization parameters. *Water Resour. Res.* **2014**, *50*, 9604–9619. [CrossRef]

151. Gleason, C.J.; Hamdan, A.N. Crossing the (Watershed) Divide: Satellite Data and the Changing Politics of International River Basins. *Geogr. J.* **2017**, *183*, 2–15. [CrossRef]

152. Hagemann, M.W.; Gleason, C.J.; Durand, M.T. BAM: Bayesian AMHG-Manning Inference of Discharge Using Remotely Sensed Stream Width, Slope, and Height. *Water Resour. Res.* **2017**, *53*, 9692–9707. [CrossRef]

153. Bonnema, M.G.; Sikder, S.; Hossain, F.; Durand, M.; Gleason, C.J.; Bjerklie, D.M. Benchmarking wide swath altimetry-based river discharge estimation algorithms for the Ganges river system. *Water Resour. Res.* **2016**, *52*, 2439–2461. [CrossRef]

154. Durand, M.; Gleason, C.J.; Garambois, P.A.; Bjerklie, D.; Smith, L.C.; Roux, H.; Rodriguez, E.; Bates, P.D.; Pavelsky, T.M.; Monnier, J.; et al. An intercomparison of remote sensing river discharge estimation algorithms from measurements of river height, width, and slope. *Water Resour. Res.* **2016**, *52*, 4527–4549. [CrossRef]

155. Sichangi, A.W.; Wang, L.; Hu, Z. Estimation of River Discharge Solely from Remote-Sensing Derived Data: An Initial Study over the Yangtze River. *Remote Sens.* **2018**, *10*, 1385. [CrossRef]

156. Chow, V.T. *Handbook of Applied Hydrology*; McGraw-Hill: New York, NY, USA, 1964.

157. Rosgen, D.L. A classification of natural rivers. *Catena* **1994**, *22*, 169–199. [CrossRef]

158. Brown, C.M.; Lund, J.R.; Cai, X.; Reed, P.M.; Zagona, E.A.; Ostfeld, A.; Hall, J.; Characklis, G.W.; Yu, W.; Brekke, L. The future of water resources systems analysis: Toward a scientific framework for sustainable water management. *Water Resour. Res.* **2015**, *51*, 6110–6124. [CrossRef]

159. Hannah, D.M.; Demuth, S.; van Lanen, H.A.J.; Looser, U.; Prudhomme, C.; Rees, G.; Stahl, K.; Tallaksen, L.M. Large-scale river flow archives: Importance, current status and future needs. *Hydrol. Process.* **2011**, *25*, 1191–1200. [CrossRef]

160. Fekete, B.M.; Looser, U.; Pietroniro, A.; Robarts, R.D. Rationale for Monitoring Discharge on the Ground. *J. Hydrometeorol.* **2012**, *13*, 1977–1986. [CrossRef]

161. Birkinshaw, S.J.; O'Donnell, G.M.; Moore, P.; Kilsby, C.G.; Fowler, H.J.; Berry, P.A.M. Using satellite altimetry data to augment flow estimation techniques on the Mekong River. *Hydrol. Process.* **2010**, *24*, 3811–3825. [CrossRef]

162. Gleason, C.J.; Wada, Y.; Wang, J. A Hybrid of Optical Remote Sensing and Hydrological Modeling Improves Water Balance Estimation. *J. Adv. Model. Earth Syst.* **2018**, *10*, 2–17. [CrossRef]

163. León, L.F.A.; Gleason, C.J. Production, Property, and the Construction of Remotely Sensed Data. *Ann. Am. Assoc. Geogr.* **2017**, *107*, 1075–1089.
164. Sneddon, C.; Fox, C. Water, Geopolitics, and Economic Development in the Conceptualization of a Region. *Eurasian Geogr. Econ.* **2012**, *53*, 143–160. [CrossRef]
165. Sneddon, C.; Fox, C. Rethinking transboundary waters: A critical hydropolitics of the Mekong basin. *Polit. Geogr.* **2006**, *25*, 181–202. [CrossRef]
166. Sneddon, C.; Fox, C. The Cold War, the US Bureau of Reclamation, and the technopolitics of river basin development, 1950–1970. *Polit. Geogr.* **2011**, *30*, 450–460. [CrossRef]
167. Dore, J.; Lebel, L. Deliberation and Scale in Mekong Region Water Governance. *Environ. Manag.* **2010**, *46*, 60–80. [CrossRef] [PubMed]
168. Ho, S. River Politics: China's policies in the Mekong and the Brahmaputra in comparative perspective. *J. Contemp. China* **2014**, *23*, 1–20. [CrossRef]

 *remote sensing*

Article

Timing of Landsat Overpasses Effectively Captures Flow Conditions of Large Rivers

George H. Allen [1,*], Xiao Yang [2], John Gardner [2], Joel Holliman [1], Cédric H. David [3] and Matthew Ross [4]

[1] Department of Geography, Texas A&M University, College Station, TX 77843, USA; joel52@tamu.edu
[2] Department of Geological Sciences, University of North Carolina at Chapel Hill, Chapel Hill, NC 27599, USA; yangxiao@live.unc.edu (X.Y.); gardj@email.unc.edu (J.G.)
[3] Jet Propulsion Laboratory, California Institute of Technology, Pasadena, CA 91109, USA; cedric.david@jpl.nasa.gov
[4] Natural Resources Ecology Laboratory, Colorado State University, Fort Collins, CO 80523, USA; matt.ross@colostate.edu
* Correspondence: geoallen@tamu.edu or georghenryallen@gmail.com

Received: 24 February 2020; Accepted: 7 May 2020; Published: 9 May 2020

Abstract: Satellites provide a temporally discontinuous record of hydrological conditions along Earth's rivers (e.g., river width, height, water quality). The degree to which archived satellite data effectively capture the overall population of river flow frequency is unknown. Here, we use the entire archives of Landsat 5, 7, and 8 to determine when a cloud-free image is available over the United States Geological Survey (USGS) river gauges located on Landsat-observable rivers. We compare the flow frequency distribution derived from the daily gauge record to the flow frequency distribution derived from ideally sampling gauged discharge based on the timing of cloud-free Landsat overpasses. Examining the patterns of flow frequency across multiple gauges, we find that there is not a statistically significant difference between the flow frequency distribution associated with observations contained within the Landsat archive and the flow frequency distribution derived from the daily gauge data ($\alpha = 0.05$), except for hydrological extremes like maximum and minimum flow. At individual gauges, we find that Landsat observations span a wide range of hydrological conditions (97% of total flow variability observed in 90% of the study gauges) but the degree to which the Landsat sample can represent flow frequency distribution varies from location to location and depends on sample size. The results of this study indicate that the Landsat archive is, on average, representative of the temporal frequencies of hydrological conditions present along Earth's large rivers with broad utility for hydrological, ecologic and biogeochemical evaluations of river systems.

Keywords: Landsat; rivers; streamflow; discharge; flow frequency; satellite revisit time; flow regime

---

1. Introduction

Rivers serve as the chief source of renewable water to humans and to freshwater habitats [1] and thus they represent an important nexus between the water cycle, civilization, and aquatic ecology. Effective water resource management depends on the ability to monitor river systems and understand how they are responding to changes in climate and land use. While hydrological simulations can provide useful data for evaluating river system dynamics, they are often unable to resolve unpredictable processes and events, leading to high uncertainty, particularly when operating over large areas [2]. Observational monitoring of rivers is therefore essential for a complete understanding of the real-world complexities of river systems.

The stream gauge is the traditional instrument used for monitoring river hydrology. Gauges provide temporally continuous data that can be highly accurate given proper gauge maintenance.

From gauge records, flow frequency distributions, or the frequency distribution of river flow through time, can be used to calculate return period and estimate future flood hazard under the assumption of stationarity [3]. However, gauges only monitor fixed points along river reaches and thus are unable to provide a comprehensive view of surface water hydrology across river networks [4]. Additionally, gauges tend to be biased in their geographic location, most often being installed along narrow, stable reaches, near populated regions, in developed countries, and on perennial river types [5]. Further, the availability of gauge data has been declining since the 1970s [6], limiting the ability to understand how changes in land use and the water cycle are impacting global rivers.

Satellite remote sensing of Earth's rivers offers an alternative, complementary approach to in situ gauging [7]. Remote sensing data are often spatially continuous and thus can be used to observe an entire river network within the constraints of a given satellite's payload and orbit. Remote sensing also enables non-intrusive measurement of rivers that avoids the inconsistencies associated with individually-maintained gauge stations. While optical remote sensing is the most popular method for studying rivers from space, it is susceptible to cloud cover, low sun angle, seasonal darkness, and nighttime conditions [8]. Other approaches like microwave remote sensing can be used to observe rivers during the day and night, and during almost all weather conditions [9–11], and high-resolution commercial imagery is emerging as a useful tool for unprecedented fine-scale monitoring of river and stream processes [12,13].

Satellite instruments cannot directly detect river discharge itself, but they can measure other properties of rivers that vary with discharge including river width [14], river height [15], and river velocity [16], as well as water quality [17]. Earth observation satellites have been operational since the early 1970s and have yielded long-term remote sensing data archives that serve as a valuable resource for monitoring rivers [18–20]. For example, remote sensing observations can be related to gauged discharge and can be used to build percentile-based rating curves [21,22]. However, the vast majority of Earth's river reaches are ungauged [23] and for these ungauged reaches, the range of hydrological conditions that satellite data cover is uncertain. A satellite's record is a discontinuous temporal sample of the overall population of hydrological conditions along rivers. However, the ability of satellite remote sensing data archives to represent flow frequency is unquantified. Does the satellite sample effectively represent flow frequency distribution in rivers?

Recent technological innovations have made this question more germane. Cloud computing image processing platforms like Google Earth Engine have substantially lowered the barriers to entry of planetary-scale analysis of entire satellite image archives [24]. These platforms have led to the development of freely-available global-scale maps of surface water occurrence and change from long-term aggregations of optical imagery [25,26] as well as open-source image processing algorithms that can be used to track changes in river morphology over decades [27]. Additionally, new global datasets of rivers including the Global River Widths from Landsat (GRWL) database [28], which contains the location and width of rivers observable by Landsat, allows for improved classification and analysis of rivers from satellites.

We are nearing the half-century anniversary of the first civilian satellite focused on Earth's land surface (ERTS-1 launched in 1972 and later renamed Landsat 1). Indeed, Landsat is the longest running Earth observation satellite program and it is a standard source of remote sensing data for land cover characterization used in hydrological applications [29]. In particular the archives of Landsat 5, 7 and 8 (hereinafter simply referred to as Landsat), which all share the same Worldwide Reference System (WRS-2), contain freely-available, cross-calibrated sensor data with compatible spatial and temporal resolutions. Landsat provides ~30 m optical and infrared observations across a 185 km swath along a sun-synchronous near-polar 16 day repeat orbit. Thus, Landsat produces consistent multispectral imagery at a high spatial resolution but at a temporally sparse interval, particularly when cloud cover prevents land surface retrieval. We focus on Landsat here because it is the most commonly used remote sensing platform for studying surface water from space [8] and it has been the basis for several popular geospatial products for studying rivers [25,28,30–32].

However, an important unknown is whether the accumulated observations of rivers from Landsat, or the Landsat sample, correspond to the true population of hydrological conditions in rivers. If the available Landsat sample adequately captures the on-the-ground frequency distribution of river flow, then Landsat-based surveys of river characteristics can be interpreted to be representative of the hydrological conditions present along Earth's rivers. Quantifying the ability of the Landsat archive to represent hydrological conditions, represented here by flow frequency, is motivated by several unexplored applications in the field of remote sensing of rivers. These include (1) using water occurrence information from the Landsat archive to estimate discharge by constructing river width-based rating curves; (2) using water occurrence information to quantify the seasonality of river and stream inundation extent, an important metric for estimating biogeochemical exchange between rivers and the atmosphere [33]; (3) using water color information to understand the variability and dynamics of river water quality [19].

If the Landsat archive is of adequate length or, in other words, if the Landsat sample size is large enough, then fundamental assumptions can be made regarding the representativeness of the Landsat record of hydrological conditions along Earth's large rivers. Thus, given the long Landsat archive (Landsat 5 launched on 1 March 1984), we hypothesize that long-term aggregations of Landsat imagery capture the flow frequency of Earth's large rivers. Here, we test this hypothesis by conducting a simple temporal sampling analysis of the United States Geological Survey (USGS) gauge records based on cloud-free overpass timing of Landsat 5, 7 and 8 (Figure 1). While a similar approach has been taken to predict the observational potential of the future Surface Water and Ocean Topography (SWOT) satellite mission [34], there has been no evaluation conducted for the widely used Landsat archive.

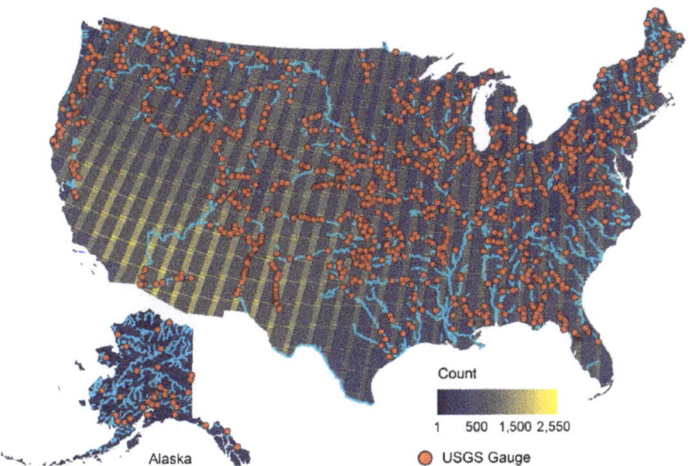

Figure 1. Map showing the United States Geological Survey (USGS) gauges used in this study and Landsat 5, 7, and 8 data availability based on the number of images with less than 30% cloud cover. Regions affected by cloud cover and low sun angle tend to have lower data availability. Rivers equal to or wider than Landsat's 30 m resolution at mean discharge shown as blue lines [28].

2. Materials and Methods

2.1. Gauge Data

We represent the true flow frequency of rivers using daily streamflow records from gauge stations in the United States (US) and operated by the USGS (Figure 1). To exclude gauges that are located on rivers too narrow to be observable by Landsat, we only consider the 1134 USGS gauges that were used to validate the Global River Widths from Landsat (GRWL) database [28]. These gauges (1) have upstream drainage areas larger than 1000 km^2; (2) have records that span at least 10 years; (3) are located within 1 km Euclidean distance of a GRWL centerline; (4) are not immediately adjacent to lakes, reservoirs or river confluences (determined using the GRWL flag field and by visually examining each gauge location [28]); (5) have in situ river width data available [35]. We only analyze discharge records occurring between 1 March 1984 and 14 August 2019, because this time period spans the range of available data from the Landsat 5, 7 and 8 missions at the time of this study's analysis. To help ensure a representative distribution of discharge at each gauge, we omit from our analysis 102 gauges that contain less than 5 years of discharge records within the 35 year study period. Additionally, because discharge conditions can be obscured by the presence of river ice [36], we exclude all gauge measurements that were taken when river ice was recorded at the gauge (USGS Quality Code set to "ice"). Similarly, we exclude discharge records that the USGS reported as provisional, estimated, underestimated, or overestimated (USGS Quality Code set to "p", "E", "<", or ">"). Combined, these exclusions remove 5.2% of the daily discharge measurements but they produce a more accurate and representative discharge record for this study. The gauges used here are located on rivers, with a median width at mean annual discharge of 76 m and a first and third quartile of 50 to 117 m, respectively, as measured by the USGS [28].

2.2. Landsat Data

To assess the ability of Landsat to capture the true flow frequency distribution, we match coincident Landsat 5, 7, and 8 overpasses with daily discharge measurements at each USGS gauge over the 35-year study period (Figure 2a,c). Note that this matching process is idealized in that the discharge value at the gauge is exactly assigned to the corresponding Landsat image. This approach effectively assumes that river discharge can be accurately estimated from Landsat width measurements, an active field of research [7,22,27,37–39]. This simplifying assumption is a necessary first step that neglects potential errors in remote sensing and discharge algorithms, but allows for our stated focus on analyzing Landsat sampling capabilities. We conduct this Landsat availability analysis using the Google Earth Engine platform [24], which hosts the entire digitized Landsat archive. Each Landsat satellite observes the same location at least once every 16 days, although in areas with frequent cloud cover, the actual interval of cloud-free observations can be much longer (Figure 1). To account for the impact of cloud cover, we also determine when an available Landsat scene is cloud-free within a 500-m radius around the gauge. We use a 500-m radius around each gauge because this distance is longer than the corresponding in situ river width at mean discharge of all but 14 gauges considered in this study [28]. We identify clouds based on the USGS Landsat Bitwise Quality Assessment (BQA) product [40]. To account for the impact of the ETM+ Scan Line Corrector failure on data quality [41], we conservatively omit all Landsat 7 observation dates after 31 May 2003, when the failure occurred. In total, the number of observations considered in this analysis after excluding clouds and problematic gauge measurements is 327,177.

2.3. Statistical Comparison at Individual Gauges

At each individual USGS gauge, we compare the flow frequency distribution of the idealized Landsat sample to the flow frequency distribution of the daily gauge record (Figure 2). The upper panel in Figure 2 shows an example of where the Landsat sample and the gauge record have flow distributions that are relatively similar, whereas the lower panel shows an example where the Landsat sample does not represent the gauged flow distribution with high fidelity. We use the non-parametric Kolmogorov–Smirnov (K-S) test [42] to characterize the statistical difference between the Landsat sample and the daily record of flow at each gauge. We use a significance level of $\alpha = 0.05$ for all statistical tests in this study.

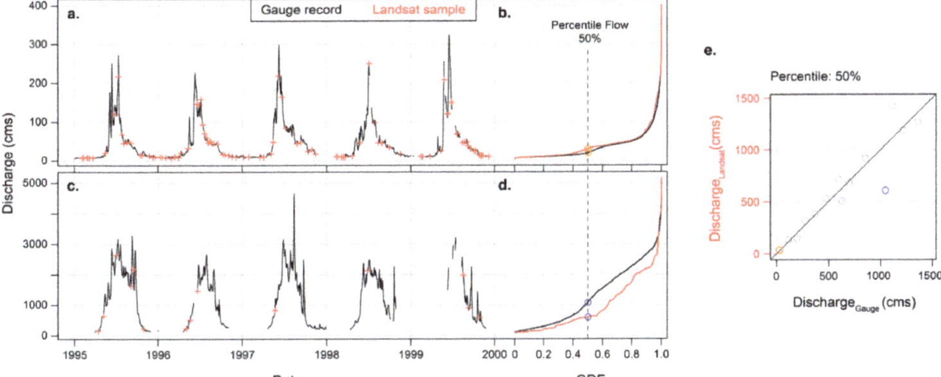

Figure 2. Hydrographs, cumulative density functions (CDFs), and a percentile plot from two example gauge stations. For visual clarity, hydrographs show a 5 year subset of the full record (1984–2019), while CDFs correspond to the full record length. Gaps in the hydrographs correspond to United States Geological Survey (USGS) flow observations that are estimated, provisional, and/or frozen. (**a**) Hydrograph of USGS Gauge 06225500 at Wind River near Crowheart, Wyoming, and (**b**) the corresponding flow frequency CDF (KSD statistic = 0.09, *p*-value < 0.001). (**c**) USGS Gauge 15129000 at Alsek River near Yakutat, Alaska, and (**d**) the corresponding flow frequency CDF (KSD statistic = 0.17, *p*-value = 0.055). (**e**) Fiftieth percentile (median) flow calculated from the full gauge record (x-axis) vs. from the Landsat sample (y-axis), with the two example gauge stations represented as colored circles. Gray circles represent median flows from other gauge stations.

Note that the K-S *p*-value is highly sensitive to sample size, causing it to be an impractical statistic for comparing across individual gauges. For example, a gauge with more Landsat observations (a larger sample size) is more likely to be considered significantly different than the same gauge with fewer observations (a smaller sample size), according to the K-S *p*-value [43,44]. Thus, the K-S *p*-value often yields contrary results, in which a large sample with a distribution that appears similar to that of its population will be considered to be significantly different (e.g., Figure 2a,b), while a small sample that appears to highly deviate from the gauge record will not be considered to be significantly different (e.g., Figure 2c,d). Due to this contradictory behavior, we do not place emphasis on the K-S *p*-value in this analysis but rather focus on the descriptive K-S D-statistic (KSD statistic), which provides a clear summary of the difference in flow frequency distributions between the Landsat sample and gauge record.

Additionally, we analyze the ability of Landsat to capture hydrological extremes at each individual gauge by determining the maximum and minimum percentile of gauged flows sampled by Landsat. We then explore potential factors affecting the ability of the Landsat sample to represent true flow frequency distributions across gauges including climate (cloudiness), watershed area and flow regime (flashiness). Flashiness is quantified according to the Richards–Baker Flashiness Index that sums the differences in daily flow divided by the total flow over a given time period [45]. This non-dimensional index is commonly used and has been observed to vary from 0 to ~1.5, although large rivers are generally less flashy and tend to exhibit index values of less than ~0.5 [34]. We correlate these factors with the KSD statistic using the Spearman rank correlation test [46] across all gauges and we examine the spatial patterns in the KSD statistic.

2.4. Statistical Comparison Across Multiple Gauges

To determine the ability of Landsat to represent river flow frequency across space, we compare the Landsat sample to the daily gauge record at different flow frequencies (Figure 2e). This approach tests whether different locations can be combined to represent flow frequency and is analogous to the classic hydrological concept of downstream hydraulic geometry [47]. In such a conceptual framework, the approach taken in Section 2.3 is equivalent to at-a-station hydraulic geometry. At each gauge, we calculate (1) discharge from the full gauge record and (2) discharge from the Landsat sample at multiple flow percentiles. Figure 2b,d shows two examples of calculating these two metrics at the 50th flow percentile (median flow). In addition to median flow, we calculate the two metrics at the following flow percentiles: 0% (minimum flow), 5%, 10%, 90%, 95%, and 100% (maximum flow).

For each of these percentiles, we compare the Landsat sample discharge to the gauge record discharge at every gauge using the non-parametric Mann–Whitney–Wilcoxon test (MWW) [48]. To avoid bias from outliers, we use the Theil–Sen median estimator [49] to derive a robust linear regression between the Landsat-sampled discharge and the gauge record discharge at each percentile. Additionally, we also calculate relative root mean square error,

$$rRMSE_j = \sqrt{\frac{1}{N}\sum_{i=1}^{N}\left(\frac{Q_{i,j}-\hat{Q}_{i,j}}{\hat{Q}_{i,j}}\right)^2},\tag{1}$$

and the relative bias,

$$rBIAS_j = \frac{1}{N}\sum_{i=1}^{N}\left(\frac{Q_{i,j}-\hat{Q}_{i,j}}{\hat{Q}_{i,j}}\right),\tag{2}$$

where N is the total number of gauges used in this study, $Q_{i,j}$ is the flow (m^3s^{-1}) from the Landsat sample distribution at a given percentile, j, at gauge i and $\hat{Q}_{i,j}$ is the flow from the gauge distribution at the same percentile and same gauge. By comparing the Landsat sample to the gauge record across all the gauges at each selected percentile, we evaluate the ability of Landsat to represent a given flow frequency through spatial averaging.

2.5. Minimum Length of Landsat Observations

We expect that the ability of satellites to effectively represent river flow frequency is related to the length of the observational archive and hence the sample size. Theoretically, longer sampling periods enable better capture of the flow frequency distribution during the sample period. To examine the effect of observation length on Landsat's capacity to represent flow frequency, we simulated different observation period lengths over which Landsat-sampled discharge from the gauge record. For each gauge in our analysis, we created 50 random permutations of a continuous temporal range with an n year duration ($n = 1, 2, 3, \ldots, 10$). To allow for random permutations of a 10 year period, we only included gauges that contained more than 15 years of continuous data (N = 927). Within each temporal range, we compared the Landsat-sampled flows and those from daily gauge records from the same period. Specifically, we calculated flow values at 0%, 1%, 5%, 50%, 95%, 99%, and 100% percentiles. From these data, we calculated the coefficient of determination (R^2), relative error metrics (rBias, rMAE, and rRMSE), and absolute error metrics (Bias, MAE, and RMSE). Together, these statistics help characterize the ability of Landsat to represent flow conditions with increasing observation duration.

3. Results

3.1. How Well Can the Landsat Archive Capture Flow Conditions at Individual Gauges?

We find that the Landsat archive contains observations corresponding to the near full range of discharge conditions for the vast majority of gauges (Figure 3a). For example, at 90% of the study gauges, the idealized Landsat sample captures at least 97% of the full range of discharge percentiles recorded by the gauge. The majority of gauges (55%) show no significant difference between the Landsat sample and the gauge record of flow according to the K-S p-value at the 95% confidence interval. However, as previously noted, the K-S p-value is exceedingly sensitive to sample size and often produces contradictory results when comparing samples of different sizes, as we do here. On the other hand, the descriptive KSD statistic, which does not exhibit this contradictory behavior, ranges from 0.016 to 0.36, with a median value of 0.083. Among the sites with no significant difference between the Landsat-sampled and gauged flow frequency, the KSD statistic ranged from 0.016 to 0.27, with a median value of 0.06. Thus, while there is considerable variation in the potential ability of Landsat to reconstruct flow frequency from gauge to gauge, the average difference between the idealized Landsat sample and the gauge flow frequency distribution is small.

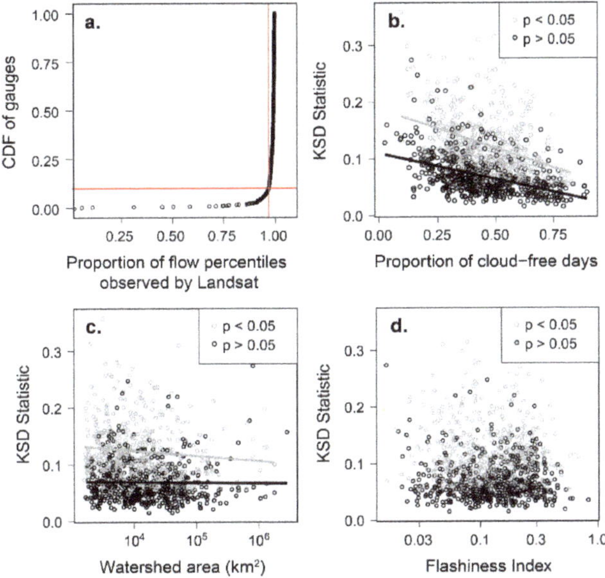

Figure 3. Individual gauge analysis, whereby each point represents metrics at a single gauge station. (**a**) Empirical CDF of the proportion of flow percentiles of the gauge record that the Landsat sample represents at each gauge. (**b**) Cloud occurrence correlates negatively with the ability of Landsat to capture flow frequency as represented by the KSD statistic. (**c**) Watershed area weakly correlates with the KSD statistic. (**d**) The Richards–Baker Flashiness Index does not correlate with the KSD statistic.

To determine potential drivers of Landsat's ability to capture river flow frequency, we examine correlations between the KSD statistic and the environmental variables of cloudiness, watershed area, and flow flashiness. We find a negative correlation ($p < 0.001$) between the proportion of cloud-free observations at a gauge and the KSD statistic (Figure 3b). This pattern is evident for locations where there is a statistically significant difference between the Landsat sample and the gauge record (gray points in Figure 3b) as well as locations that exhibit no significant difference (black points in Figure 3b; Spearman rank correlation coefficients of r = −0.40 and r = −0.47, respectively). We also find a weak negative correlation between the KSD statistic and watershed area (r = −0.09 and −0.16; $p = 0.02$ and $p < 0.001$; Figure 3c) for locations with significant difference and locations with a significant difference, respectively. Conversely, we find no significant correlations between flow flashiness and the KSD statistic at sites with statistically significant differences (r = 0.012, $p = 0.78$) nor at sites with statistically insignificant differences (r = 0.018, $p = 0.66$; Figure 3d). Landsat-observable rivers are large and span a relatively narrow range of the Richards–Baker Flashiness Index, from 0.016 to 0.84, compared to small streams that can vary up to 1.5 [34]. The lack of correlation between flashiness and the KSD statistic implies that flow regimes on large Landsat-observable rivers do not affect Landsat's ability to capture flow frequency distribution. We find no readily apparent spatial patterns in the D-statistic (see Figure S1 for an interactive map showing flow frequency distributions of the Landsat sample and the gauge record for each stream gauge). While more sophisticated statistical approaches could be employed to predict locations where Landsat best captures river flow frequency, this task is beyond the scope of this study.

3.2. How Well Can the Landsat Archive Capture Flow Conditions across Multiple Gauges?

Comparing the Landsat sample to the gauge record across all gauge stations reveals consistent patterns between a given flow percentile and the ability of Landsat to potentially represent flow at that percentile (Figure 4). For a wide range of flows (i.e., the 1% through the 95% percentiles in Figure 4), we find no statistically significant difference between the Landsat sample and the full gauge record according to the MWW test. Thus, except for extreme hydrological conditions like maximum and minimum flow, the idealized Landsat sample is not statistically different from the daily flow measured by multiple gauges (α = 0.05). Error metrics tend to be highest at extreme flow conditions and lowest at median flow. For intermediate percentiles that show no statistically significant difference, rRMSE values range from 12% to 78% and rBIAS values range from −6.4% to 8%.

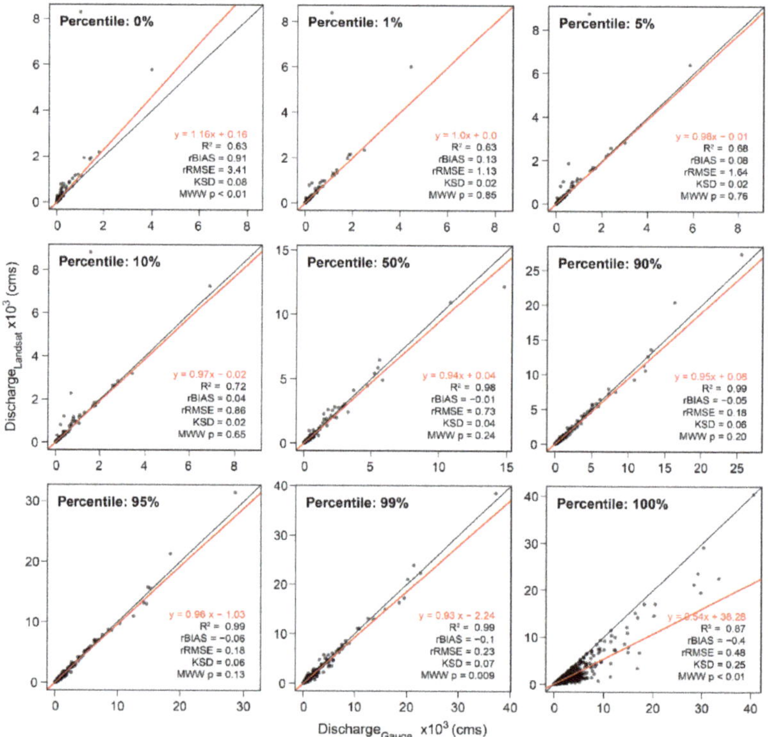

Figure 4. Full gauge record flow percentiles vs. Landsat-sampled flow percentiles for each USGS gauge studied (N = 1134). The black line represents the 1:1 line and the red line represents the Theil–Sen median estimator best fit (equation shown in red). R^2: coefficient of determination; rBias: relative bias; rRMSE: relative root mean square error; KSD: Kolmogorov–Smirnov D-statistic; MWW p: Mann–Whitney–Wilcoxon *p*-value.

Examining the error statistics of the extreme flow percentiles yields additional insights. At minimum flow, the Landsat sample always either matches or overestimates minimum flow, as seen by the points always being above the 1:1 line in the 0% percentile panel of Figure 4, resulting in a high positive relative bias of 200%. Conversely, at maximum flow, the Landsat sample either equals or underestimates maximum flow, producing a negative relative bias of −31%. These underestimates tend to increase in magnitude with increasing discharge (errors exhibit heteroscedasticity), resulting in a Theil–Sen median estimator that significantly deviates from unity at the 100% percentile (red line). These patterns also persist at the 99% percentile, in which the Landsat sample tends to underestimate discharge, albeit to a lesser degree than at maximum flow.

3.3. What Is the Minimum Length of Landsat Observations Needed to Represent Flow Frequency?

Our findings confirm that long periods of Landsat observation correspond with an improved ability of the Landsat archive to contain observations that can effectively represent river flow frequency. Generally, after 3 years of Landsat observation, all percentiles except for minimum and maximum flow are in close agreement with the reference values derived from the gauge record, with R^2 values above 0.9 (Figure 5a). Moreover, with the exception of minimum and maximum flow, we find a general trend of decreasing relative error statistics (rBias, rMAE, and rRMSE; Figure 5b) and absolute error statistics (Bias, MAE, RMSE; Figure 5c) with increasing duration of observation. Thus, as satellite observation duration increases, the distribution of flow frequency observed by Landsat expectedly converges with the flow frequency of the daily gauge record, except for extreme flows like minimum and maximum discharge.

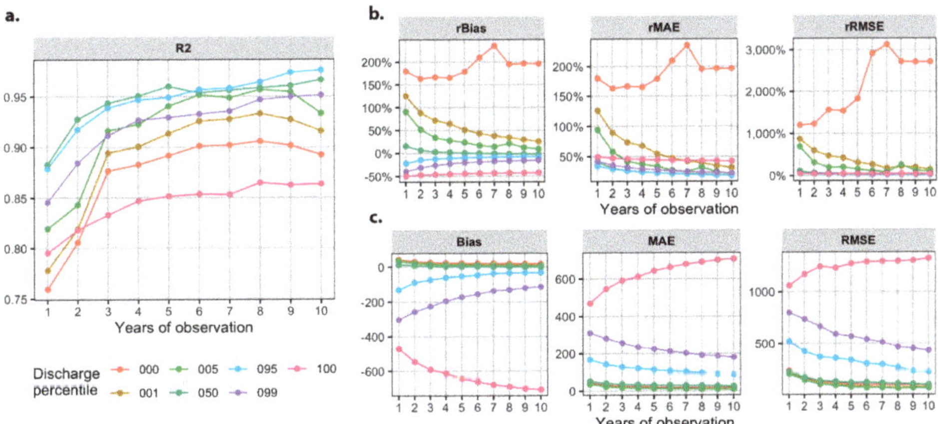

Figure 5. Landsat's representation of flow conditions with increasing observation duration. (**a**) Variations in R^2 value with increasing observation duration. (**b**) Relative metrics (rBias, rMAE, rRMSE) comparing gauge daily records and Landsat-sampled gauge records. (**c**) Absolute error metrics (Bias, MAE, RMSE, unit: cms) comparing gauge daily records and Landsat-sampled gauge records. Median statistical values plotted for each percentile.

Additionally, our results reveal that the degree of increasing similarity between the Landsat sample and the gauge record flow frequency distributions strongly depends on the flow percentile being studied. With increasing observation duration, smaller percentile flows generally show a more dramatic improvement in performance relative to larger flows. For example, Figure 5a shows a more substantial increase in R^2 for low flow percentiles (0%, 1%, and 5% flow) at a 3 year duration relative to the larger flow percentiles. This pattern also persists for the relative error metrics whereby the decreasing trend is greatest in smaller percentiles, with the exception of minimum and maximum flow. We note that low flow frequencies correspond to higher relative errors compared to higher flow frequencies, likely because of their smaller denominators in Equations (1) and (2). Conversely, high flow frequencies correspond to higher absolute error metrics relative to lower flow frequencies and the non-linear gap between the percentile curves may result from the positive skewness of flow frequency distributions in most rivers.

4. Discussion

4.1. Interpretations of Primary Findings

Our results indicate that Landsat can effectively capture river flow frequency over large spatial areas given an adequate duration of observation and accurate remote sensing and discharge algorithms. Although Landsat's sampling capability varies at individual gauge sites, we find that spatially averaging over multiple gauges enables effective representation of flow frequency, with the exception of extremely high and low flows (Figure 4). Landsat more aptly captures flow frequency at the lowest flows rather than at the highest flows, likely because of the positive skew characteristic of flow frequency distributions (e.g., Figure 2b). Small changes in flow frequency at the highest flows represent large variations in discharge that are difficult to capture due to the 16 day repeat orbit of Landsat. Cloudy conditions often accompany high flows, which further inhibits Landsat's ability to characterize the frequency of high flow events. The duration of observation also plays an important role in the ability of Landsat to capture river flow frequency. As expected, a longer observation duration will produce a better representation of river flow frequency, albeit with diminishing returns (Figure 5). Several error metrics initially improve rapidly with time until approximately a 3 year duration, after which they improve more gradually. Thus, we suggest that at least 3 years of Landsat observations should be aggregated before they adequately contain observations representative of river flow frequency.

4.2. Implications for River Remote Sensing Applications

River flow frequency analysis is a key tool for a variety of hydrological applications including flood hazard and risk evaluations, hydraulic engineering, and water resources management [50,51]. Flow frequency is also related to a river's water quality and distribution of freshwater habitats [52–54]. While satellite remote sensing cannot measure discharge directly, it can measure other attributes of rivers that scale with discharge including river morphology and water quality [47,55]. For example, Landsat can observe surface water inundation extent from which river surface area and width can be extracted, which both scale with discharge [27,56–58].

This study's findings indicate river water occurrence data derived from long-term aggregations of Landsat observations correspond to the flow frequency of Earth's large rivers. For example, on average, median river width derived from long-term temporal composites of classified Landsat data [25,26] corresponds to median river flow. Our results also indicate that these same relationships can be extended to a wide range of flow frequencies, except for extremely high and low flows. This result has key utility for developing percentile-based width rating curves for estimating discharge [14,22] or for estimating variability in river surface area [59]. We emphasize that at any single given location, these relationships do not necessarily apply but rather that these relationships are valid when averaging over space. Thus, applying at-a-station hydraulic geometry [47] at individual single cross-sections solely from Landsat water occurrence data may often be invalid. However, our results suggest that developing at-a-station hydraulic geometry relationships across multiple cross-sections over a large area is valid for non-extreme flow frequencies and given an adequate Landsat sample size with potential implications for remote sensing of discharge approaches [15,37–39].

Our results also have implications for riparian ecology and river water quality applications. Flow is a "master" variable in river ecology and water quality [54] and, like river width and surface area, Landsat can also measure water quality parameters [31,60]. Specifically, Landsat imagery is commonly paired with in situ water samples to derive empirical relationships with optically-active constituents such as suspended sediment concentrations [61,62], chlorophyll-a [60,63], and colored dissolved organic matter (CDOM) [64,65]. While these water quality parameters generally vary with discharge, the relationship between river flow and water quality varies. Suspended sediment generally increases non-linearly with flow [66] but chlorophyll-a and CDOM can increase, decrease, or vary independently of flow depending on river size, season, and watershed properties [67–70]. Understanding flow conditions captured by satellite observations is therefore important for deriving representative surveys of river water quality measurements. The most extreme flow events are rarely observed by Landsat, but Landsat observes a wider range of flows (97% of flow percentiles at 90% of gauges) than well-designed water quality field sampling programs which sample 80% of flow percentiles at best [71]. Thus, as remote sensing of water quality methods typically relies on in situ field measurements, it is critical that field sampling programs collect measurements at high and low flows to match the wide range of hydrological conditions captured by Landsat observations.

4.3. Limitations and Future Directions

While the results of this study are encouraging for river remote sensing applications, our study does have limitations. First, we assume that daily discharge measured from the USGS study gauges represents the true flow frequency of rivers, but river gauges are biased in their placement and fluctuations in river discharge often can occur over subdaily timescale [5,72]. Related to this point, our results may differ in regions outside the US, which have dissimilar conditions. The gauge stations used in this study measure the majority (52%) of US Landsat-observable river reaches but only 6.6% of Earth's observable rivers are located within the US, according to the GRWL database [28]. Future work could explore this uncertainty by using an international gauge database [73] or a global hydrological simulation [74] to sample streamflow worldwide. Second, we emphasize that Landsat will not necessarily produce a representative sample of flow at any given single location along a river network. However, our results indicate that, given an adequate period of observation, spatially-averaged Landsat observations can capture flow frequency distribution over large spatial areas. Further, Landsat does not adequately capture minimum and maximum flow conditions in most locations. Third, while this study does not assume stationarity, we emphasize that river flow frequency is non-stationary such that flow frequencies derived over one time period cannot necessarily be used to infer flow frequencies over another period [3]. Finally, this study does not consider the uncertainty associated with the Landsat remote sensing measurements themselves. So, while we find that the timing of Landsat observations is adequate to capture river flow frequency, the ability of Landsat to accurately measure the parameter of interest (e.g., river width, suspended sediment concentration, discharge) itself remains unconstrained here. Constraining this uncertainty is application and algorithm specific [7,20] and beyond the scope of this analysis. Additionally, determining the drivers of the heterogeneity in Landsat's ability to represent flow frequency from location to location (Figure S1) is a recommended topic for further research.

Similar approaches to this study may be used to understand river sampling capabilities of satellite missions other than Landsat. Satellite programs with optical sensors and shorter revisit times, such as Sentinel 2 (2–5 days) and Planet (~1 day), likely capture river flow frequency over a shorter study duration, although this advantage of more frequent retrieval will be bottlenecked by persistent cloud cover in some regions. Other sensor technologies enable remote sensing of rivers over a broader array of atmospheric and solar illumination conditions. Indeed, thermal, passive microwave, radar and lidar remote sensing have distinct advantages over optical remote sensing and can provide alternative observations of river systems. Similar to this study, a recent analysis found that 3 years of SWOT data were sufficient to represent the flow frequency distribution over the Mississippi River basin [34]. SWOT will have a longer repeat orbit (21 days) and narrower swath width (100 km) than Landsat but these attributes are counterbalanced by the ability of its Ka-band radar instrument to collect surface returns during cloudy and nighttime conditions [11].

5. Conclusions

This study's findings show that the Landsat archive can effectively represent river flow frequency over large areas given an adequate period of observation and accurate remote sensing and discharge algorithms. At individual locations, the ability of Landsat to capture flow frequency in large rivers is positively correlated with the cloud occurrence, weakly correlated with watershed area, and does not correlate with flow flashiness (Figure 3; Figure S1). While the Landsat record captures a wide range of flow conditions (97% of the flow percentiles at 90% of sites), its ability to capture the flow frequency distribution varies widely from location to location (KSD statistic ranges from 0.016–0.36). This implies that, at any single site along a river, the Landsat archive cannot be assumed to adequately capture flow frequency. Nevertheless, we find that spatially averaging over multiple locations effectively enables representation of hydrological conditions at a given flow frequency, with the exception of hydrological extremes like maximum and minimum flow ($\alpha = 0.5$; Figure 4). Such an averaging process could hence be used to improve existing Landsat-based discharge algorithms. Landsat can better capture flow frequency at the lowest flows than at the highest flows likely because of the positive skew characteristic of river flow frequency distributions. We also find that a longer Landsat observation time period positively impacts the representation of river flow frequency, albeit with diminishing returns with increasing observation time (Figure 5). We suggest that, on average, a minimum of 3 years of aggregated observations is necessary if Landsat is used to reconstruct flow frequency distribution. We emphasize that this analysis only considers the timing of Landsat observations in relation to river flow frequency and does not consider the ability of Landsat to accurately measure the parameter of interest (e.g., river width, suspended sediment, discharge). We also note that the gauges used in this analysis only measure a relatively small portion of the global observable river network, solely located within the United States. Regardless, the results of this study support the hypothesis that long-term aggregations of Landsat data can be used to capture the flow frequency of Earth's large rivers. Thus, Landsat-based surveys of river characteristics can be interpreted to be representative of the hydrological conditions present along Earth's large rivers, with wide-ranging utility for river hydrology, water quality, and ecology.

Supplementary Materials: The following are available online at http://www.mdpi.com/2072-4292/12/9/1510/s1, Figure S1: Interactive map showing values of the KSD statistic over each gauge and the flow frequency distribution of the Landsat sample and the gauge record. The data produced in this study are openly available under a Creative Commons Attribution License at https://doi.org/10.5281/zenodo.3817346. The software used to analyze the data and produce the figures is available at https://github.com/geoallen/ROTFL/ under a Berkeley Software Distribution 3-Clause License.

Author Contributions: Conceptualization, G.H.A.; methodology, all authors; formal analysis, G.H.A., X.Y., J.G., and M.R.; data curation, J.H. and G.H.A.; writing—original draft preparation, G.H.A., X.Y., J.G., J.H., C.H.D., and M.R.; writing—review and editing, G.H.A., X.Y., J.G., J.H., C.H.D., and M.R.; visualization, G.H.A., M.R., X.Y., and J.G.; supervision, G.H.A. All authors have read and agreed to the published version of the manuscript.

Funding: This research was funded by NASA THP grant #NNH17ZDA001N. X.Y. was funded by a subcontract from the SWOT Project Office at the NASA/Caltech Jet Propulsion Laboratory to Tamlin M. Pavelsky and J.G. was funded by NSF-EAR Postdoctoral Fellowship grant #1806983. C.H.D. was supported by the Jet Propulsion Laboratory, California Institute of Technology, under a contract with NASA.

Acknowledgments: We thank Cassandra Nickles and Edward Beighley for providing helpful feedback during the initial phase of this project.

Conflicts of Interest: The authors declare no conflict of interest. The funders had no role in the design of the study; in the collection, analyses, or interpretation of data; in the writing of the manuscript, or in the decision to publish the results.

References

1. Vorosmarty, C.J.; McIntyre, P.B.; Gessner, M.O.; Dudgeon, D.; Prusevich, A.; Green, P.; Glidden, S.; Bunn, S.E.; Sullivan, C.A.; Liermann, C.R.; et al. Global threats to human water security and river biodiversity. *Nature* **2010**, *467*, 555–561. [CrossRef] [PubMed]
2. Wood, E.F.; Roundy, J.K.; Troy, T.J.; van Beek, L.P.H.; Bierkens, M.F.P.; Blyth, E.; de Roo, A.; Döll, P.; Ek, M.; Famiglietti, J.; et al. Hyperresolution global land surface modeling: Meeting a grand challenge for monitoring Earth's terrestrial water. *Water Resour. Res.* **2011**, *47*. [CrossRef]
3. Milly, P.C.D.; Betancourt, J.; Falkenmark, M.; Hirsch, R.M.; Kundzewicz, Z.W.; Lettenmaier, D.P.; Stouffer, R.J. Stationarity Is Dead: Whither Water Management? *Science* **2008**, *319*, 573. [CrossRef] [PubMed]
4. Alsdorf, D.; Bates, P.; Melack, J.; Wilson, M.; Dunne, T. Spatial and temporal complexity of the Amazon flood measured from space. *Geophys. Res. Lett.* **2007**, *34*, L08402. [CrossRef]

5. Park, C.C. World-wide variations in hydraulic geometry exponents of stream channels: An analysis and some observations. *J. Hydrol.* **1977**, *33*, 133–146. [CrossRef]

6. Gudmundsson, L.; Do, H.X.; Leonard, M.; Westra, S. The Global Streamflow Indices and Metadata Archive (GSIM)—Part 2: Quality control, time-series indices and homogeneity assessment. *Earth Syst. Sci. Data* **2018**, *10*, 787–804. [CrossRef]

7. Gleason, J.C.; Durand, T.M. Remote Sensing of River Discharge: A Review and a Framing for the Discipline. *Remote Sens.* **2020**, *12*, 1107. [CrossRef]

8. Huang, C.; Chen, Y.; Zhang, S.; Wu, J. Detecting, Extracting, and Monitoring Surface Water from Space Using Optical Sensors: A Review. *Rev. Geophys.* **2018**, *56*, 333–360. [CrossRef]

9. Twele, A.; Cao, W.; Plank, S.; Martinis, S. Sentinel-1-based flood mapping: A fully automated processing chain. *Int. J. Remote Sens.* **2016**, *37*, 2990–3004. [CrossRef]

10. Lan Anh, D.T.; Aires, F. River Discharge Estimation based on Satellite Water Extent and Topography: An Application over the Amazon. *J. Hydrometeorol.* **2019**, *20*, 1851–1866. [CrossRef]

11. Biancamaria, S.; Lettenmaier, D.P.; Pavelsky, T.M. The SWOT Mission and Its Capabilities for Land Hydrology. *Surv. Geophys.* **2016**, *37*, 307–337. [CrossRef]

12. Cooley, W.S.; Smith, C.L.; Stepan, L.; Mascaro, J. Tracking Dynamic Northern Surface Water Changes with High-Frequency Planet CubeSat Imagery. *Remote Sens.* **2017**, *96*, 1306. [CrossRef]

13. Vanderhoof, K.M.; Burt, C. Applying High-Resolution Imagery to Evaluate Restoration-Induced Changes in Stream Condition, Missouri River Headwaters Basin, Montana. *Remote Sens.* **2018**, *10*, 913. [CrossRef]

14. Smith, L.C.; Pavelsky, T.M. Estimation of river discharge, propagation speed, and hydraulic geometry from space: Lena River, Siberia. *Water Resour. Res.* **2008**, *44*, W03427. [CrossRef]

15. Tourian, M.J.; Schwatke, C.; Sneeuw, N. River discharge estimation at daily resolution from satellite altimetry over an entire river basin. *J. Hydrol.* **2017**, *546*, 230–247. [CrossRef]

16. Kääb, A.; Altena, B.; Mascaro, J. River-ice and water velocities using the Planet optical cubesat constellation. *Hydrol. Earth Syst. Sci.* **2019**, *23*, 4233–4247. [CrossRef]

17. Onderka, M.; Pekárová, P. Retrieval of suspended particulate matter concentrations in the Danube River from Landsat ETM data. *Sci. Total Environ.* **2008**, *397*, 238–243. [CrossRef]

18. Smith, L.C. Satellite remote sensing of river inundation area, stage, and discharge: A review. *Hydrol. Process.* **1997**, *11*, 1427–1439. [CrossRef]

19. Topp, N.S.; Pavelsky, M.T.; Jensen, D.; Simard, M.; Ross, R.V.M. Research Trends in the Use of Remote Sensing for Inland Water Quality Science: Moving Towards Multidisciplinary Applications. *Water* **2020**, *12*, 169. [CrossRef]

20. Piégay, H.; Arnaud, F.; Belletti, B.; Bertrand, M.; Bizzi, S.; Carbonneau, P.; Dufour, S.; Liébault, F.; Ruiz-Villanueva, V.; Slater, L. Remotely sensed rivers in the Anthropocene: State of the art and prospects. *Earth Surf. Process. Landf.* **2020**, *45*, 157–188. [CrossRef]

21. Tourian, M.J.; Sneeuw, N.; Bárdossy, A. A quantile function approach to discharge estimation from satellite altimetry (ENVISAT). *Water Resour. Res.* **2013**, *49*, 4174–4186. [CrossRef]

22. Pavelsky, T.M. Using width-based rating curves from spatially discontinuous satellite imagery to monitor river discharge. *Hydrol. Process.* **2014**. [CrossRef]

23. Pavelsky, T.M.; Durand, M.T.; Andreadis, K.M.; Beighley, R.E.; Paiva, R.C.D.; Allen, G.H.; Miller, Z.F. Assessing the potential global extent of SWOT river discharge observations. *J. Hydrol.* **2014**, *519 Pt B*, 1516–1525. [CrossRef]

24. Gorelick, N.; Hancher, M.; Dixon, M.; Ilyushchenko, S.; Thau, D.; Moore, R. Google Earth Engine: Planetary-scale geospatial analysis for everyone. *Big Remote Sensed Data Tools Appl. Exp.* **2017**, *202*, 18–27. [CrossRef]

25. Pekel, J.-F.; Cottam, A.; Gorelick, N.; Belward, A.S. High-resolution mapping of global surface water and its long-term changes. *Nature* **2016**, *540*, 418–422. [CrossRef]

26. Pickens, A.H.; Hansen, M.C.; Hancher, M.; Stehman, S.V.; Tyukavina, A.; Potapov, P.; Marroquin, B.; Sherani, Z. Mapping and sampling to characterize global inland water dynamics from 1999 to 2018 with full Landsat time-series. *Remote Sens. Environ.* **2020**, *243*, 111792. [CrossRef]

27. Yang, X.; Pavelsky, T.M.; Allen, G.H.; Donchyts, G. RivWidthCloud: An Automated Google Earth Engine Algorithm for River Width Extraction from Remotely Sensed Imagery. *IEEE Geosci. Remote Sens. Lett.* **2019**, 1–5. [CrossRef]

28. Allen, G.H.; Pavelsky, T.M. Global extent of rivers and streams. *Science* **2018**. [CrossRef]

29. Lettenmaier, D.P.; Alsdorf, D.; Dozier, J.; Huffman, G.J.; Pan, M.; Wood, E.F. Inroads of remote sensing into hydrologic science during the WRR era. *Water Resour. Res.* **2015**, *51*, 7309–7342. [CrossRef]

30. Jones, W.J. Improved Automated Detection of Subpixel-Scale Inundation—Revised Dynamic Surface Water Extent (DSWE) Partial Surface Water Tests. *Remote Sens.* **2019**, *11*, 374. [CrossRef]

31. Ross, M.R.V.; Topp, S.N.; Appling, A.P.; Yang, X.; Kuhn, C.; Butman, D.; Simard, M.; Pavelsky, T.M. AquaSat: A Data Set to Enable Remote Sensing of Water Quality for Inland Waters. *Water Resour. Res.* **2019**, *55*. [CrossRef]

32. Yamazaki, D.; Ikeshima, D.; Sosa, J.; Bates, P.D.; Allen, G.H.; Pavelsky, T.M. MERIT Hydro: A High-Resolution Global Hydrography Map Based on Latest Topography Dataset. *Water Resour. Res.* **2019**, *55*, 5053–5073. [CrossRef]

33. Raymond, P.A.; Hartmann, J.; Lauerwald, R.; Sobek, S.; McDonald, C.; Hoover, M.; Butman, D.; Striegl, R.; Mayorga, E.; Humborg, C.; et al. Global carbon dioxide emissions from inland waters. *Nature* **2013**, *503*, 355–359. [CrossRef] [PubMed]

34. Nickles, C.; Beighley, E.; Zhao, Y.; Durand, M.; David, C.; Lee, H. How Does the Unique Space-Time Sampling of the SWOT Mission Influence River Discharge Series Characteristics? *Geophys. Res. Lett.* **2019**, *46*, 8154–8161. [CrossRef]

35. Allen, G.H.; Pavelsky, T.M. Patterns of river width and surface area revealed by the satellite-derived North American River Width data set. *Geophys. Res. Lett.* **2015**, *42*, 395–402. [CrossRef]

36. Yang, X.; Pavelsky, T.M.; Allen, G.H. The past and future of global river ice. *Nature* **2020**, *577*, 69–73. [CrossRef]

37. Gleason, C.J.; Smith, L.C. Toward global mapping of river discharge using satellite images and at-many-stations hydraulic geometry. *Proc. Natl. Acad. Sci. USA* **2014**, *111*, 4788–4791. [CrossRef]

38. Feng, D.; Gleason, C.J.; Yang, X.; Pavelsky, T.M. Comparing Discharge Estimates Made via the BAM Algorithm in High-Order Arctic Rivers Derived Solely from Optical CubeSat, Landsat, and Sentinel-2 Data. *Water Resour. Res.* **2019**, *55*, 7753–7771. [CrossRef]

39. Durand, M.; Gleason, C.J.; Garambois, P.A.; Bjerklie, D.; Smith, L.C.; Roux, H.; Rodriguez, E.; Bates, P.D.; Pavelsky, T.M.; Monnier, J.; et al. An intercomparison of remote sensing river discharge estimation algorithms from measurements of river height, width, and slope. *Water Resour. Res.* **2016**, *52*, 4527–4549. [CrossRef]

40. Foga, S.; Scaramuzza, P.L.; Guo, S.; Zhu, Z.; Dilley, R.D.; Beckmann, T.; Schmidt, G.L.; Dwyer, J.L.; Joseph Hughes, M.; Laue, B. Cloud detection algorithm comparison and validation for operational Landsat data products. *Remote Sens. Environ.* **2017**, *194*, 379–390. [CrossRef]

41. Maxwell, S.K.; Schmidt, G.L.; Storey, J.C. A multi-scale segmentation approach to filling gaps in Landsat ETM+ SLC-off images. *Int. J. Remote Sens.* **2007**, *28*, 5339–5356. [CrossRef]

42. Massey, F.J. The Kolmogorov-Smirnov Test for Goodness of Fit. *J. Am. Stat. Assoc.* **1951**, *46*, 68–78. [CrossRef]

43. Fasano, G.; Franceschini, A. A multidimensional version of the Kolmogorov–Smirnov test. *Mon. Not. R. Astron. Soc.* **1987**, *225*, 155–170. [CrossRef]

44. Crutcher, H.L. A Note on the Possible Misuse of the Kolmogorov-Smirnov Test. *J. Appl. Meteorol.* **1975**, *14*, 1600–1603. [CrossRef]

45. Baker, D.B.; Richards, R.P.; Loftus, T.T.; Kramer, J.W. A new flashiness index: Characteristics and applications to midwestern rivers and streams. *JAWRA J. Am. Water Resour. Assoc.* **2004**, *40*, 503–522. [CrossRef]

46. Spearman, C. The Proof and Measurement of Association between Two Things. *Am. J. Psychol.* **1904**, *15*, 72–101. [CrossRef]

47. Leopold, L.B.; Maddock, T., Jr. *The Hydraulic Geometry of Stream Channels and Physiographic Implications*; US Government Printing Office: Washington, DC, USA, 1953; p. 57.

48. Mann, H.B.; Whitney, D.R. On a Test of Whether one of Two Random Variables is Stochastically Larger than the Other. *Ann. Math. Stat.* **1947**, *18*, 50–60. [CrossRef]

49. Sen, P.K. Estimates of the Regression Coefficient Based on Kendall's Tau. *J. Am. Stat. Assoc.* **1968**, *63*, 1379–1389. [CrossRef]

50. Chow, V.T.; Maidment, D.R.; Mays, L.W. *Applied Hydrology*; McGraw-Hill Series in Water Resources and Environmental Engineering; McGraw-Hill: New York, NY, USA, 1988; ISBN 0-07-010810-2.

51. Rao, A.R.; Hamed, K.H. *Flood Frequency Analysis*, 1st ed.; CRC Press: Boca Raton, FL, USA, 2000; ISBN 978-0-8493-0083-7.

52. Bunn, S.E.; Arthington, A.H. Basic Principles and Ecological Consequences of Altered Flow Regimes for Aquatic Biodiversity. *Environ. Manag.* **2002**, *30*, 492–507. [CrossRef]

53. Nilsson, C.; Renöfält, B.M. Linking Flow Regime and Water Quality in Rivers. *Ecol. Soc.* **2008**, *13*, 18. [CrossRef]

54. Poff, N.L.; Allan, J.D.; Bain, M.B.; Karr, J.R.; Prestegaard, K.L.; Richter, B.D.; Sparks, R.E.; Stromberg, J.C. The natural flow regime. *BioScience* **1997**, *47*, 769–784. [CrossRef]

55. Rijn, L.C. van Sediment Transport, Part II: Suspended Load Transport. *J. Hydraul. Eng.* **1984**, *110*, 1613–1641. [CrossRef]

56. Hou, J.; van Dijk, A.I.J.M.; Renzullo, L.J.; Vertessy, R.A.; Mueller, N. Hydromorphological attributes for all Australian river reaches derived from Landsat dynamic inundation remote sensing. *Earth Syst. Sci. Data* **2019**, *11*, 1003–1015. [CrossRef]

57. Isikdogan, F.; Bovik, A.; Passalacqua, P. RivaMap: An automated river analysis and mapping engine. *Big Remote Sensed Data Tools Appl. Exp.* **2017**, *202*, 88–97. [CrossRef]

58. Pavelsky, T.M.; Smith, L.C. RivWidth: A Software Tool for the Calculation of River Widths from Remotely Sensed Imagery. *Geosci. Remote Sens. Lett. IEEE* **2008**, *5*, 70–73. [CrossRef]

59. Allen, G.H.; Yang, X.; Lin, P.; Pan, M.; Holliman, J.; Yamazaki, D.; Liu, S.; Raymond, P.A. Seasonal variations in global river and stream inundation extent. In Proceedings of the Multisource Remote Sensing of Rivers, Lakes, Reservoirs, and Wetlands, San Francisco, CA, USA, 11 December 2019.

60. Ritchie, J.C.; Zimba, P.V.; Everitt, J.H. Remote sensing techniques to assess water quality. *Photogramm. Eng. Remote Sens.* **2003**, *69*, 695–704. [CrossRef]

61. Mertes, L.A.; Smith, M.O.; Adams, J.B. Estimating suspended sediment concentrations in surface waters of the Amazon River wetlands from Landsat images. *Remote Sens. Environ.* **1993**, *43*, 281–301. [CrossRef]

62. Ritchie, J.C.; Cooper, C.M.; Yongqing, J. Using Landsat multispectral scanner data to estimate suspended sediments in Moon Lake, Mississippi. *Remote Sens. Environ.* **1987**, *23*, 65–81. [CrossRef]

63. Kuhn, C.; de Matos Valerio, A.; Ward, N.; Loken, L.; Sawakuchi, H.O.; Kampel, M.; Richey, J.; Stadler, P.; Crawford, J.; Striegl, R. Performance of Landsat-8 and Sentinel-2 surface reflectance products for river remote sensing retrievals of chlorophyll-a and turbidity. *Remote Sens. Environ.* **2019**, *224*, 104–118. [CrossRef]

64. Brezonik, P.L.; Olmanson, L.G.; Finlay, J.C.; Bauer, M.E. Factors affecting the measurement of CDOM by remote sensing of optically complex inland waters. *Remote Sens. Environ.* **2015**, *157*, 199–215. [CrossRef]

65. Griffin, C.G.; McClelland, J.W.; Frey, K.E.; Fiske, G.; Holmes, R.M. Quantifying CDOM and DOC in major Arctic rivers during ice-free conditions using Landsat TM and ETM+ data. *Remote Sens. Environ.* **2018**, *209*, 395–409. [CrossRef]

66. Asselman, N.E.M. Fitting and interpretation of sediment rating curves. *J. Hydrol.* **2000**, *234*, 228–248. [CrossRef]

67. Creed, I.F.; McKnight, D.M.; Pellerin, B.A.; Green, M.B.; Bergamaschi, B.A.; Aiken, G.R.; Burns, D.A.; Findlay, S.E.; Shanley, J.B.; Striegl, R.G. The river as a chemostat: Fresh perspectives on dissolved organic matter flowing down the river continuum. *Can. J. Fish. Aquat. Sci.* **2015**, *72*, 1272–1285. [CrossRef]

68. Dolph, C.L.; Hansen, A.T.; Finlay, J.C. Flow-related dynamics in suspended algal biomass and its contribution to suspended particulate matter in an agricultural river network of the Minnesota River Basin, USA. *Hydrobiologia* **2017**, *785*, 127–147. [CrossRef]

69. Lucas, L.V.; Thompson, J.K.; Brown, L.R. Why are diverse relationships observed between phytoplankton biomass and transport time? *Limnol. Oceanogr.* **2009**, *54*, 381–390. [CrossRef]

70. Moatar, F.; Abbott, B.W.; Minaudo, C.; Curie, F.; Pinay, G. Elemental properties, hydrology, and biology interact to shape concentration-discharge curves for carbon, nutrients, sediment, and major ions. *Water Resour. Res.* **2017**, *53*, 1270–1287. [CrossRef]

71. Hooper, R.P.; Aulenbach, B.T.; Kelly, V.J. The National Stream Quality Accounting Network: A flux-based approach to monitoring the water quality of large rivers. *Hydrol. Process.* **2001**, *15*, 1089–1106. [CrossRef]

72. Allen, G.H.; David, C.H.; Andreadis, K.M.; Hossain, F.; Famiglietti, J.S. Global Estimates of River Flow Wave Travel Times and Implications for Low-Latency Satellite Data. *Geophys. Res. Lett.* **2018**, *45*, 7551–7560. [CrossRef]

73. Do, H.X.; Gudmundsson, L.; Leonard, M.; Westra, S. The Global Streamflow Indices and Metadata Archive (GSIM)-Part 1: The production of a daily streamflow archive and metadata. *Earth Syst. Sci. Data* **2018**, *10*, 765–785. [CrossRef]

74. Lin, P.; Pan, M.; Beck, H.E.; Yang, Y.; Yamazaki, D.; Frasson, R.; David, C.H.; Durand, M.; Pavelsky, T.M.; Allen, G.H.; et al. Global Reconstruction of Naturalized River Flows at 2.94 Million Reaches. *Water Resour. Res.* **2019**. [CrossRef]

Article

Using Small Unmanned Aircraft Systems for Measuring Post-Flood High-Water Marks and Streambed Elevations

Brandon T. Forbes [1,*], Geoffrey P. DeBenedetto [2], Jesse E. Dickinson [1], Claire E. Bunch [1] and Faith A. Fitzpatrick [3]

[1]   U.S. Geological Survey Tucson, Tucson, AZ 85719, USA; jdickins@usgs.gov (J.E.D.); cebunch@usgs.gov (C.E.B.)
[2]   U.S. Geological Survey Flagstaff, Flagstaff, AZ 86001, USA; gdebened@usgs.gov
[3]   U.S. Geological Survey Middleton, Middleton, WI 53562, USA; fafitzpa@usgs.gov
*   Correspondence: bforbes@usgs.gov

Received: 14 February 2020; Accepted: 29 April 2020; Published: 1 May 2020

Abstract: Floods affected approximately two billion people around the world from 1998–2017, causing over 142,000 fatalities and over 656 billion U.S. dollars in economic losses. Flood data, such as the extent of inundation and peak flood stage, are needed to define the environmental, economic, and social impacts of significant flood events. Ground-based global positioning system (GPS) surveys of post-flood high-water marks (HWMs) and topography are commonly used to define flood inundation and stage, but can be time-consuming, difficult, and expensive to conduct. Here, we demonstrate and test the use of small unmanned aircraft systems (sUAS) and close-range remote sensing techniques to collect high-accuracy flood data to define peak flood stage elevations and river cross-sections. We evaluate the elevation accuracy of the HWMs from sUAS surveys by comparison with traditional GPS surveys, which have acceptable accuracy for many post-flood assessments, at two flood sites on two small streams in the U.S. Mean elevation errors for the sUAS surveys were 0.07 m and 0.14 m for the semiarid and temperate sites, respectively; those values are similar to typical errors when measuring HWM elevations with GPS surveys. Results demonstrate that sUAS surveys of HWMs and cross-sections can be an accurate and efficient alternative to GPS surveys; we provide insights that can be used to decide whether sUAS or GPS techniques will be most efficient for post-flood surveying.

Keywords: flooding; high-water marks (HWMs); small unmanned aircraft systems (sUAS); drone; photogrammetry; hydraulic modeling; aerial photography; surveying; hydrology; inundation

---

1. Introduction

Floods are largely uncontrollable phenomena that threaten infrastructure, property, and human life around the globe. Floods affected approximately two billion people from 1998–2017, causing over 142,000 fatalities and over 656 billion U.S. dollars in economic losses [1]. In Houston, Texas, for example, widespread inundation over large urban areas caused by Hurricane Harvey in 2017 resulted in damage totaling 130 billion U.S. dollars [2]. As large floods have become more frequent and damaging over the past century in the U.S. and around the globe [3,4], flood-magnitude data are critical for future flood prediction, understanding the risks to human health and safety, design of flood protection and critical infrastructure, and planning efforts to reduce the magnitude of losses due to flooding.

Flood magnitudes can be characterized by the peak flood discharge, peak stage (maximum water elevation during the flood), and by the extent of inundation (the margin of the inundated area). Widespread flood extents are commonly assessed from high-water marks (HWMs) that indicate the highest elevation reached by the water surface during a flood and measurements of the terrain surface that had been inundated [5]. HWMs are commonly the primary evidence of peak-flood stage and are left by many different types of floods in river channels, floodplains, and coastlines. Common types of HWMs left by floods are debris lines, wash lines, cut lines, and mud lines (Figure 1) [6]. Once HWMs have been measured, their elevations and locations are combined with terrain data, commonly cross-sectional surveys or digital terrain models (DTMs), and provide measures of the depth, extent, and cross-sectional area of floods.

Figure 1. Examples of different types of high-water marks (HWMs) that have the potential to be identified within high-resolution orthophotographs collected using unmanned aircraft systems (sUAS). Panel (**A**) shows a wash line formed when material is washed down the bank and/or laid over, panel (**B**) shows a mud line deposited by the flood, panel (**C**) shows a debris line deposited by the flood, panel (**D**) shows a cut line caused by erosion. [6].

Elevations and locations of HWMs can be identified and measured by various approaches, the most common being field reconnaissance and ground-based surveying. Typically, ground-based measurements are collected by survey-grade global positioning system (GPS) equipment soon after the floods recede. The inundation extent is often inferred by assuming that the HWMs represent the peak flood stage over the affected area [6]. This is done by overlaying the HWMs onto DTMs using geographic information system (GIS) software to produce maps of flood extent. However, point measurements of HWMs do not always provide sufficient resolution to accurately define a continuous inundation surface throughout the flooded area. Similarly, the horizontal and vertical resolutions of existing DTMs are generally too coarse, commonly greater than 1 m, and may not appropriately represent terrain features.

Alternatively, surveys using small unmanned aircraft systems (sUAS) conducted after floods have occurred can quickly collect data across relatively large affected areas (hundreds of square meters to multiple square kilometers) and provide orthophotographs (mosaicked images that have a spatial reference) with sufficient resolution to allow visual identification of HWMs [7]. Imagery data collected in a specific manner employing photogrammetric techniques using sUAS can be post-processed to generate accurate DTMs which can be used to map the location and elevation of the visually identified HWMs and the terrain features that were inundated. Post-flood HWMs and terrain data can be collected quickly and efficiently soon after a flood and before the HWMs naturally deteriorate. The sUAS provides the mobility to rapidly cover areas from more than one operating location in a short time, which can dramatically increase HWM data collection efficiency, the number of HWMs surveyed, and the safety of field personnel [8,9]. A further benefit of the sUAS surveys, relative to ground-based survey methods, is that high-spatial density data are generated to represent flood peak stages within sUAS data, which can then be used for flood mapping and calibration of hydraulic and inundation models.

Here, we demonstrate and test the use of sUAS to characterize flood events on two small stream channels in the United States. The primary objectives are to demonstrate the sUAS survey techniques for locating and obtaining HWM elevations and inundated area to evaluate the accuracy of (1) DTM elevations derived from photogrammetric techniques and (2) HWM locations and elevations that were visually identified from orthophotographs and mapped using the DTM generated from the sUAS survey. The sUAS HWM elevations are evaluated by comparing them to ground-truth elevation data collected using GPS methods at the HWM locations and inundated terrain within the same stream channel. The orthophotographs generated from the sUAS are produced at fine enough resolution that individual HWMs can be identified at a specific pixel in the photograph, which then represents the location of the HWM. Elevations and coordinates of HWMs at those pixels are compared against GPS data to determine the accuracy and repeatability of this technique.

First, we describe the two field sites, Underwood Creek in Wisconsin, USA, and Brawley Wash in Arizona, USA. Then the GPS survey and sUAS survey methods, and the methods and results of the comparison of the HWM elevations from the GPS and sUAS methods are described. Finally, we describe field conditions that could limit the success of these techniques and discuss best practices and lessons learned for future post-flood sUAS HWM data collection. Our focus on examining the accuracy of using sUAS for post-flood data collection builds upon the work by Diakakis et al. [10]. Our assessment of increasing data collection efficiency in a post-flood environment complements the work of Smith et al. [11].

2. Materials and Methods

2.1. Site Descriptions

In this paper, we discuss the analysis of two small river reaches, Underwood Creek in Wisconsin, and Brawley Wash in Arizona, USA, to demonstrate and test the application of sUAS HWM survey techniques. Both sites experienced flooding in 2018 and have the requisite datasets: HWMs surveyed by GPS and high-resolution aerial photographs collected by sUAS using photogrammetric techniques in which HWMs were visible. Although these rivers and flood events are relatively small, the techniques demonstrated here can be scaled up and deployed to quantify the extent of HWMs typically found in much larger magnitude floods.

Our first study area was Underwood Creek, U.S. Geological Survey (USGS) gage 04087088, near the town of Wauwatosa in southeastern Wisconsin. This perennial stream drains a suburban area of 46.9 km^2 and flows are influenced by local urban runoff from impervious surfaces [12]. The channel gradient is low (approximately 0.0025), and the main channel is incised and meanders through the floodplain. Tall grasses, cattails, and shrubs are prevalent in the channel and floodplain as seen in Figure 2. The flood being analyzed occurred on August 20 and 21, 2018 (Figure 3), inundating a heavily vegetated and terraced floodplain. Sixteen HWMs were surveyed using GPS surveying equipment on September 14, 2018 [13], explained in detail below, all of which were used for comparison with the sUAS survey. The sUAS survey was completed on August 23, 2018 [14], and cross-sections in support of hydraulic modeling were surveyed on July 17, 2019 [15] (Figure 4). Channel change is not suspected in the time between the surveys due to the well-established vegetation at the site persisting throughout the year. The peak discharge for the flood event on August 20, 2018, was 29.2 m^3/s, derived from the stage-discharge rating [12].

Figure 2. Photograph of Underwood Creek collected by the sUAS on August 23, 2018 [14]. Notice the meandering channel, tall grass, cattail, and shrubs on the floodplain. The direction of flow is from the bottom of the image to the top.

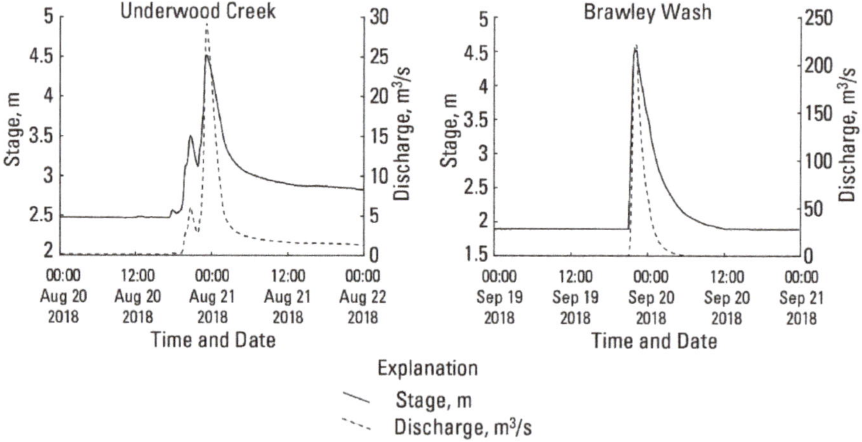

Figure 3. Stage and discharge hydrograph of the flood event at Underwood Creek on August 20 and 21, 2018 [12] (**left**), and Brawley Wash, Arizona, on September 19, 2018 [16] (**right**).

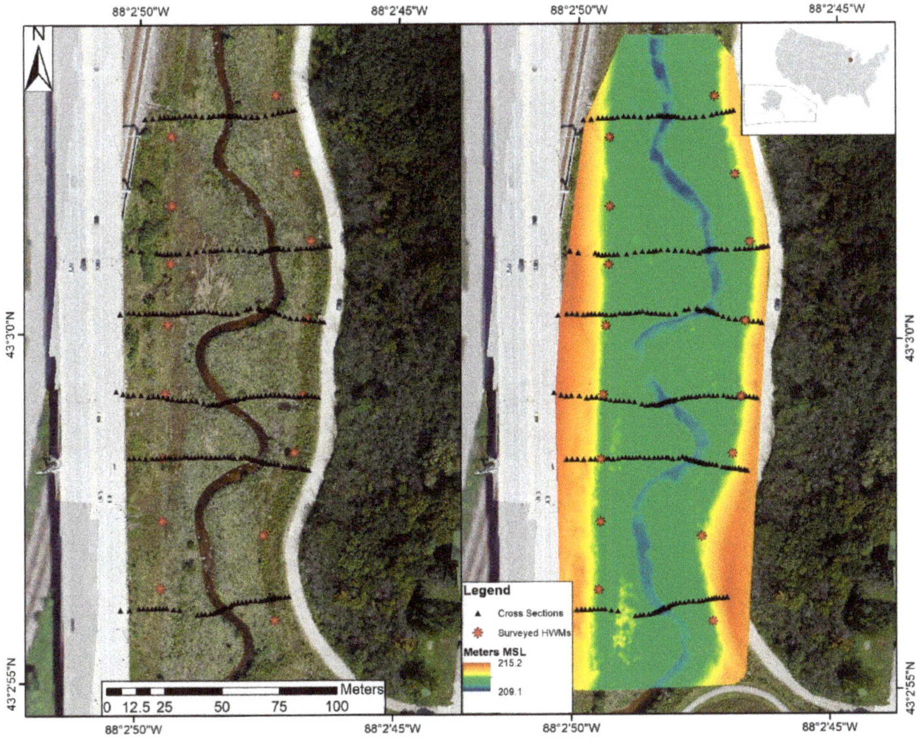

Figure 4. HWMs [13] and cross-sectional terrain data [15] collected by GPS (**left**) and the digital terrain model (DTM) generated from sUAS data [14] (**right**) at Underwood Creek. Flow direction is from bottom to top of the graphic (south to north).

Our second study area was Brawley Wash, USGS gage 09487000, near the town of Three Points, Arizona. The site has a drainage area of 2010 km^2 [16] and is a sand-bedded ephemeral wash that flows only in response to heavy rain events that cause flash floods (Figure 5). The channel gradient is low, about 0.005, and the gage site often requires indirect measurements to quantify peak flows because of the flashy flows commonly observed in the region. Dense shrub and tree vegetation are present on the right lower and mid-overbank areas of this straight continuous channel, with intermittent shrub vegetation on the left lower overbank. The flow event studied was contained in the primary floodplain of the main channel and occurred on September 19, 2018 (Figure 3). The peak discharge of 220 m^3/s [16] was calculated using traditional indirect discharge measurement techniques, which has a 5 to 10-year peak flood recurrence interval [17] (Figure 3). 78 HWMs were surveyed on the left bank of the stream in adittion to seven cross-sections using GPS equipment on October 1, 2018 [18], all of which were used for comparison with the sUAS survey (Figure 6). High-water marks were not found on the right bank during the field campaign. The reach was surveyed with sUAS on January 31, 2019 [19].

Figure 5. Photograph of Brawley Wash, a sandy ephemeral channel with dense shrub vegetation, primarily mesquite, palo verde, and desert broom on the banks.

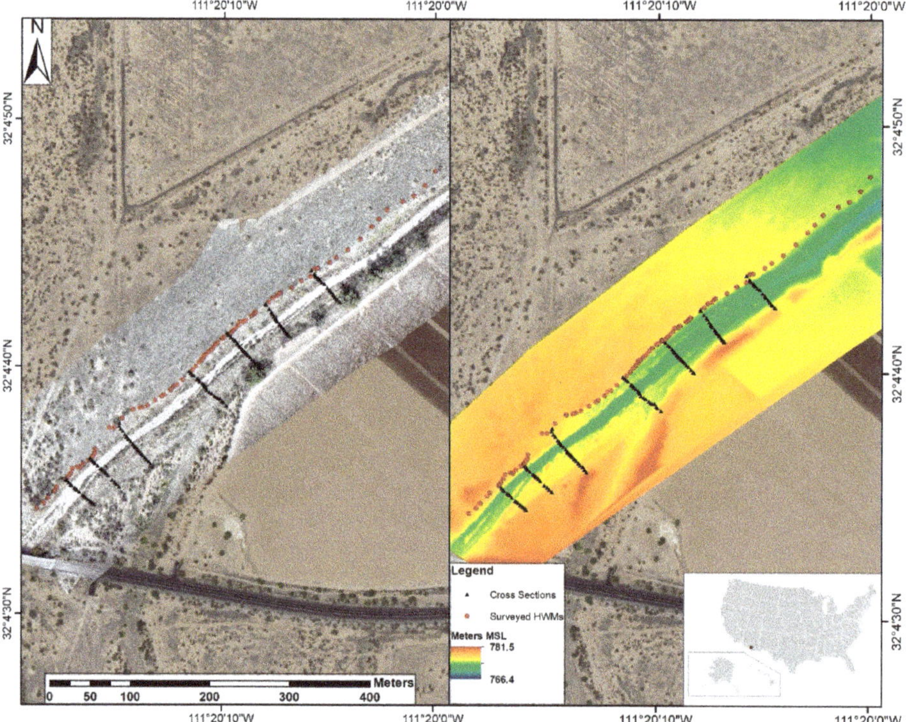

Figure 6. HWMs and cross-sectional terrain data collected by GPS [18] and the orthophotograph collected by sUAS [19] (**left**) and the DTM generated from sUAS data [19] (**right**) at Brawley Wash. Flow direction is from the bottom left to the top right of the graphic (southwest to northeast).

2.2. Ground-Based GPS Data Collection

The GPS surveys at both sites were conducted following USGS data collection standards [20]. The survey at Underwood Creek utilized a cellular-enabled Real-Time Network where the Leica GS14 GPS unit accessed multiple base stations in the region to find the absolute position of points during the survey. The GPS survey conducted at Brawley Wash was done using a static solution with the base station setup over a single point during the entire data collection. The data collected by the base station was sent to the Online Positioning User Service (OPUS) to tie in the position of the survey to NAD83 and NAVD88 datums [21]. The GPS-surveyed points have two different types of uncertainty associated with them, absolute position and model composition. Absolute position is associated with the accuracy of the base location, cell towers or GPS base station with OPUS correction, and generally are 0.02 m horizontal and 0.04 m vertical. This accuracy addresses how a model will plot on a map or in GIS and does not affect the relative positions between points. The accuracy of surveyed points in relation to each other, composition, is more important for generating a comparative model and establishing representative geometries and is what is important in this research. The uncertainty of composition for these models is generally 0.015 m both horizontally and vertically [20].

Uncertainty about the flood stage and extent is influenced by the equipment and survey techniques, as well as the conditions that form and represent the HWMs in the field. In addition to the minimal uncertainty caused by the GPS survey equipment, HWMs do not always represent a perfect measure of the water surface at its peak and HWMs rarely exhibit a smooth surface in situ. HWMs are formed by the interaction and deposition of material at the interface between the flowing water surface and the river's bank, bridges or other structures, or stable vegetation, such as large trees. At this interface, the elevation of the HWMs can be affected by turbulence, surging flow, local influence from vegetation and channel-bed material, drawdown effects, pileup, among other factors which all commonly cause noise in the water-surface elevation. The HWMs can also degrade soon after the water recedes due to wind, rain, gravity, and cleanup efforts. Depending on where and when the HWM is measured, the survey data can show variability in the elevation of the measured surface in the reach. Under standard USGS HWM procedures, HWMs are assigned an uncertainty value to represent the vertical range of peak water surface that could be described by the HWM in the field [6]. The HWMs at both study locations are considered to be of excellent or good quality with a vertical uncertainty of 0.015 to 0.03 m [6].

Interpreting a continuous representation of the peak water surface elevation using the GPS HWM data is performed throughout the reaches at both study locations to properly compare data collected from the sUAS. This is done by using all the GPS point data collected in the field, coupled with survey uncertainty and HWM quality, to define a smoothed water surface elevation across the entire length of the reaches. In the two test cases here, the HWMs were of good to excellent quality, so all HWMs were weighted equally when establishing the spatial trend of the water surface. For each test site, a Locally Weighted Scatterplot Smoothing (LOWESS) curve [22] was fit to the GPS surveyed HWM data as a function of downstream distance to act as the measured water surface trend along each reach. Establishing a trend in the water surface that best fits HWM point data is common practice when conducting this type of data collection, where outliers are commonly ignored [5,6]. The water-surface trend is used in many different modeling approaches, many of which are sensitive to the overall slope of the water surface and are critical to categorize correctly [23]. HWM elevation data collected from the sUAS were compared to the LOWESS curves drawn at each study site.

2.3. sUAS Photogrammetric Data Collection and Processing

Overlapping color images of sharp focus and appropriate exposure were collected using the sUAS at both of the study locations. At Underwood Creek, data were collected with a DJI Inspire quadcopter and a Sony NEX5 camera. At Brawley Wash, data collection was done using a 3DR Solo micro-quadcopter and a fixed custom-mounted Ricoh GRII camera. Basic flight planning and command-and-control software were used to plan and execute area coverage, manage flight lines, and acquire photos with 60% photographic overlap, ensuring good stereo-photograph geometry for photogrammetric processing. Underwood Creek was flown at approximately 75 m above ground level under partially cloudy skies; the 0.163 km^2 area was captured with 289 images. Brawley Wash was flown at approximately 90 m above ground level on a clear sunny day; the 0.278 km^2 area was captured with 317 images.

Photographs at both sites that were collected by sUAS were processed in Agisoft Metashape [24] to develop a three-dimensional point cloud model, a DTM, and an orthorectified mosaicked image. Processing techniques and objectives may be referenced from the USGS National UAS Projects Office publication, Unmanned Aircraft Systems Data Post-Processing, Structure-From-Motion

Photogrammetry [25]. Gradual selection techniques to statistically reduce model errors were used to establish tie-points that had a base-to-height ratio of 1:2.3 (reconstruction uncertainty = 10), a standard deviation of between 3-4 (projection accuracy 3-4), and a compositional pixel accuracy of better than 0.2 pixels (reprojection error 0.2–0.14). When the models are developed, ground control points are heavily weighted the ground targets that were surveyed by the GPS equipment with 0.02 m positional accuracy). Camera positions measured with GPS sensors onboard the sUAS were processed with looser tolerance (~10–30 m) as forward motion, capture and write times, and collection in ellipsoid height all negatively affect image position data compared to the GPS data. Once acceptable common imagery tie-points were established, data were processed to generate products including a three-dimensional point cloud model, a DTM, and an orthorectified mosaicked image.

The next step in the photogrammetric data processing procedure was the development of bare-earth DTMs. Raw, point-cloud data generated by prior steps include vegetation due to the capture of photographic pixels on vegetated surfaces visible from the sUAS's bird's-eye view. By contrast, GPS surveys of HWMs measure the position of the HWM on the bank's bare-earth surface, beneath any vegetation. In order to obtain comparable data sets, near bare-earth DTMs needed to be processed and produced for both study sites. Classification tools within Agisoft Metashape [24] and Global Mapper [26] were used to build the bare-earth DTMs. Classification steps included removing outlier data, noise filtering, color identification, and geometric classification tools to remove or classify points that represent features that are not ground. Challenges in vegetation classification include plant material that has a similar color to ground surfaces (dormant grasses), vegetation that obscures all view of the ground surface when viewed from above, and oversimplification of vertical surfaces below dense vegetation in the channels in areas where banks are steep and undercut. Ultimately, a combination of automated and hand-editing techniques was used to produce the bare-earth DTMs used in the subsequent analyses.

2.4. Evaluating sUAS Surveys for Collection of Post-Flood Data

The data collected by sUAS were first examined to determine if HWMs were clearly visible in the orthophotographs, as some HWMs can be faint, small, and difficult to distinguish. To accurately map HWM elevations, HWMs must be clearly identifiable within the orthophotograph, making crisp and clear imagery imperative. The orthophotographs were examined to determine if HWMs were visible in both stream reaches when zooming in to the photographs using GIS software.

2.4.1. Comparing GPS Surveyed HWM Elevations to DTM Elevations

The accuracy of the sUAS surveys for obtaining flood peak-stage and channel elevation was evaluated first by comparing sUAS elevations with those measured by GPS surveys. The elevations from the GPS surveys were treated as the control or "true" values for the comparisons. The first comparison evaluated the DTM elevations generated by the sUAS survey and photogrammetric data at the exact locations where the GPS points were collected. Using Global Mapper GIS software [26], the orthophotograph was draped over the DTM, and the GPS points were rendered onto the orthophotograph and DTM. The positions of the GPS surveyed HWMs and channel topography were duplicated in the GIS software by creating point features. The point locations were then assigned the underlying sUAS generated DTM elevation values and corresponding elevations were compared.

2.4.2. Comparing GPS Surveyed Cross Sections to DTM Elevations at Surveyed Points

The next comparison evaluated the elevations of cross-sections surveyed in each channel. This comparison took the same approach when evaluating the HWMs compared to the DTM elevations described in the previous section but was done using the surveyed cross-section points. This analysis is particularly important because an accurate cross-sectional area is important in hydraulic analysis, such as multiplying the area by the mean velocity in meters per second to obtain a discharge in cubic meters per second. Additionally, when cross-sections are surveyed in situ, field staff make informed decisions about what points to survey at critical inflections in the channel that best characterize the overall shape of the channel, ignoring vegetation and objects that are perceived to be mobile, like floating vegetation debris or trash litter accumulated by flood inflow. When comparing cross-sections in the GIS environment, points were picked directly under the surveyed points just like the prior HWM analysis, benefitting from the field staff's decision making. GIS also allows an analyst to pull points under a line by a set number or regular interval, or inflection points, but these methods may over-generalize the channel, dramatically changing the true cross-sectional area. These techniques do not account for the influence of objects that should not be included such as vegetation artifacts or debris mats or litter.

2.4.3. "Blind" Hydrologist Selected HWMs from sUAS-based GIS Products

The next comparison assessed error associated with visual identification of HWMs using the orthophotograph and the measurement of HWM elevations obtained from the DTMs, all of which were done using the remotely-sensed data products through GIS. For this comparison, GPS survey elevations were not available at every location along the reach to serve as "true" elevation comparisons. Instead, the HWM elevations were compared to an estimated flood-peak surface profile using a LOWESS fit [22] of the GPS elevations in the downstream direction. In this evaluation, three separate hydrologists identified HWMs from the sUAS data along the river reach to test repeatability and the variability of different interpretations of HWM locations from orthophotographs. The three trials were done without access to the control data set, so all HWMs and elevations were identified directly from the sUAS data. The selected HWMs were plotted in a downstream distance along the river's thalweg, and the results of the three trials were compared to the LOWESS curves generated from GPS survey point data that describe average water surface slopes at each study location.

2.4.4. LOWESS Curve Fit to Hydrologist Selected HWMs compared to GPS LOWESS Curve Fit

The final comparison assessed the difference between a LOWESS curve of the visually selected HWM elevations and the LOWESS curve of the GPS-measured, or true, HWM elevations. Hydraulic modeling often requires a best-fit line to represent the slope of the water's surface during a flood event commonly profiled through the collection of HWM point data. HWM data are typically noisy and contain outliers, as the water surface is rarely a smooth linear plane at the peak of floods, requiring a smoothed curve to be developed to represent the overall trend in the water's surface. Traditional GPS field surveys often generate points that are used to profile the continuous water surface, averaging a line between points with weighting dependent upon point quality, with inflections at the different cross-section locations. This study used a LOWESS curve that fits a curve to the HWMs, creating a continuous water surface that may better represent the dynamic conditions in the reaches. A locally weighted and smoothed line or curve better represents data used in hydraulic modeling when compared to the likely noisy point data collected by surveyed HWMs. LOWESS curves were compared for the two data sets, the "true" GPS field survey data compared to the "blind" sUAS-based GIS selected data by the three hydrologists which represent the final comparison.

For each comparison discussed above, the mean error (ME) and mean absolute error (MAE) of the residual differences of the two separate measures were calculated. The ME statistic was used to assess the potential vertical bias in the sUAS survey that may result from the effects of vegetation. Vegetation in the photogrammetric data collected using sUAS could result in higher measurements of elevation in the DTM, relative to bare-earth elevations. The MAE statistic was used to compare the variability between the measured and true values to gauge the differences between the control and sUAS data. The assumption was that the lower the MAE or magnitude of error, the higher the confidence in the water surface profile established along the river's banks.

3. Results

3.1. HWM Identification in the sUAS Orthophotographs

At Underwood Creek, HWMs were identified along the margin of the flood on both banks by the color change caused by vegetation that had been laid down during the event and in some locations, by debris lines. To aid identification at Underwood Creek, high-visibility paint was applied by the GPS survey team at multiple locations to identify HWM locations along the river's banks (Figure 7). This technique proved valuable to aid HWM identification in the orthophotograph collected using sUAS. At Brawley Wash, HWMs were identified as flattened vegetation coupled with almost continuous debris and wash line along the left bank (Figure 8). The HWMs were easily identifiable in the sUAS generated orthophotographs at both study sites.

Figure 7. Clipped orthophotograph showing Underwood Creek with zoomed inlay (upper right); notice vegetation along the sides of the channel, and two roads (light gray) on both sides of the channel. The yellow boxes indicate the extent of the HWMs indicated by high visibility paint marks but also flattened vegetation and debris lines. The downstream direction is toward the top of the photograph.

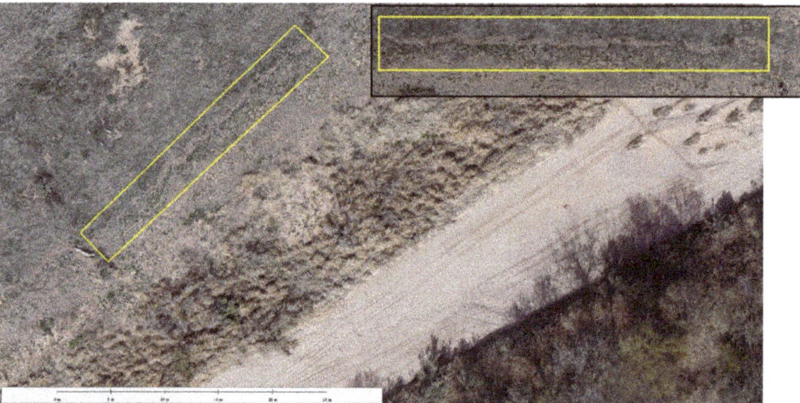

Figure 8. Clipped orthophotograph with a zoomed inlay of HWM (upper right) showing the Brawley Wash stream channel (light tan) and vegetation along the sides of the channel. The channel is dry in the imagery. The box brackets a portion of the debris line indicated by the HWM left by the flooding event. The stream flows from the lower left to the upper right of the frame.

3.2. Terrain Modeling Results

Point cloud data were classified into ground and vegetation classes, hand edited to best represent a near bare-earth model, and then used to render a triangulated irregular network (TIN) surface for each study site. The TIN model creates a continuous surface and interpolates to fill in areas of sparse data because vegetation has been removed. A DTM was constructed from the TIN models with a 3-cm pixel resolution to create an elevation background data set at the averaged positional accuracy of the survey-grade GPS data collected at ground targets. While the compositional accuracies of the GPS data and the sUAS derived models were around 1.5 cm or better, the positional (absolute) accuracies of base points and each respective survey point was 2 cm horizontal and 4 cm vertical. High compositional accuracy is important for individual models as the goal is the best representation of the natural environment, especially when those models are driving further analysis such as hydraulic modeling. Though the models are generally more accurate in composition than the positional accuracy of the GPS, using the GPS accuracy for resolution ensures appropriate confidence is given to any individual part of the model.

Holes in the terrain data caused by the removal of vegetation were interpolated and filled by converting sUAS collected points to a TIN and then to a DTM. This is a common occurrence when doing vegetation removal using photogrammetric data because the model only contains data from what is visible from above. The Underwood Creek DTM had a noisy overbank surface on the floodplain due to the dense, partly laid-over vegetation present on the banks. These conditions effectively created a DTM that was slightly higher than the actual ground surface by approximately 0.4 m. The DTM for Brawley Wash characterized the upper overbank conditions well, based on comparison with GPS survey data in that area where vegetation was mostly grasses, but the steep primary channel banks and mid-height vegetation at the overbank transitions proved problematic, causing both oversimplifications of the bank transitions and residual vegetative noise. The model containing vegetation compared to the resultant near bare-earth DTM from Underwood Creek and Brawley Wash can be seen in Figure 9 below.

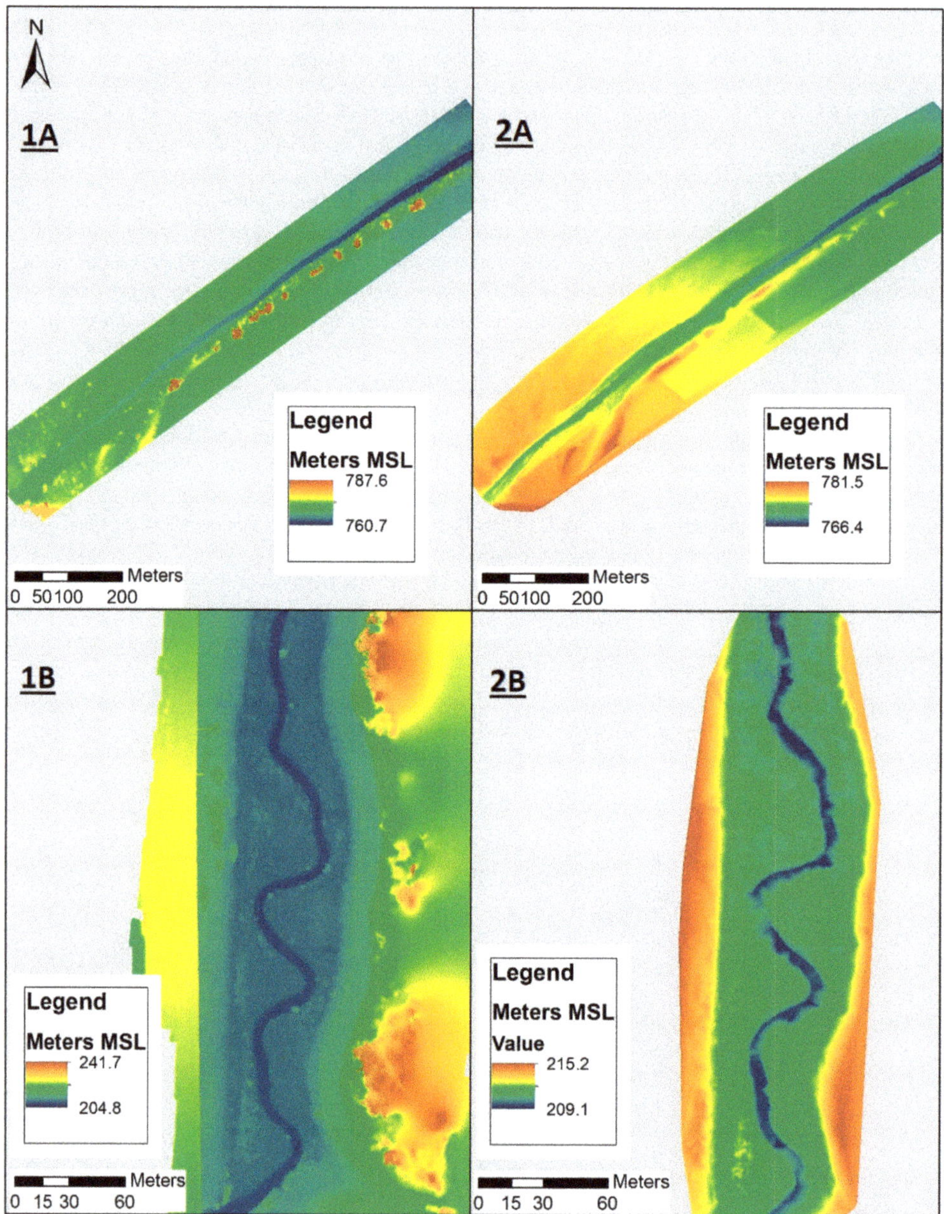

Figure 9. Full (**1A** and **1B**) and near bare-earth (**2A** and **2B**) DTMs at Brawley Wash (**1A** and **2A**) and Underwood Creek (**1B** and **2B**). Notice the absence of vegetation on the river's banks in the near bare-earth models on the right. This can also be seen in the total elevation ranges for all the maps, the near bare-earth DTM is a smaller range in both cases.

3.3. Evaluating the DTM and sUAS Elevations

The first comparison between GPS surveyed points and data derived from sUAS compares the GPS surveyed HWMs to the elevations of the DTM at the same location on the river's banks as described in

Section 2.4.1. In theory, the elevation of the GPS collected points and the elevation of the DTM at that location should be very close, within the compositional error of the surface model and GPS equipment, about 0.01 to 0.04 m since they are collocated in the same position. The horizontal position of the GPS points was plotted on top of the DTM elevation sampled at that same location, and elevations from the GPS survey and the sampled DTM location are compared. The ME and the MAE were calculated to compare the control data to the sUAS data. The relative ME and the relative MAE were calculated to compare the overall trend in the water surface in the reach, a critical parameter when conducting the hydraulic analysis. The results of this comparison can be seen in Figure 10. Some variability between the control and sUAS datasets that can be seen in Table 1.

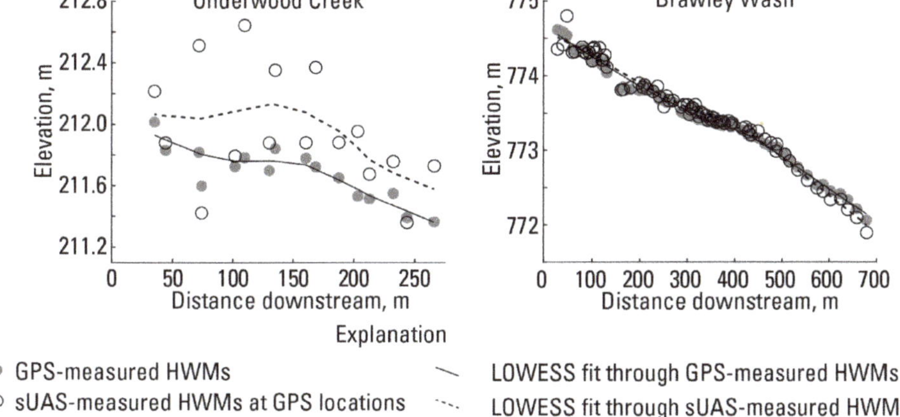

Figure 10. Figure comparing the DTM elevations from the sUAS and the GPS elevations at identical locations of all the HWMs in Underwood Creek (**left**) and Brawley Wash (**right**) as described in Section 2.4.1. LOWESS fits through the DTM and GPS elevations show the overall change in water-surface slopes in the stream reaches and how they compare.

Table 1. Errors of the HWM elevations from the sUAS compared to elevations estimated by a LOWESS fit through GPS HWM elevations along downstream transects in Underwood Creek. Elevations are compared at identical horizontal locations. ME is the mean error of elevations in each transect, MAE is the mean absolute error, and n is the number of elevations used to compute the error. The relative errors are the ME and MAE divided by the upstream and downstream elevation drop in the LOWESS elevations.

Site	Point Error					Curve Error				
	ME (m)	MAE (m)	Relative ME (-)	Relative MAE (-)	n	ME (m)	MAE (m)	Relative ME (-)	Relative MAE (-)	n
Underwood	0.26	0.33	46.7%	58.7%	14	0.28	0.28	49.8%	49.8%	14
Brawley	0.02	0.07	1.0%	2.8%	78	0.01	0.05	0.3%	2.1%	78

This comparison shows the difference in performance between a fully vegetated site, (Underwood Creek) and a site with little vegetation (Brawley Wash) when using sUAS mapping techniques. Underwood Creek showed an overall ME of 0.26 m, whereas Brawley Wash showed a ME of only 0.02 m. This positive bias between the true and sUAS measured elevations is assumed to be attributed to the vegetation located in the reach that caused the DTM to be erroneously high in elevation. In contrast, Brawley Wash, with little to no vegetation on the banks near the HWMs, showed an overall ME much closer to the true elevation measured by the GPS. This trend can also be seen in the relative ME of the overall trend in the water surface being 46.7% at Underwood Creek and just 1.0% at Brawley Wash.

A LOWESS curve was fit to describe both the GPS and sUAS elevations at each surveyed HWM, and the results of their comparisons are similar to the point data comparison discussed above.

Underwood Creek sUAS data ME was 0.28 m higher than the GPS survey, which is assumed to be due to the effects of vegetation, but had a similar downstream trend of water-surface slope as the GPS data, and Brawley Wash sUAS data ME was just 0.01 m (Table 1). Using point data to interpret the water surface profile along the river reaches is important for hydraulic modeling and inundation mapping. The downstream slope is an important component in these applications and can have a large effect on the modeling result, especially in low-gradient river systems.

The second comparison between GPS surveyed points and data derived from sUAS compared GPS surveyed cross-section elevations with elevations of the DTM at the same locations as described in Section 2.4.2. This was done in the same manner as the HWM comparison described above, where the position of the GPS surveyed point was used to compare the ground measurement against the DTM elevation at the same location. The results of this analysis can be seen in Figure 11. Some sections compare well, for example, sections 1–3 at Brawley Wash are all less than 0.2 m ME (Table 2). Sections 1 and 3 are less than 0.29 m ME at Underwood Creek (Table 2) even though the banks are covered with dense vegetation which in many places obscures the ground.

Other sections have fair results when streambed elevations collected from GPS and sUAS are compared. Sections 4 and 7 at Brawley Wash compare within 0.41 m ME and sections 5 and 6 at Underwood Creek were within 0.46 m ME. This is again believed to be an artifact of the dense vegetation present on the channel banks, which limits the ability of the sUAS to photograph the ground surface. As mentioned above, water can cause issues for photogrammetric data processing techniques, and in this case, the actual bottom elevation of the channel was not measured at some locations in the sUAS terrain data, although the difference is quite small, about 0.20 m ME.

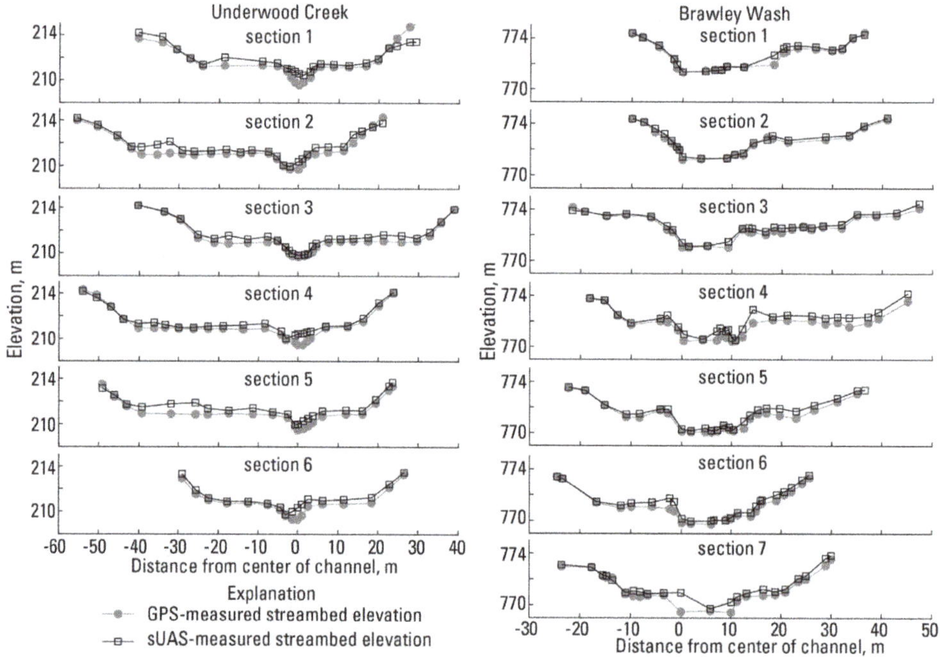

Figure 11. Figure comparing DTM elevations from the sUAS and GPS elevations along channel sections at identical locations at Underwood Creek (left) and Brawley Wash (right) as described in Section 2.4.2.

Table 2. Errors of the DTM elevations from the sUAS in comparison with GPS-measured elevations at identical streambed locations along cross-sections in Underwood Creek and Brawley Wash. ME is the mean error of elevations in each transect, MAE is the mean absolute error, and n is the number of elevations used to compute the error.

Section	Underwood Creek			Brawley Wash		
	ME (m)	MAE (m)	n	ME (m)	MAE (m)	n
1	0.26	0.45	22	0.14	0.17	18
2	0.37	0.39	25	0.11	0.14	18
3	0.29	0.29	25	0.18	0.19	24
4	0.34	0.37	23	0.41	0.41	24
5	0.46	0.46	20	0.25	0.25	22
6	0.41	0.41	15	0.28	0.28	22
7	–	–	–	0.32	0.34	20

3.4. Evaluating the sUAS Surveys for Identification and Mapping of HWMs

The third and final comparison assesses differences in the HWM water surface trend based on blindly selected HWMs with the GPS measured HWMs as described in Sections 2.4.3 and 2.4.4. The HWMs were chosen using the orthophotograph and mapped using the DTM by three different hydrologists at both sites. Because this is a subjective process that may include operator error, the exact same analysis was performed independently by each of the three hydrologists. Hydrologists were instructed to use only the orthophotograph and hydraulic judgment to select HWMs that they felt best represented the HWM profile in the reach. Hydrologist #1 chose 18 HWMs at Underwood Creek and 25 HWMs at Brawley Wash, Hydrologist #2 chose 19 and 20, respectively, and Hydrologist #3 chose 28 and 42, respectively. The results of these trials compared with the 16 and 78 HWMs surveyed using GPS are shown in Figure 12.

The first evaluation of these trials assesses differences between the LOWESS fits of the GPS and the sUAS-derived HWM point data as described in Sections 2.4.3 and 2.4.4. At Underwood Creek, the three hydrologist-chosen HWM elevations had ME = 0.05 m or less and MAE = 0.20 m or less (Table 3) compared to the LOWESS fit of the control data. The relative ME compares the overall trend in the water surface in the reach to the control data; results for Hydrologist #1 and #3 show low bias with relative ME of 2.2% and 0.3% respectively. Results for Hydrologist #2 show a larger bias of -8.1% in the water surface trend comparison. The relative MAE shows a range of 22.6% to 35.7%.

The next evaluation at Underwood Creek compares the LOWESS fits of both the GPS and sUAS measured HWM data as described in Section 2.4.4. The ME of the three hydrologist-chosen data set compared to the control data were -0.04 m to 0.01 m. The MAE comparing the data sets was 0.03 to 0.09 m. Promising results were also found when comparing the relative error between the hydrologist-chosen HWMs and the control data. Trials had relative ME of −7.9 to 1.1% and had relative MAE of 4.5 to 16.1% Table 3.

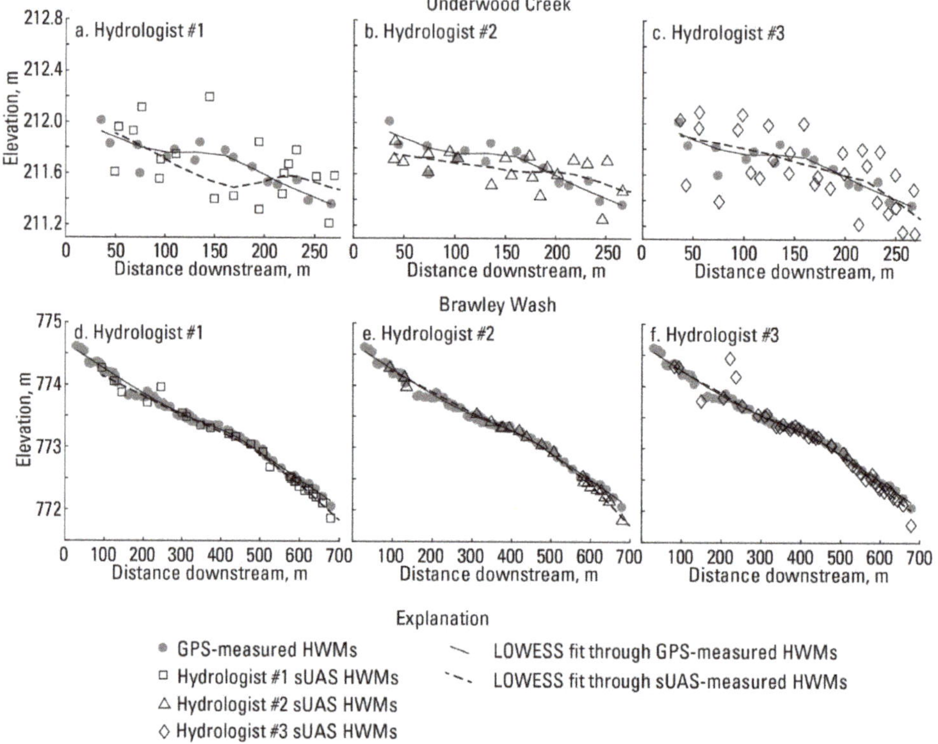

Figure 12. Comparison of HWM elevations from the GPS survey and the sUAS derived DTM as described in Sections 2.4.3 and 2.4.4. Three hydrologists independently identified HWM locations and derived elevations from the sUAS orthophotos and DTMs. The GPS elevations at the same HWM locations were estimated from a LOWESS fit to the GPS data. LOWESS fits through the sUAS HWM elevations (dashed line) and GPS HWM elevations (solid line) show estimated water-surface slopes at times of high water.

Table 3. Error in sUAS-derived HWM elevations, relative to ground-truth elevations estimated by a LOWESS fit through GPS-surveyed HWM elevations, for Underwood Creek. Elevations are compared at identical horizontal locations. The relative errors are the ME and MAE divided by difference between the most upstream and downstream elevations of the LOWESS curve. (ME, mean error, MAE, mean absolute error; n, sample size).

Underwood Creek	Point Error					Curve Error				
	ME (m)	MAE (m)	Relative ME (-)	Relative MAE (-)	n	ME (m)	MAE (m)	Relative ME (-)	Relative MAE (-)	n
Hydrologist #1	0.01	0.20	2.2%	35.7%	18	−0.02	0.09	−3.4%	16.1%	18
Hydrologist #2	−0.05	0.13	−8.1%	22.6%	19	−0.04	0.09	−7.9%	15.9%	19
Hydrologist #3	0.002	0.18	0.3%	32.4%	28	0.01	0.03	1.1%	4.5%	28

At Brawley Wash, the three hydrologist-chosen HWM elevations had less than ME = -0.06 m and MAE = 0.09 m error compared to the LOWESS fit of the control data as described in Section 2.4.3 (Table 4). This shows that the sUAS-measured data compared well to the control and that the data had minimal scatter because the MAE was low. The relative ME compares the overall trend in the water surface in the reach to the control data, results show an ME of −2.4 to −0.6% and had MAE of 2.3 to 3.6% (Table 4).

The next evaluation at Brawley Wash compared the LOWESS fits of both the GPS and sUAS-measured HWM data as described in Section 2.4.4. The ME of the three hydrologist-chosen data set compared to the control data were −0.07 to −0.03 m. The MAE comparing the data sets was within 0.04 to 0.07 m. Promising results were also found when comparing the relative error between the hydrologist chosen HWMs and the control data. Trials had relative ME of −3.1 to −1.2% and relative MAE of 1.6 to 3.1% (Table 4). This result shows that the data measured using only sUAS fits the control data well with minimal scatter and provided an accurate measure of the peak water surface slope in the reach.

Table 4. Error in sUAS-derived HWM elevations, relative to ground-truth elevations estimated by a LOWESS fit through GPS-surveyed HWM elevations, for Brawley Wash. Elevations are compared at identical horizontal locations. The relative errors are the ME and MAE divided by the difference between the most upstream and downstream elevations of the LOWESS curve. (ME, mean error; MAE, mean absolute error; n, sample size).

Brawley Wash	Point Error					Curve Error				
	ME (m)	MAE (m)	Relative ME (-)	Relative MAE (-)	n	ME (m)	MAE (m)	Relative ME (-)	Relative MAE (-)	n
Hydrologist #1	−0.06	0.09	−2.4%	3.6%	25	−0.07	0.07	−3.1%	3.1%	25
Hydrologist #2	−0.03	0.05	−1.4%	2.3%	20	−0.04	0.05	−1.7%	2.0%	20
Hydrologist #3	−0.01	0.09	−0.6%	3.6%	42	−0.03	0.04	−1.2%	1.6%	42

4. Discussion

The results presented here show that locating and measuring HWMs using sUAS and photogrammetric techniques is possible, and for the cases examined, proved accurate compared to data obtained using traditional GPS survey methods. The noteworthy finding in this research was the results of three comparisons at both study sites demonstrated the ability of flood hydrologists to successfully locate and measure HWMs accurately using only the products collected by the sUAS. The results from the three trials performed independently by three hydrologists also showed that skilled identification and interpretation of HWMs within the sUAS data resulted in the best comparison to the control data.

The most important findings were that the hydrologist-selected marks provided the best results in this analysis. The information presented in Figure 10 and Table 1 shows the results of the HWM location surveyed with the GPS and compares it to the elevation of the DTM at that same location. The results of this analysis at Underwood Creek were fair, ME = 0.26 m and MAE = 0.33 m (Table 1), but that amount of uncertainty would not be considered highly accurate for hydraulic modeling applications. However, the results of the hydrologist-selected HWMs at Underwood Creek showed better results, ME = −0.05 and MAE = 0.20 (Table 3). This shows that the style in which HWMs are surveyed using ground-based methods compared to sUAS identification might be different. When field teams collect HWMs, they have the opportunity to survey HWMs using the GPS in locations that may not be visible from above, which was the case at Underwood Creek. The HWMs selected by the three hydrologists were only HWMs that were visible from above and were likely HWMs that were located in areas free of vegetation and on the ground, not located underneath vegetation. This produced a better result in all cases and shows that even sites with heavy vegetation can be analyzed to produce accurate measurements of HWM using sUAS techniques.

Because the study plots at Underwood Creek and Brawley Wash had such different conditions, particularly the type of HWMs left by the flood and the amount of vegetation present in the channel, this study has shown conditions where techniques using sUAS would be promising, and conditions where photogrammetric data collection might not yield adequate results. As discussed, vegetation is a limitation for this type of data collection because it obstructs the view of the ground surface from the camera. An inconsistent bare-earth model results in areas within the DTM being artificially elevated. Targeting areas away from vegetation where the HWMs reside at or near the ground surface will produce the most defensible results when using these techniques, as was shown in comparison three where the hydrologists selected marks that were visible from above and likely on the ground.

Our results provide guidance for planning sUAS photograph collection campaigns to document flooding extent. Important factors to consider when planning sUAS flights include coverage area, flight altitude, resolution of the camera, and desired accuracy of the resulting HWM survey. Data collection and ideal pixel resolution can be guided by the resolution necessary to positively identify HWMs. Whether the flood evidence is a debris line, sediment color change, laid-down vegetation, or a cut-line in a bank, 4–6 pixels are needed to discretely identify that object or change in surface representing the HWM. For this study, the derived orthophotograph for each project area provided a single measurable photographic product at 2.33 cm^2 pixel size or smaller which provided the resolution required to positively identify the HWMs at both study locations, despite their geographic and geomorphic differences. If the HWMs investigated are less apparent than discussed in this research, then higher resolution orthophotographs would be needed and could be achieved with better camera sensors or by flying closer to the surface (lower altitude). In the case of fine seed lines, positive results might not be possible from sUAS collection if the HWM cannot be clearly identified in the orthophotograph.

Sound close-range photogrammetry techniques and a strong understanding of digital camera setup are critical for collecting accurate DTMs using sUAS. The near-nadir view used for data collection provides the most accurate compositional data but is limited in profiling complex vertical features. In areas of obscuring or overhanging vegetation, data collection at 10–15 degrees oblique in addition to nadir may help fill in ground surfaces and help with point classification, reducing data gap sizes. Shutter speeds too slow for the flight height or speed will create motion blur. Inappropriate aperture settings can over or under-expose imagery and potentially sacrifice depth of field. ISO settings too high (over 800) reduce crispness and clarity in the pixels and may create difficulties identifying features. These and other camera-specific settings need to be considered together to create clear and crisp photographs to feed photogrammetry analysis and produce accurate DTMs.

Point cloud processing algorithms are constantly advancing in modeling and GIS software, including the software packages used for this study, Metashape [24] and Global Mapper [26]. As photogrammetric data sets become increasingly common, the software packages are improving, furthering the ability to classify and edit photogrammetric data. Additional research is needed to establish more robust techniques for creating bare-earth models from photogrammetric data sets. Ultimately, the more accurate the bare-earth model, the more accurate measurements of HWMs in vegetated areas will be when using the sUAS techniques presented here.

Though out of reach for most potential users due to cost, commercial sUAS size aerial light detection and ranging (lidar) sensors are increasing in availability while decreasing in price. Lidar is the industry standard for generating point cloud models that can penetrate vegetation. This allows for not only a bare-earth model by only using the lowest or last returns, but also models of vegetation that can be analyzed, potentially providing useful information when looking for changes in channel roughness. Additionally, scanning sensors that mount to sUAS do not rely on rotating mirrors, motorized parts, and are generally lighter weight are becoming available and will lead to more accurate terrain data. Though these sensors currently are lacking in data density and point accuracy, this advancement represents progress in the field. The increasing availability of lidar means the subjective and possibly inaccurate process of classifying photogrammetric point clouds in order to make a bare-earth DTM could become unnecessary. Available aerial lidar models will also create comparative data sets to further determine utilities and deficiencies of photogrammetric data and is an opportunity for further research.

Identifying and measuring HWMs using sUAS has implications for data collection efficiency and personnel safety. To document widespread flooding, such as storm surge or inundation due to extreme precipitation, runoff, and surge events, survey teams must travel to many different locations to document HWMs with traditional methods. If operated from an area with a clear and safe viewshed, sUAS can access large areas and are not inhibited by localized ground conditions or hazards. This allows for imagery and topographic data collection over a much larger area and with more complete spatial coverage compared to traditional data collection methods. Additionally, sUAS can collect data in areas that are inaccessible after natural disasters due to inundated roads, closed bridges, or other obstacles, while the range of sUAS surveys reduces the number of locations and time spent traveling. sUAS data collection after floods can be completed in less time, with less physical exertion by personnel, and with less exposure to hazards.

Traditional surveys often require extensive hiking over rugged terrain with dense vegetation and difficult stream crossings to locate HWMs and survey cross-sections. Complete topographic data throughout a reach can be collected efficiently using sUAS, whereas traditional surveys provide limited cross-section data that are time-consuming and often challenging to collect. When collecting flood data with field crews and GPS survey equipment, the data collected are limited to the conditions seen on the ground at the time of the survey. Additionally, cross-sectional data is collected in parallel strips at a distance sufficient to characterize slope and fall between locations. This type of data collection benefits from significant field experience and might be limited by access, time, topography, and equipment requirements. sUAS collection can survey a complete inundation area rapidly without the need for complete physical access. Additionally, the models generated might be used for countless analyses with a wide variety of hydraulic software packages to best fit the complex conditions seen during flood events. Cross-sections may be rendered from DTMs at any interval or distance and are not limited to access. Further, complete DTMs may be used in two-dimensional hydraulic models to represent the complex terrain inundated, not limited by the uncertainty and assumptions between individual cross-sections. A more complete understanding of the natural environment and the complex interactions occurring during floods may be gained from more complete and holistic models.

5. Conclusions

We demonstrate that post-flood HWMs can be reasonably (within acceptable error) identified and mapped using sUAS-collected imagery processed using photogrammetric techniques. These techniques can increase data collection efficiency after natural disasters and major flooding while also keeping field personnel in potentially safer positions. The first important finding is that when the photogrammetric data sets are collected with appropriate ground control, the DTMs allow HWMs to be identified with low positional bias compared to the GPS-surveyed HWMs. Secondly, for the two cases presented, HWMs are clearly visible in orthophotographs with a typical resolution for sUAS-derived products. HWMs that were clearly visible include wash lines, debris lines, and mud lines, as well as HWMs depicted by vegetation that has been laid over by the flood.

Vegetation is an important factor to consider when deciding whether to apply these sUAS-based HWM survey techniques. To accurately identify HWMs and map terrain elevation, images collected by the sUAS must contain pixels that represent the ground surface. In areas where tree canopy and thick vegetation dominate, aerial images may only capture pixels of the vegetation and therefore the terrain products that are produced do not accurately represent the elevation of the land surface in those areas. This can be seen in the results when comparing the two study sites investigated here with points collected with GPS surveying equipment. The results show Brawley Wash was mainly free of vegetation at the HWMs and closely reproduced the GPS surveyed data, within 0.02 m ME, compared to Underwood Creek where vegetation affected the accuracy of the bare-earth model at the HWMs and produced a 0.26 m ME.

Vegetation has similar effects on the bare-earth model when comparing measurements of the cross-sections in both of the study locations. Both of the reaches contained dense stands of vegetation in places that limited the ability of the sUAS to photograph the ground surface at these locations. This caused varying success when comparing GPS measured cross-sections to sUAS-measured sections. Cross-sections 1 and 3 at Underwood Creek and 1–3 at Brawley Wash compare well to the GPS-collected data. However, cross-sections 5 and 6 at Underwood Creek and 7 at Brawley Wash show deviation from the actual ground-surface measurements. Cross-section data collection for hydraulic modeling requires specific locations of sections based upon channel slope and width. If sections cannot be located in areas free of vegetation where the DTM is proven accurate, field measurements might still be required to achieve accurate measures of the land surface.

sUAS-based surveying is proving to be a promising tool in post-flood data collection. Advancements in flight time and lift capabilities, coupled with the miniaturization of sensors such as high-resolution cameras and aerial lidar sensors, could increase sUAS collection in this field. Further analysis is needed to compare lidar data to photogrammetrically derived models, but ultimately a combination of lidar and stereo photography might provide the best representation of the natural environment with and without vegetation for complete hydraulic models. Still, depending on the absolute accuracy needed and the implications of the resultant data (i.e., flood inundation mapping for legal designation) ground-truthed data might be most appropriate. With an understanding of the available tools, their benefits and limitations, managers are empowered to tailor a response effort to the known field conditions, the legal implications of the data collected, and need for field personnel efficiency and safety.

Author Contributions: Project conceptualization: B.T.F., G.P.D. Methodology: B.T.F., G.P.D. Validation: B.T.F., G.P.D., C.E.B.; Analysis: B.T.F., G.P.D., J.E.D., C.E.B. Investigation: B.T.F., G.P.D., J.E.D., C.E.B. Data curation: B.T.F., G.P.D., F.A.F. Writing—original draft: B.T.F., G.P.D., J.E.D., C.E.B.; Writing—review and editing: B.T.F., G.P.D., J.E.D., C.E.B., F.A.F.; Visualization: B.T.F., G.P.D., J.E.D., C.E.B.; Funding acquisition: B.T.F., G.P.D. All authors have read and agreed to the published version of the manuscript.

Funding: The data collection an Underwood Creek was funded with support from the Great Lakes Restoration Initiative with support from the U.S. Army Corp of Engineers. The data collection at Brawley Wash was funded by the Arizona Department of Transportation and Pima County Flood Control. Analysis and publication costs were funded with support from the U.S. Geological Survey Hydrologic Remote Sensing Branch.

Acknowledgments: The authors would like to thank the Arizona Department of Transportation and Pima County Flood Control District for supporting this, and many other data collection efforts around Arizona. The authors would also like to thank the U.S. Geological Survey Hydrologic Remote Sensing Branch, the U.S. Geological Survey National Unmanned Aircraft Systems Project Office, and the U.S. Geological Survey Extreme Hydrologic Events Team for their ongoing support. Any use of trade, firm, or product names id for descriptive purposes only and does not imply endorsement by the U.S. Government.

Conflicts of Interest: The authors declare no conflict of interest.

References

1. Wallemacq, P.; UNISDR; CRED. *Economic Losses, Poverty and Disasters 1998–2017*; UNISDR: Brussels, Belgium, 2018. [CrossRef]
2. NOAA National Centers for Environmental Information (NCEI). U.S. Billion-Dollar Weather and Climate Disasters. Available online: https://www.ncdc.noaa.gov/billions/ (accessed on 28 January 2020).
3. Pielke, J.; Downton, M.W. Precipitation and Damaging Floods: Trends in the United States, 1932–97. *J. Clim.* **2000**, *13*, 3625–3637. [CrossRef]
4. Milly, P.C.D.; Wetherald, R.T.; Dunne, K.A.; Delworth, T.L. Increasing Risk of Great Floods in a Changing Climate. *Nature* **2002**, *415*, 514–517. [CrossRef] [PubMed]
5. Benson, M.A.; Dalrymple, T. General Field and Office Procedures for Indirect Measurements: U.S. Geological Survey Techniques of Water-Resources Investigations, Book 3, Chap. Al, 30. 1967. Available online: http://pubs.er.usgs.gov/publication/twri03A1 (accessed on 28 January 2020).
6. Koenig, T.A.; Bruce, J.L.; O'Connor, J.; McGee, B.D.; Holmes, R.R.; Hollins, R.J.; Forbes, B.T.; Kohn, M.S.; Schellekens, M.F.; Martin, Z.W.; et al. Identifying and Preserving High-Water Mark Data. In *U.S. Geological Survey Techniques and Methods, Book 3, Chap. A24*; U.S. Geological Survey: Reston, VA, USA, 2016. [CrossRef]
7. Murphy, R.R. Use of Small Unmanned Aerial Systems for Emergency Management of Flooding. Federal Highway Administration Tech Brief. 2019. Available online: https://www.fhwa.dot.gov/uas/resources/hif19019.pdf (accessed on 24 March 2020).
8. Whitehead, K.; Hugenholtz, C. Remote sensing of the environment with small unmanned aircraft systems (UASs), part 1: A review of progress and challenges. *J. Unmanned Veh. Syst.* **2014**, *2*, 69–85. [CrossRef]

9. Whitehead, K.; Hugenholtz, C.H.; Myshak, S.; Brown, O.; LeClair, A.; Tamminga, A.; Barchyn, T.E.; Moorman, B.; Eaton, B. Remote Sensing of the Environment with Small Unmanned Aircraft Systems (UASs), part 2: Scientific and Commercial Applications. *J. Unmanned Veh. Syst.* **2014**, *2*, 86–102. [CrossRef]

10. Diakakis, M.; Andreadakis, E.; Nikolopoulos, E.I.; Spyrou, N.I.; Gogou, M.E.; Deligiannakis, G.; Katsetsiadou, N.K.; Antoniadis, Z.; Melaki, M.; Georgakopoulos, A.; et al. An Integrated Approach of Ground and Aerial Observations in Flash Flood Disaster Investigations. The Case of the 2017 Mandra Flash Flood in Greece. *Int. J. Disaster Risk Reduct.* **2019**, *33*, 290–309. [CrossRef]

11. Smith, M.W.; Carrivick, J.L.; Hooke, J.; Kirkby, M.J. Reconstructing Flash Flood Magnitudes Using "Structure-from-Motion": A Rapid Assessment Tool. *J. Hydrol.* **2014**, *519*, 1914–1927. [CrossRef]

12. U.S. Geological Survey. 04087088 Underwood Creek at Wauwatosa, WI. Available online: https://waterdata.usgs.gov/wi/nwis/inventory/?site_no=04087088&agency_cd=USGS (accessed on 28 January 2020).

13. Haserodt, M.J. 04087088-Underwood Creek at Wauwatosa, WI–2018/09/14 GPS Survey. In *U.S. Geological Survey Data Release*; U.S. Geological Survey: Reston, VA, USA, 2020. [CrossRef]

14. EarthExplorer. Available online: https://earthexplorer.usgs.gov/ (accessed on 19 March 2020).

15. Prokopec, J.G.; Lund, J.W. 04087088–Underwood Creek at Wauwatosa, WI–2019/07/17 GPS Survey. In *U.S. Geological Survey Data Release*; U.S. Geological Survey: Reston, VA, USA, 2020. [CrossRef]

16. U.S. Geological Survey. 09487000 Brawley Wash near Three Points, AZ. Available online: https://waterdata.usgs.gov/az/nwis/inventory/?site_no=09487000&agency_cd=USGS (accessed on 28 January 2020).

17. Paretti, N.V.; Kennedy, J.R.; Turney, L.A.; Veilleux, A.G. Methods for Estimating Magnitude and Frequency of Floods in Arizona, Developed with Unregulated and Rural Peak-Flow Data through Water Year 2010. In *U.S. Geological Survey Scientific Investigations Report 2014-5211*; U.S. Geological Survey: Reston, VA, USA, 2014. [CrossRef]

18. Bunch, C.E.; Forbes, B.T.; DeBenedetto, G.P. 09487000–Brawley Wash near Three Points, AZ–2018/09/19 GPS Survey. In *U.S. Geological Survey Data Release*; U.S. Geological Survey: Reston, VA, USA, 2019. [CrossRef]

19. U.S. Geological Survey. 09487000–Brawley Wash near Three Points, AZ–2019/01/31 UAS Data. Available online: https://www.sciencebase.gov/catalog/item/5d3f3216e4b01d82ce8d91f6 (accessed on 12 February 2020).

20. Rydlund, P.H.; Densmore, B.K. Methods of Practice and Guidelines for Using Survey-Grade Global Navigation Satellite Systems (GNSS) to Establish Vertical Datum in the United States Geological Survey. In *U.S. Geological Survey Techniques and Methods, Book 11, Chap. D1*; U.S. Geological Survey: Reston, VA, USA, 2012. [CrossRef]

21. National Geodetic Survey. OPUS: Online Positioning User Service. Available online: https://www.ngs.noaa.gov/OPUS/about.jsp (accessed on 28 January 2020).

22. Burkey, J. LOWESS, Locally Weighted Scatterplot Smoothing for Linear and Non-Linear Data (Enhanced). Available online: https://www.mathworks.com/matlabcentral/fileexchange/22470-lowess-locally-weighted-scatterplot-smoothing-for-linear-and-non-linear-data-enhanced (accessed on 28 January 2020).

23. Dalrymple, T.; Benson, M.A. Measurement of Peak Discharge by the Slope-Area Method. In *U.S. Geological Survey Techniques of Water-Resources Investigations, Book 3, Chap. A2*; U.S. Geological Survey: Reston, VA, USA, 1967. [CrossRef]

24. *Agisoft Metashape*, version 1.5.2; Agisoft LLC: St. Petersburg, Russia, 2019.

25. U.S. Geological Survey National Unmanned Aircraft Systems Office. Unmanned Aircraft Systems Data Post-Processing: Structure-from-Motion Photogrammetry. Available online: https://uas.usgs.gov/nupo/pdf/PhotoScanProcessingDSLRMar2017.pdf (accessed on 28 January 2020).

26. *Global Mapper*, version 20.0; Blue Marble Geographics: Hallowell, MA, USA, 2018.

Article

Longitudinal, Lateral, Vertical, and Temporal Thermal Heterogeneity in a Large Impounded River: Implications for Cold-Water Refuges

Francine H. Mejia [1],*, Christian E. Torgersen [1], Eric K. Berntsen [2], Joseph R. Maroney [2], Jason M. Connor [2], Aimee H. Fullerton [3], Joseph L. Ebersole [4] and Mark S. Lorang [5]

1   U.S. Geological Survey, Forest and Rangeland Ecosystem Science Center, Cascadia Field Station, Seattle, WA 98195-2100, USA; ctorgersen@usgs.gov
2   Kalispel Tribe of Indians Natural Resources Department, Usk, WA 99180, USA; eberntsen@knrd.org (E.K.B.); jmaroney@knrd.org (J.R.M.); jconnor@knrd.org (J.M.C.)
3   Fish Ecology Division, Northwest Fisheries Science Center, National Marine Fisheries Service, National Oceanic and Atmospheric Administration (NOAA), Seattle, WA 98112, USA; aimee.fullerton@noaa.gov
4   U.S. Environmental Protection Agency, Center for Public Health and Environmental Assessment, Pacific Ecological Systems Division, Corvallis, OR 97333, USA; Ebersole.Joe@epa.gov
5   FreshwaterMap, Bigfork, MT 59911, USA; mark@freshwatermap.com
*   Correspondence: fmejia@usgs.gov

Received: 15 February 2020; Accepted: 26 April 2020; Published: 28 April 2020

Abstract: Dam operations can affect mixing of the water column, thereby influencing thermal heterogeneity spatially and temporally. This occurs by restricting or eliminating connectivity in longitudinal, lateral, vertical, and temporal dimensions. We examined thermal heterogeneity across space and time and identified potential cold-water refuges for salmonids in a large impounded river in inland northwestern USA. To describe these patterns, we used thermal infrared (TIR) imagery, in situ thermographs, and high-resolution, 3-D hydraulic mapping. We explained the median water temperature and probability of occurrence of cool-water areas using generalized additive models (GAMs) at reach and subcatchment scales, and we evaluated potential cold-water refuge occurrence in relation to these patterns. We demonstrated that (1) lateral contributions from tributaries dominated thermal heterogeneity, (2) thermal variability at confluences was approximately an order of magnitude greater than of the main stem, (3) potential cold-water refuges were mostly found at confluences, and (4) the probability of occurrence of cool areas and median water temperature were associated with channel geomorphology and distance from dam. These findings highlight the importance of using multiple approaches to describe thermal heterogeneity in large, impounded rivers and the need to incorporate these types of rivers in the understanding of thermal riverscapes because of their limited representation in the literature.

Keywords: water temperature; salmonids; Pend Oreille River; thermal infrared (TIR); acoustic Doppler current profiler (ADCP); channel bathymetry; flow velocity; cold-water refuge; dam; remote sensing

---

1. Introduction

In rivers and streams where humans have altered the thermal regime, not only does summer water temperature frequently exceed the thermal tolerance of native salmonids, thus influencing their distribution [1], but these human modifications can amplify or dampen spatial and temporal variation [2]. Specifically, dams and diversions can greatly modify riverine thermal regimes by restricting or eliminating connectivity in longitudinal, lateral, vertical, and temporal dimensions [3,4]. Large dams can be intentionally managed to release cold water from deep reservoirs to provide habitat for salmonid fisheries, whereas releases from any size dam and diversion can increase downstream temperatures by releasing warm water directly from the reservoir surface [5].

The thermal effects of dams have been studied for many years [6,7], and it is recognized that these effects depend on the landscape position of the dam, dam operation, water release depth, and geomorphic setting [5]. Direct modifications to the thermal regime are made by upstream water releases and dam operations. When taking into consideration the geomorphic setting, confined valleys are likely to be more susceptible to changes in water temperature due to water releases than are unconfined valleys, which have greater surface water-groundwater exchange and greater hydrologic storage [8]. Indirect modifications of the thermal regime are those that affect the processes of heat exchange between water and the environment [9]. These indirect modifications to the thermal regime occur when dams alter sediment transport and channel geometry and fragment riparian and riverine habitat [8,10–12]. In impounded rivers, flow velocity, water depth, and the extent of channel wetted perimeter are decoupled (spatially and temporally) from the river's geomorphic setting due to dam operations [8]. Water releases from the dam that have a limited sediment load and excess energy may incise the channel bed and coarsen the channel bed material [11]. However, dam effects on flow and sediment are also dependent on the geomorphic setting where lateral and vertical movement of flow and sediment are constrained by valley shape [8]. Changes in water depth and flow velocity are likely greatest in confined valleys, whereas changes in the extent of wetted perimeter and channel width may be largest in unconfined valleys, such as in the Colorado, Green, and Missouri rivers [8,13].

The mechanisms by which channel geometry (i.e., water depth, wetted perimeter, stream width, and cross-sectional area) indirectly affect water temperature are complex. In confined sections of a river where the river is narrow and deep, two main mechanisms may be at play, depending on flow velocity. At high flow velocities, water may be cooler due to mixing and topographic shading, but at low velocities, thermal stratification may occur in the summer when surface water temperature is higher than deeper water. In contrast, in unconfined sections, where changes in wetted perimeter and channel width are largest, warmer water temperatures are the result of a larger surface area exposed to solar radiation [14]. Rivers with dense submerged aquatic vegetation (SAV) may also experience thermal stratification, as SAV steepens vertical temperature gradients and decreases water velocity [15].

Tributaries and groundwater inputs can also have an impact on the thermal regime and its variability by creating discontinuities in the longitudinal profile regardless of whether a river or stream is dammed or not [16–18]. The extent of the influence of these tributaries and groundwater inputs on the physical characteristics of the main stem river depend on their size relative to the main stem, morphology or junction angle, and dam operations [8,19–24].

Although dam-induced modifications to a river's thermal regime can influence the quantity and distribution of cold-water refuges for salmonids, cold-water refuge quantity and distribution in large impounded rivers have not been well studied. Previous studies on the spatial and temporal variability of water temperature and cold-water refuges have focused on small- to medium-size streams [16,25–29] and some large unregulated rivers [18,30,31]. In contrast, studies on large regulated rivers have mostly emphasized thermal variability changes [8,32–34] or their cold-water refuge distribution [35–37], but not both.

In the Pend Oreille River, adult bull trout (*Salvelinus confluentus*) and westslope cutthroat trout (*Oncorhynchus clarkii lewisi*) experience summer water temperatures that exceed their upper tolerances, 18–20 °C [38–42]. Bull trout and westslope cutthroat trout in Pend Oreille River telemetry studies have been observed behaviorally thermoregulating in cold tributary confluences and areas with groundwater inputs (E.K. Berntsen, J.R. Maroney, and J. Connor, Kalispel Tribe of Indians, personal observation). We examined thermal heterogeneity across space and time and identified potential cold-water refuges for adult salmonids in the Pend Oreille River, a large impounded river with an epilimnetic-release dam in the inland northwestern USA. We had three specific objectives:

(1) Compare longitudinal patterns of water temperature derived from thermal infrared imagery to instream measurements in the water column.

(2)   Assess spatial (lateral, longitudinal, and vertical) and temporal thermal heterogeneity and potential cold-water refuges for salmonids at multiple scales.

(3)   Evaluate statistical associations between hydrologic and geomorphic variables and occurrence of cool-water areas and median water temperature.

We hypothesized that (1) water temperature in the main stem of the Pend Oreille River increases as water moves downstream, but lateral contributions from tributaries create discontinuities in the longitudinal profile; (2) the confluence area formed by the main stem Pend Oreille River and cold tributaries provide potential cold-water refuges for cold-water fishes; and (3) in the summer, thermal stratification from various sources (i.e., seeps, groundwater inputs, low water velocities, and submerged aquatic vegetation) creates vertical temperature gradients that isolate cool water near the river bottom to potentially provide refuge for fish.

2. Materials and Methods

2.1. Study Area

The study area encompassed the Box Canyon reservoir section of the Pend Oreille River, an 88.5-km portion of the river between the town of Ione, Washington, and Albeni Falls Dam in Idaho (USA) (Figure 1).

The reservoir's surface area varies between 2800 and 3600 ha, depending on water elevation and flow. Flows are maintained for maximum power production below approximately 1980 m^3 s^{-1}. If flows exceed this level, the reservoir operates in free-flow mode without power production. The Pend Oreille River, a sixth order stream, is the fourth-largest tributary of the Columbia River and drains an area of 62,678 km^2 with a mean annual discharge of 709 m^3 s^{-1} for the period from 1904 to 2015. Seasonal weather patterns throughout the study area are typical of the Northern Rockies ecoregion with hot, dry summers and cold, relatively wet winters. During the Ice Age, the Pend Oreille Lobe of the Glacial Lake Missoula formed the Pend Oreille River. The glacial Lake Missoula, part of the Cordilleran Ice Sheet, extended south and covered the valley and the surrounding low hills with glacial and alluvial material deposited during the retreat of ice creating an alluvial valley. Alluvial valleys can range from confined (floodplain width/channel width of less than 2) to unconfined (floodplain width/channel width greater than 4). Because landscape metrics such as confinement and tributary confluences are strongly associated with the density and occurrence of cold-water refuges [22], describing these landscape metrics in the Pend Oreille River is important for understanding watershed-scale relationships. The Pend Oreille River has two large, moderately confined sections: (1) Above Box Canyon dam, the first section between river kilometer (rkm) 53 to rkm 93, and (2) below Albeni Falls Dam between rkm 131 to rkm 144. In contrast, the most unconfined section is between rkm 94 and rkm 121 [43]. Calispel Creek and Le Clerc Creek (C6) are the largest tributaries within the study area (Figure 1).

The Pend Oreille River within the study area is constrained by two main roads and levees where most of the original forests have been converted to agricultural or residential land. Additionally, cottonwood galleries have declined because of limited regeneration and replacement due to flow management. In contrast, there are numerous seasonally flooded wetlands located near the Kalispel Tribe reservation. In addition, flow is managed by the U.S. Army Corps of Engineers (ACOE) for flood control, and comparisons between 1998 and 2015 aerial imagery revealed that there have not been significant changes in channel morphology. The entire Pend Oreille River is considered impaired for water temperature in the states of Idaho and Washington because water temperatures exceed the water quality standards criterion of 20 °C at numerous locations in the river. However, cold-water refuges present within the Pend Oreille have been observed to be used by salmonids during the summer when water temperatures are above their optimum (E.K. Berntsen, J.R. Maroney, and J. Connor, unpublished data).

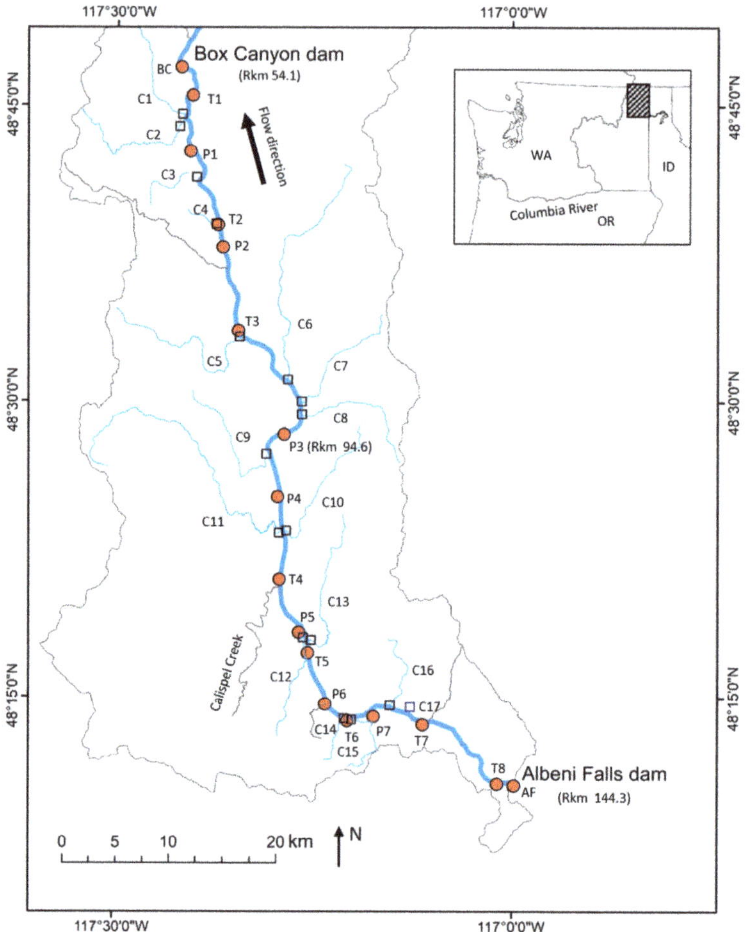

Figure 1. Pend Oreille River study area in eastern Washington and northern Idaho, USA. Orange circles are main stem sites labeled as either P (peaks) or T (troughs), and hollow squares are tributary confluences (C) included in the study. Gray tributaries were sampled but not included in analysis because of missing data. Black arrow describes direction of flow from south to north.

2.2. Water Temperature Data

2.2.1. Airborne Thermal Infrared (TIR) Imagery

To characterize longitudinal thermal heterogeneity of the Pend Oreille River, we used airborne thermal infrared imagery (TIR) collected in 2001. TIR surveys were conducted on 16 August 2001 between the hours of 14:00 and 17:00 [44]. The data were collected using a TIR imager with a wavelength range of 8-12 microns [45]. The sensor was mounted to the underside of a helicopter. The helicopter was flown longitudinally along the river channel with the sensor and camera in a vertical or near vertical position at 1372 m above ground level. Flight altitudes were determined based on the river width and floodplain characteristics. The altitude selected allowed for an image with a ground width of 483 m and a pixel size of 0.8 m. Ground-truth measurements of instream temperature at the time of the survey indicated that radiant temperatures were accurate to within 0.5 °C.

To compare spatial patterns of instream water temperatures to TIR data, we selected a day in 2017 that was similar in air temperature and discharge to the conditions when the thermal imagery was collected in 2001. Air temperatures on 22 August 2017 ranged from 11.0 °C at midnight to near 33.0 °C at 15:00 at the Tacoma Creek station, a Remote Automated Weather Station (RAWS, https://raws.dri.edu/index.html, accessed on 25 October 2017). Mean air temperatures from 14:00 to 17:00 were near 32.0 °C. Air temperatures reported by Watershed Sciences [44] on 16 August 2001 at 14:00 to 17:00 during the TIR survey ranged from 31.5 °C to 32.2 °C. The Tacoma Creek station in 2001 recorded a minimum air temperature of 14.0 °C and a maximum air temperature of 33.9 °C at 16:00. Discharge was 229 m^3 s^{-1} in 2001 and 215 m^3 s^{-1} in 2017.

2.2.2. Instream Measurement

We deployed instream Hobo pendant data loggers (Onset Computer Corporation, Bourne, Massachusetts, USA; ± 0.5 °C) and recorded water temperatures at 1-h intervals at 17 sites in the main stem (labeled as either a peak, P, or a trough, T) and at 22 confluences (labeled as C) from 17 July to 14 September 2017 to describe the longitudinal and lateral thermal heterogeneity of the Pend Oreille River from rkm 53 to rkm 144 (i.e., Box Canyon Reservoir) (Figure 1 and Table 1). At each site, loggers were positioned approximately 30 cm from the bed and 30 cm from the surface of the river to characterize vertical thermal heterogeneity. Each main stem logger was placed at a major temperature deviation (i.e., a peak [n = 7] or a trough [n = 8]) in the 2001 thermal infrared longitudinal profile based on the smoothed trend from the 10-km moving average [26]. We also deployed loggers near Albeni Falls and Box Canyon dams.

Table 1. Site identification and name for instream loggers and their corresponding river kilometer (rkm) location and depth at deployment. Main stem sites are labeled as either P (peaks) or T (troughs) and tributary confluences are labeled as C.

Site ID	Name	Rkm	Site Type	Depth (m) at Deployment
BC	BC	54.1	Main stem near dam	15.8
T1	T1	56.7	Main stem, trough	13.8
C1	Cedar Cr.	59.0	Confluence	0.9
C2	Big Muddy Cr.	59.3	Confluence	5.0
P1	P1	62.7	Main stem, peak	11 ?
C3	Renshaw Slough	65.9	Confluence	1.6
C4	Tiger Slough	70.6	Confluence	8.1
T2	T2	70.7	Main stem, trough	9.8
P2	P2	72.8	Main stem, peak	8.2
T3	T3	81.0	Main stem, trough	7.3
C5	Ruby Cr.	81.6	Confluence	0.5
C6	Le Clerc Cr.	88.3	Confluence	0.6
C7	Middle Cr.	90.7	Confluence	0.6
C8	Mill Cr.	91.9	Confluence	0.6
P3	P3	94.6	Main stem, peak	7.4
C9	Cusick Cr.	97.3	Confluence	0.6
P4	P4	101.6	Main stem, peak	1.4
C10	Cee Cee Ah Cr.	104.8	Confluence	1.2
C11	Tacoma Cr.	104.9	Confluence	1.8
T4	T4	109.3	Main stem, trough	4.0
P5	P5	114.9	Main stem, peak	3.8
C12	Davis Cr	115.6	Confluence	0.8
C13	Skookum Cr.	116.1	Confluence	0.8
T5	T5	117.3	Main stem, trough	4.5
P6	P6	122.4	Main stem, peak	5.1
C14	Kent Cr.	124.5	Confluence	0.7
T6	T6	124.9	Main stem, trough	6.1
C15	McCloud Cr.	125.1	Confluence	0.4
P7	P7	127.2	Main stem, peak	4.9
C16	Indian Cr.	129.1	Confluence	0.5
C17	Trout ponds	131.0	Confluence	0.4
T7	T7	132.6	Main stem, trough	3.7
T8	T8	142.5	Main stem, trough	4.4
AF	AF	144.3	Main stem near dam	6.1

Characterization of the longitudinal thermal profile is important because not all rivers warm as their water moves downstream from the headwaters to the mouth, and their profile can inform how to implement long-term monitoring programs and climate-adaptation strategies [26]. Additionally, in slow-moving rivers like the Pend Oreille in the summer months, the surface temperature may not be indicative of the overall water temperature or of cold-water areas near the river bottom due to vertical stratification.

2.3. Characterization of Riverine Landscape

2.3.1. Geographical Information System

We used publicly available spatial data layers including land use [46], geology [22,47], and tributaries [17] as explanatory variables in our statistical analysis to assess associations with cool-water areas in the Pend Oreille River. The specific variables in our analyses were land use (% developed land), aquifer permeability (high versus low), degree of confinement (m m^{-1}), and tributaries (area in km^2 and number).

Land Use

We grouped land use into two categories (percent developed or percent undeveloped) from the 2010 data produced by the Washington Department of Ecology (https://ecology.wa.gov/Research-Data/Data-resources/Geographic-Information-Systems-GIS/Data, accessed on 16 March 2018). Of the 84 land use categories, seven were described as nondeveloped land, and the remaining 77 categories were classified as developed land. To estimate percent developed land we generated points every 1 km on the centerline of the Pend Oreille River from rkm 53 to the Albeni Falls dam (rkm 144). We buffered the Pend Oreille River shoreline at 1 km and generated Thiessen polygons using every other river kilometer point that then were intersected with the buffered river layer to demarcate river segments. These river segments were bisected to create 1-km river segments within which we estimated the percent contribution of developed land on each bank.

Aquifer Permeability

We used aquifer permeability data for the Columbia River basin (https://catalog.data.gov/dataset/pacific-northwest-pnw-hydrologic-landscape-hl-polygons-and-hl-code, accessed on 19 October 2017). These data are based on similarities in lithology. They estimate hydraulic conductivity value, identify areas of low or high aquifer permeability, and describe locations where shallow subsurface groundwater mixes with deep groundwater flows and where the loss or gain of water through groundwater fluxes may be important [48].

Valley Width and Confinement

Confinement was calculated as the ratio of floodplain width to bankfull channel width. Confined streams had ratios of less than 2, moderately unconfined streams had ratios between 2 and 4, and streams with ratios greater than 4 were considered unconfined [49]. We mapped confinement using the NetMap floodplain width layer [50] and bankfull channel width (m) data obtained from Beechie and Imaki [43]

Tributaries

We used the NetMap stream layer to count the number of tributaries per kilometer and to determine the watershed area of each tributary to infer tributary size (km^2). The NetMap stream layer is derived from digital elevation models (DEM) ranging in resolution from 10 m to 1 km. This layer was produced using a consistent method to extract a synthetic routed and attributed network including drainage area and drainage area per contour length [50].

2.3.2. Three-Dimensional Hydraulic Mapping and Channel Morphology

To characterize the Pend Oreille River morphology, Freshwater Map, Inc. (https://freshwatermap. com) conducted 3-D bathymetric mapping of 81 km of the Pend Oreille River from the Albeni Falls Dam to the Box Canyon Dam. Eight, one-person pontoon boats deployed Teledyne, RiverPro acoustic Doppler current profilers (ADCP, Poway, California) with fully integrated Hemisphere, Vector V102GPS compass units deployed from catamaran rafts while floating on the river to measure water depth and velocities and current vectors. Mapping was conducted in two separate surveys: On 11–14 September 2017, and on 26 March 26–1 April 2018. Sampling dates in March were chosen because SAV was not present at that time. In shorelines areas, technicians stayed in deeper water to avoid SAV. SAV can also cause incorrect readings of the streambed with ADCP.

Water Depth and Velocity

To measure depth, velocity, and discharge in the September 2017 survey, the ADCP team made at least two separate data collection runs each day. The first run covered one half of the river channel to a midpoint of the day's collection objective (river left). The second half of the river channel (river right) was measured as the team returned to the day's measurement starting point. The team maintained a preset distance apart and semi-staggered positioning. River discharge at the Newport U.S. Geological Survey (USGS) gage 12395500 during the survey ranged from 150 m^3 s^{-1} to 385 m^3 s^{-1}. For the March 2018 survey, depth, velocity, and discharge data were collected on the southern 48 km of the river, and the eight-person team made four passes each day over 8 km for a total 32 passes each day for 6 days. River discharge ranged from 694 m^3 s^{-1} to 745 m^3 s^{-1}. For both sample periods, the measurement error was within 1% of the actual flow and depths being sampled.

The survey team also measured river discharge at 20 transect locations along the river, once at the beginning of each day and once at the end. Discharge measurements were replicated 5 to 8 times at each location following standard USGS protocols for discharge measurements, resulting in a mean variance in discharge measurements less than 1% for most locations and less than 3% for all, following the methods of Lorang and Tonolla [51] and Marotz and Lorang [52]. A DEM that included bathymetry tied to publicly available lidar (light detection and ranging) data was produced from the water depth data. The DRIFTER routine in River Analyzer was used to create bottom boundary and top velocity maps.

Transect Velocity and Discharge Calculations

The reservoir was divided into 50,000 cross-channel slices using the SLICER routine in River Analyzer. Each slice was 12 m apart, spanning from river left to river right, and contained data from 32 longitudinal transects composed of 10-cm velocity bin ensembles extending from the surface to the bottom minus a 10-cm blanking distance at the surface and a variable bottom blanking distance of 10 cm to 50 cm depending on variable signal attenuation factors. Velocity ensembles covered depths from 30 cm to 8 m resulting in 8000 to 10,000 velocity data points that went into each slice discharge calculation. Wetted perimeter was calculated based on bottom ping data which defined the bathymetry that gave the distance along the bottom as the wetted perimeter for each slice. Hydraulic radius for each slice was derived by summing the cross-sectional area for each transect slice and dividing by the wetted perimeter.

We also calculated the Froude number (Fr), a dimensionless value that describes the energetic state of the water column and values greater than 1 represent supercritical flow:

$$\mathrm{Fr} = V/(gD)^{0.5} \tag{1}$$

where V is the average slice water velocity (m s^{-1}), D is the hydraulic depth (cross-sectional area/top width, in meters), and g is the acceleration of gravity (9.8 m s^{-2}). For our case we wanted to quantify the relative energetic state of the water column as a potential proxy for potential water column mixing.

Lastly, continuous transects along the river, each 100-m wide and perpendicular to the river centerline, were created to summarize depth and velocity (i.e., mean, standard deviation, median, 25th quantile, 75th quantile, minimum, and maximum). Each of these transect polygons were associated with Universal Transverse Mercator (UTM) coordinates of the point where the river centerline intercepted the midline of the perpendicular transect, a rectangular polygon with 100 m width.

Submerged Aquatic Vegetation

SAV is a dominant feature of this river and can influence sediment dynamics, water velocity, and thermal gradients [15]. To characterize SAV distribution, we used data mapped by The Pend Oreille County Public Utility District in 1997. We recognize that although these data may not reflect current conditions, they are representative of 2001 conditions when the TIR data were collected. For our analysis, we only used the areas classified as densely vegetated regardless of plant species composition, and we did not consider unvegetated areas. The midpoint of these polygons was then linearly referenced to the nearest river kilometer.

2.4. Data Analysis

2.4.1. Spatial and Temporal Analysis

To describe spatial and temporal thermal heterogeneity of the Pend Oreille River, we used digital hydrography data (mixed-scale hydrography, version 3.1, or "MSHv3.1") at 1:24,000 scale or finer for the state of Washington from the StreamNet Project, Pacific States Marine Fisheries Commission as the map template in a geographical information system (GIS) (https://www.streamnet.org/gisdata/map_data_base/Hydrort_MSHv31_September2012.xml, accessed on 13 October 2016). We then reduced the thermal and environmental ancillary data (e.g., flow velocity, depth) to one-dimensional positions, expressing these geographic positions as measurements along the river's path in relation to the confluence with the Columbia River [53]. We calculated minimum, average, maximum, daily minimum, and daily maximum differences between surface and bottom water temperatures, and the consecutive 7-day moving average of daily maximum (7DADM) temperature for all sites. We calculated the 7DADM because this metric incorporates the average of maximum temperatures that fish are exposed to over a week and is used in water quality standards to protect against acute effects such as mortality and migration blockage [54]. Confluences were considered stratified when differences between hourly surface and bottom water temperatures were equal or greater than 2 °C, and we estimated the percentage of time that confluences were stratified.

To determine whether main stem or tributary confluences could potentially be used as cold-water refuges by adult salmonids, we used the period from 9 August to 13 September 2017. Sites were considered potential cold-water refuges based on Greer et al. [55]: (1) Ambient refuges, defined by differences of 1 °C or 2 °C, and (2) physiological refuges, defined as the physiological threshold for migrating adult salmonids (18 °C) [54–57]. More specifically, bottom water temperatures in the main stem can potentially function as cold-water refuges when the difference between the surface water temperature and bottom water temperature was greater than 1 °C or 2 °C for at least an hour a day, or when bottom 7DADM water temperatures were lower than 18 °C. Confluences can be potential cold-water refuges when their 7DADM temperatures are 1 °C or 2 °C cooler than the 7DADM main stem temperature, or the 7DADM confluences temperature were below 18 °C. In the Pend Oreille River, acoustic-tagged salmonids were observed near shore and at shallow depths, but detection in deeper areas was difficult because of the loss of signal (J. Connor, unpublished data). In contrast, in the main stem Columbia River, tagged bull trout used deep (12.6 m on average), slow habitats [58]. We characterized the temporal variability and persistence (i.e., the number of times a potential cold-water refuge at a given location was observed) [18] for each of the thermal criteria outlined above. Finally, we estimated the spacing between all persistent physiological cold-water refuges and compared them to the spacing of physiological cold-water refuges present at least 50% of the time to determine (1) how far salmonids would have to travel before they experienced thermal relief during their migration, and (2) how often salmonids may use these ambient cold-water refuges as stepping stones because the distances between physiological cold-water refuges may exceed the swimming abilities of trout under these stressful conditions [59].

2.4.2. Statistical Analysis

We calculated z-scores for all the channel morphology metrics measured or estimated during 3-D mapping to normalize measures for temporal variation among the two surveys. Z-score distributions all have the same mean and standard deviation, and individual scores from different distributions can be directly compared. The complete suite of variables for which we calculated z-scores included surface width (m), cross-sectional area (m^2), wetted perimeter (m), hydraulic radius (m), water column velocity (ms^{-1}), maximum water depth (m), mean water depth (m), and Froude number.

Generalized Additive Modeling

We explained the probability of occurrence of cool water areas (i.e., troughs in the longitudinal profile) and the median water temperature from the TIR data using generalized additive models (GAMs). Generalized additive models were selected because of their ability to compute nonlinear relationships between the response and the set of explanatory variables [60]. The response variables were (1) the presence of troughs, defined as areas of cooler water than the median radiant water temperature of the river, and (2) the median radiant temperature. The GAM analysis uses smoothers to describe the empirical relationships between explanatory variables and response variables and, therefore, does not assume a specific relationship. We used the GAM function in the Mixed GAM Computation Vehicle with Automatic Smoothness Estimation (MGCV) package of the statistical program R [61,62]. For the probability of occurrence model and the median radiant water temperature, we used a binomial distribution with a logit-link function and a Gaussian distribution with a log function, respectively, to determine whether there were significant relationships between the response and explanatory variables.

Because rivers are hierarchical systems, we considered two spatial scales and selected explanatory variables that characterized temperature patterns at each scale: Reach (100 m to 1 km), and subcatchment (1 km to tens of kilometers) (Table 2).

Table 2. Explanatory variables evaluated for associations with median radiant water temperature and the probability of occurrence of cool-water areas.

Variable	Units/Category	Data Type	Description
Reach scale			
Cross sectional area	z-scores	continuous	Measured with acoustic Doppler current profilers (ADCP)
Maximum water depth	z-scores	continuous	Measured with ADCP
Degree of confinement	m m^{-1}	continuous	Measured with remote sensing
Dense SAV	km^2	categories	Sum of all dense submerged aquatic vegetation patches regardless of plant species composition.
Subcatchment scale			
Land use	%	continuous	Measured with remote sensing as percent developed land.
Aquifer permeability,	high/low	category	Locations where shallow subsurface groundwater meets deep groundwater flows
Size of tributaries	km^2	continuous	Watershed size
Distance from Albeni Falls dam	km	continuous	Distance from Albeni Falls dam to Box Canyon dam (lower boundary)

We checked for concurvity using the MGCV package [63]. Concurvity is the existence of nonlinear dependencies among predictor variables [64], which we minimized by removing variables with a Pearson correlation greater than 0.40. Presence of concurvity in the data may lead to poor parameter estimation and loss of interpretation as the effect of a predictor on the response variable may change depending on the order of the predictors in the model if concurvity is present [64].

To automate model selection from the global model for each response, we used the Multi-Model Inference (MuMIn) R package [65]. We used Akaike's Information Criterion (AIC) with the small-sample bias adjustment (AICc) and calculated Akaike weights to assess the relative fit of each candidate model and compared candidate models for the same data with the best-fitting models with the lowest AICc [66]. We then selected the models that had an AICc difference less than 2. Models with a difference less than 2 were considered to have a substantial level of empirical support for the most plausible candidate model [66]. Finally, we selected the most parsimonious model from the models that had AICc difference less than 2.

3. Results

3.1. Comparisons between TIR and Instream Measurements

The TIR data indicated a downstream warming trend, with several peaks and troughs along the longitudinal thermal profile (Figure 2).

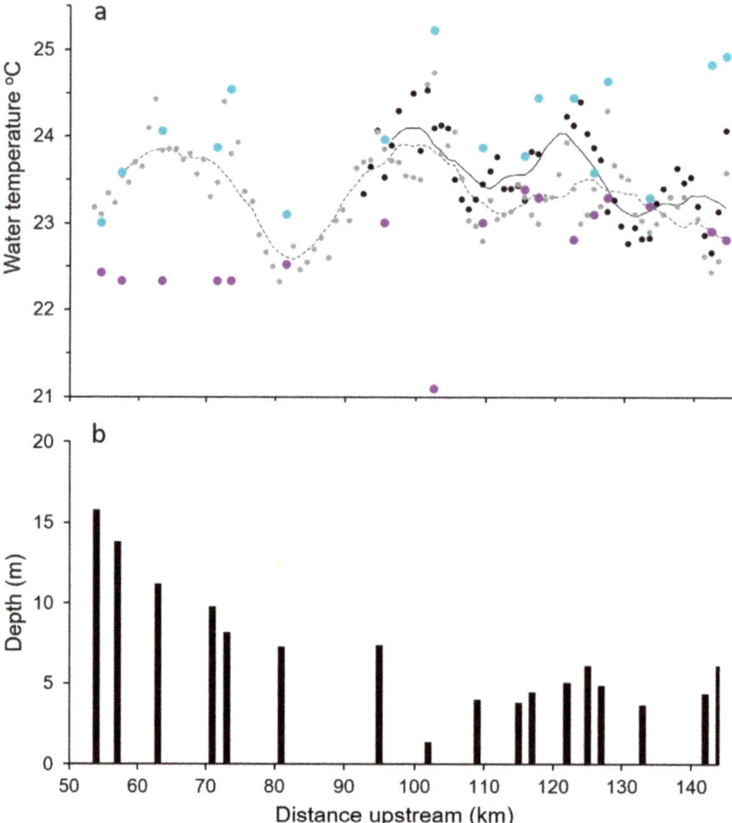

Figure 2. Longitudinal thermal profile of the Pend Oreille River based on airborne thermal infrared (16 August 2001) and instream (22 August 2017) temperature data (**a**). Black circles and gray circles represent radiant water temperature for the right and left banks, respectively. Dotted and smoothed lines are 10-km centered moving averages. Purple circles and cyan circles represent bottom daily maximum water temperature and surface daily maximum water temperature, respectively. Water depth at deployment for all instream water temperature main stem sites (**b**).

The median radiant temperatures were 23.4 °C for the right bank and 23.6 °C for the left bank, with standard deviations of ±0.50 and ±0.48, respectively. We found that the 22 August 2017 instream surface water temperature lowest trough and highest peak coincided with the TIR lowest trough and highest peak. The lowest TIR trough was 22.3 °C and was located near river kilometer (rkm) 80 in a narrow and deep section of the river and near a cold tributary, Ruby Creek (C5, 17.9 °C). The troughs were generally associated with either cold tributaries or springs and seepages. In contrast, the largest TIR peak was 24.7 °C. This section of the river is wide and shallow and has warm water inputs from Cusick Creek (C9, 28.3 °C) and a large side channel (25.7 °C).

The TIR sensor only detects thermal radiation emitted from the upper 0.1 mm of the water surface, but there were numerous instances where thermal stratification may have occurred. Examples of midstream stratification were observed intermittently (e.g., rkm 57 to rkm 80, rkm 88 to 102, and rkm 122 to rkm 126) and in tributary and side channel junctions where slow-moving water cooler than the main stem generally sank to the bottom when it mixed with the warmer main stem water. The instream measurements confirm some of this stratification; for example, in midstream near rkm 102 on 22 August 2017, maximum surface water temperature was 25.2 °C and minimum bottom water temperature was 21.1 °C, a 4.1 °C difference between surface and bottom water temperatures.

A total of 25 tributaries and side channels were mapped during the TIR surveys, and of those, only six were cooler than the mean temperature of the Pend Oreille River. Four of these six cooler tributaries were found between rkm 81 and rkm 93, near where the largest TIR trough was observed. However, the bottom instream loggers deployed at confluences in 2017 revealed that most of the tributaries sampled were cooler than the main stem Pend Oreille River and had strong vertical temperature gradients at their confluences (Figure 3). The confluences of tributaries such as Renshaw Creek (C3) and Tiger Slough (C4) had average temperature differences of 9.5 °C and 11.9 °C between the surface and bottom, respectively. Streams such as Skookum Creek (C13) and Indian Creek (C16) were well mixed.

3.2. Characterization of Spatial and Temporal Thermal Heterogeneity

3.2.1. Longitudinal

We did not observe a warming trend in the 7DADM surface or bottom water temperature along the longitudinal profile, nor when we averaged the entire period of collection (17 July to 14 September), by each month. However, the daily maximum water temperature data showed several peaks and troughs along the longitudinal profile that the daily minimum water temperature did not (Figures 4a and S1). The range of mean bottom water temperature along the longitudinal profile was 21.2–23.0 °C, and the range of mean surface water temperature was 22.7–23.4 °C. The minimum (19.6 °C) and maximum (28.8 °C) surface water temperature and minimum (19.0 °C) bottom water temperature were all observed at P4, whereas the maximum bottom water temperature (25.4 °C) was observed at P7.

3.2.2. Lateral: Tributary Confluences

Of the 22 confluences we surveyed, 17 were retained for analysis because they did not have any missing data (Figure 3). Contrary to the limited spatial longitudinal variability in the main stem, spatial thermal variability at the confluences was approximately an order of magnitude greater than the main stem. Overall, the range of water temperatures for all confluences during the entire period varied considerably. Surface temperatures had a wider range (7.9–32.3 °C) than bottom temperatures (7.9–27.4 °C). C17 was the most variable confluence at the surface (Δ24.4 °C) and C14 was the most variable confluence at the bottom (Δ16.9 °C), whereas the least variable confluence was C4 for both surface and bottom (Δ2.7 °C). The 7DADM for the surface water temperature ranged from 18.8 °C (C13) to 30.2 °C (C17), and the 7DADM bottom water temperature ranged from 12.4 °C (C4) to 24.5 °C (C14). As expected, the 7DADM for the bottom water temperature was cooler than the surface water 7DADM by at least 1 °C degree at all confluences (except at C13, 18.7 °C).

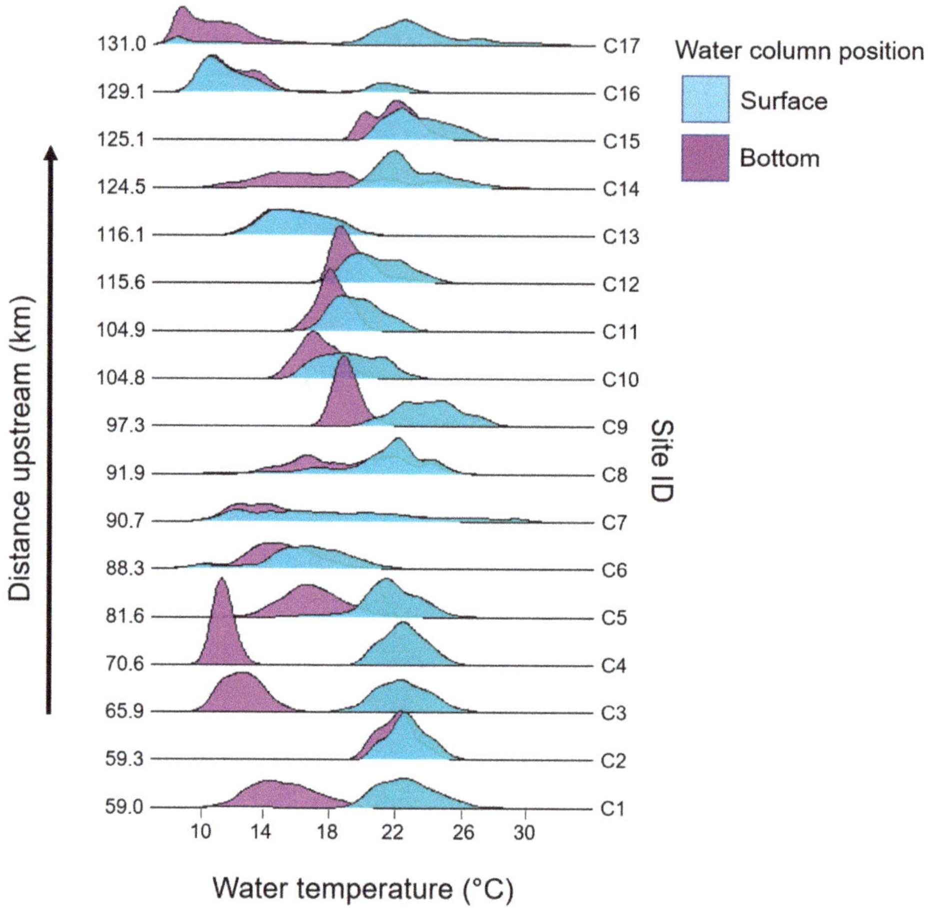

Figure 3. Water temperatures in tributary confluences (rows) measured from 17 August to 13 September 2017 with instream temperature loggers. Bottom (purple) and surface (cyan) hourly water temperature distributions illustrate the extent of vertical thermal stratification.

3.2.3. Vertical: Water Column

Differences between daily minimum bottom and daily minimum surface water temperatures in the main stem were less than 0.5 °C, with daily minimum water temperatures near 20.3 °C (SD ± 0.24 to 0.36 °C) except for P4, where minimum bottom and surface water temperatures were 19.0 °C and 19.6 °C, respectively. Maximum daily surface water temperatures in the main stem were on average 1.2 °C warmer than maximum daily bottom water temperature. Furthermore, hourly differences between surface and bottom water temperature ranged from −0.2 °C to 4.8 °C: 17 sites had differences greater than 1 °C, and 13 sites had differences greater than 2 °C (Figures 4b and S1).

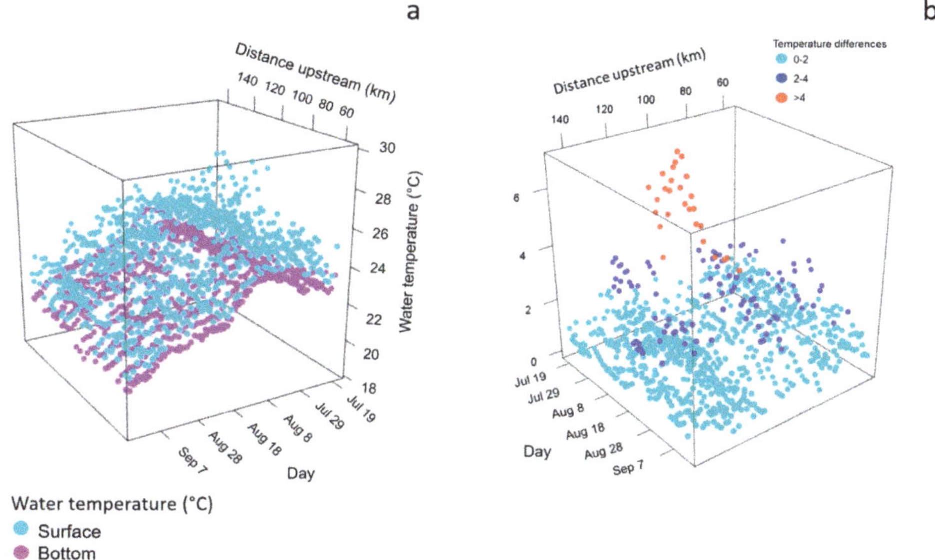

Water temperature (°C)
Surface
Bottom

Figure 4. Patterns of longitudinal thermal heterogeneity (**a**) of the Pend Oreille River measured with instream maximum and minimum temperature loggers deployed from 17 July to 14 September 2017, and (**b**) maximum differences between surface water temperature and bottom water temperature observed from instream temperature loggers deployed along the longitudinal profile (**b**). See Supplementary Material S2 for R code to generate graphics for interactive visualization in three-dimensional space.

Unlike the main stem where water temperature differences between the surface and bottom were small, temperature differences at the confluences in the tributaries were more pronounced. Hourly differences between surface and bottom water temperature ranged from −11.1 °C to 19.2 °C, which was nearly four times greater than the vertical variability in the main stem. Furthermore, we identified five confluences that had daily minimum bottom water temperatures of at least 1 °C cooler than the daily minimum surface water temperature, with the largest hourly differences observed for the entire period at C3 (12.4 °C) and C4 (15.2 °C). We identified eight confluences that had periods where hourly surface water was at least 1 °C cooler than bottom water. These occurrences were most pronounced at C7 (11.1 °C), C8 (4.3 °C), and C14 (4.6 °C) in late August and September when dense SAV patches trapped warm water near the streambed while surface water cooled at night (Figure S1).

3.2.4. Temporal

We initiated data collection on 8 July 2017, but we restricted our data analysis to the relatively warmer 17 August to 13 September 2017 period for comparison among sites due to missing data. Water temperature in the main stem (surface and bottom) was above 20 °C except at one site (P4) during our last week of sampling in September. Diel thermal patterns in the main stem were relatively uniform along the longitudinal profile (Figure 5), and typically water was warmest from late morning to late afternoon, with considerable variation from day to day. These diel patterns were also more pronounced at the surface than at the bottom (Figure 5).

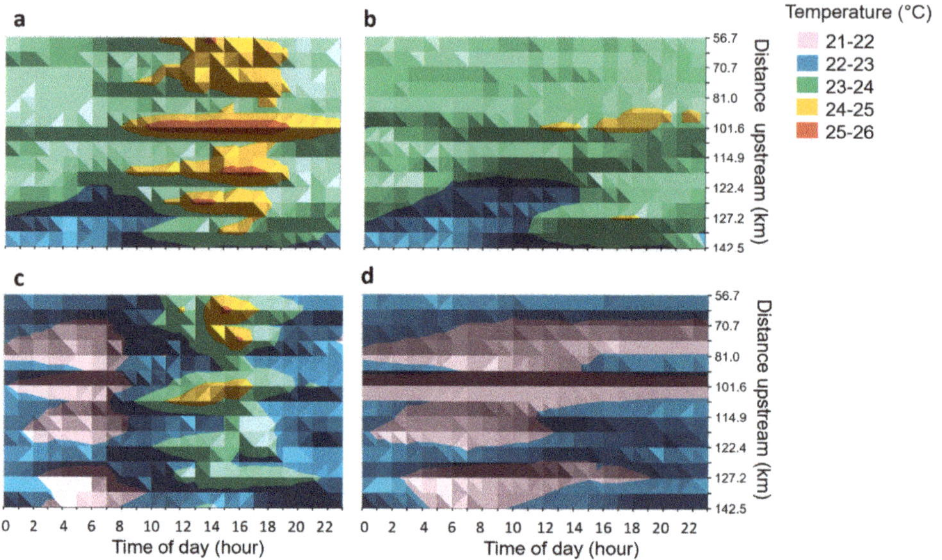

Figure 5. Hourly water temperature near the surface (**a**) and bottom (**b**) for 23 July 2017 and surface (**c**) and bottom (**d**) for 27 August 2017 along the longitudinal profile of the Pend Oreille River. Values were interpolated between sites and times. Y-axes indicate river kilometer locations and are not scaled linearly. These days were selected because both days had a maximum air temperature of 33.3 °C but they had different diel patterns. On 27 August, air temperature was 3 °C cooler in the morning than on 23 July.

Diel patterns at confluences, like the main stem, revealed peaks in mid to late afternoon and troughs late at night or early morning for both surface and bottom thermographs. Over the course of the summer, most of the confluences water temperature decreased, but some were highly variable (Supplementary Material S2).

3.2.5. Implications for Potential Cold-Water Refuges

The presence of a potential cold-water refuge depended on the definition used. We did not observe main stem water temperatures cooler than 18 °C during our study, so none of the main stem sites were classified as a potential physiological cold-water refuge. The percentage of time during which the main stem sites met the 1 °C ambient criterion ranged from 14% (P5) to 100% (P4). However, when we used the 2 °C ambient criterion, 5 sites (T1, T3, P3, P5, and T53) were not considered potential ambient cold-water refuges, an additional 5 sites were considered potential ambient cold-water refuges less than 10% of the time (BC, T4, P6, T7 and AF), and one site (P4) decreased slightly to 98% (Table 3).

Table 3. Seven-day average daily maximum (7DADM) water temperature at main stem sites where instream loggers were deployed at the surface and on the bottom of the river. Cold-water refuge persistence (percentage of time) is presented for two ambient criteria based on whether the difference between the surface water temperature and bottom water temperature was greater than 1 °C or 2 °C for at least an hour a day.

Site	Main Stem Water Temperature 7DADM (°C)		Cold-Water Refuge Persistence (%)	
	Bottom	*Surface*	*2 °C Criterion*	*1 °C Criterion*
BC	24.2	24.4	3	19
T1	24.1	25.3	0	74
P1	24.2	25.8	22	78
T2	24.2	24.9	17	69
P2	24.1	25.3	14	61
T3	24.0	24.6	0	20
P3	24.3	24.7	0	22
P4	21.9	25.4	98	100
T4	24.1	24.6	3	29
P5	24.5	25.0	0	14
T5	24.2	25.1	0	34
P6	24.1	25.0	6	54
T6	24.3	24.7	14	54
P7	24.5	25.0	20	66
T7	24.4	25.1	3	31
T8	24.2	25.1	11	51
AF	24.2	24.9	9	57

Five confluences were classified as potential physiological cold-water refuges over 96% of the time when bottom 7DADM water temperatures were below 18 °C (Table 4), whereas only two confluences never experienced temperatures below this physiological criterion (C2, C9). The remaining 10 confluences were classified as potential cold-water refuges between 10% and 62% of the time. Over twice as many confluences (11) were always classified as potential ambient cold-water refuges in relation to the main stem, and only one confluence, C2, was never a potential cold-water refuge regardless of the ambient criterion used (2 °C versus 1 °C). The remaining six confluences were potential cold-water refuges 52–96% of the time when using the 1 °C criterion or 31–86% of the time when using the more restrictive 2 °C criterion (Table 3). Overall, persistence of potential cold-water refuges was associated with the difference between main stem 7DADM bottom water temperature and confluence 7DADM bottom water temperature regardless of criterion (Spearman r = 0.89, 0.81, 0.82, Figure 6). The spacing along the longitudinal profile among persistent physiological cold-water refuges formed by confluences was greater than 2 km but less than 41 km (median = 7 km). When we considered physiological cold-water refuges that were present at least 50% of the time, the spacing (distance between nearest neighbors) was reduced to less than 18 km, but the median distance increased from 7 to 11 km.

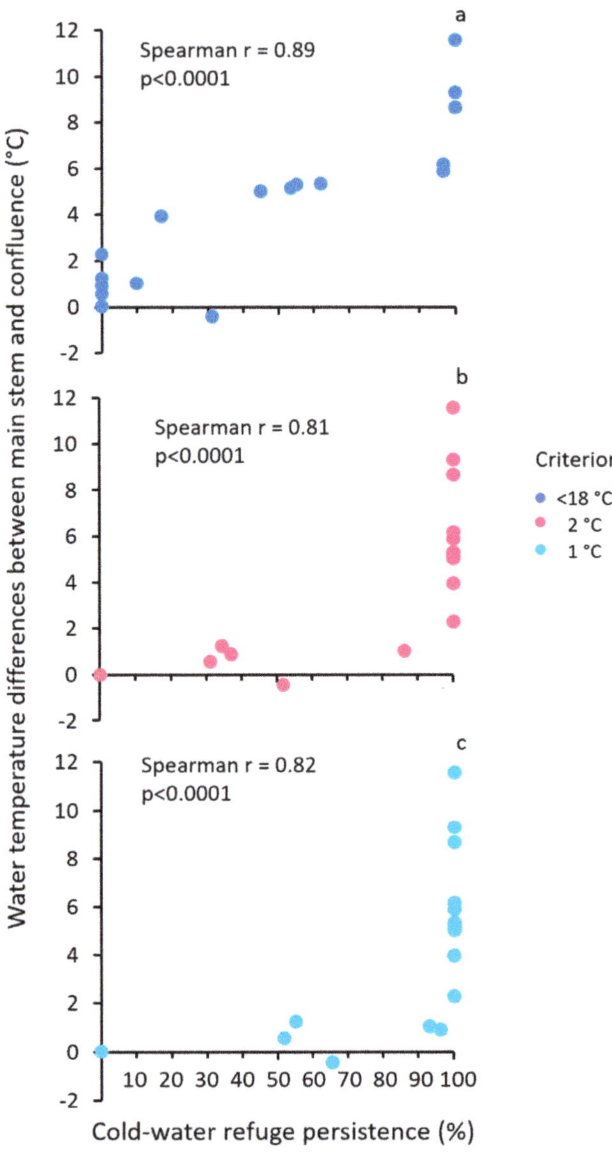

Figure 6. Relationships between persistence of cold-water refuges and bottom water temperature differences (7DADM) between the main stem and confluences for three different cold-water refuge criteria: (**a**) Physiological, <18 °C, (**b**) ambient 2 °C, and (**c**) ambient 1 °C.

Table 4. Seven-day average daily maximum (7DADM) water temperature at main stem and confluence sites where instream loggers were deployed at the bottom of the river. Cold-water refuge persistence (percentage of time) is presented for ambient and physiological criteria based on whether the 7DADM confluence temperatures were (1) 1 or 2 °C cooler than the 7DADM main stem temperature, or (2) below 18 °C. Stratification is defined as the percentage of time when hourly differences between surface and bottom water temperatures are equal or greater than 2 °C.

Site	Bottom Water Temperature 7DADM (°C)		Cold-Water Refuge Persistence (%)			Stratification (%)
	Confluence	*Main stem*	*Criterion = <18 °C*	*Criterion = 2 °C*	*Criterion = 1 °C*	
C1	18.8	24.0	53	100	100	97
C2	24.0	24.0	0	0	0	1
C3	14.7	24.0	100	100	100	100
C4	12.4	24.0	100	100	100	100
C5	20.1	24.0	17	100	100	95
C6	18.1	24.0	97	100	100	36
C7	23.0	24.0	10	86	93	43
C8	22.8	24.0	0	35	55	25
C9	20.8	21.7	0	37	96	70
C10	18.5	23.8	62	100	100	39
C11	18.8	23.8	45	100	100	32
C12	21.7	24.0	0	100	100	33
C13	18.7	24.0	55	100	100	0
C14	24.5	24.0	31	52	66	77
C15	23.7	24.3	0	31	52	36
C16	18.1	24.3	97	100	100	14
C17	15.6	24.3	100	100	100	87

3.3. Characterization of Riverine Landscape

3.3.1. GIS Data

Land use (% developed land), aquifer permeability (high versus low), degree of channel confinement (m m^{-1}), and tributary contributions (area in km^2 and number of tributaries) were used to characterize the riverine landscape in the Pend Oreille River (Figure 7). The percent developed land along the Pend Oreille River ranged from 3% near Box Canyon dam to 75% near the city of Cusick at rkm 101. Three areas with a high percentage of developed land (peaks) were observed: One near the town of Ione, Washington; another at the largest thermal peak near the town of Cusick; and a third peak near the town of Oldtown, Idaho, near Albeni Falls dam. Aquifer permeability was low in most of the Pend Oreille River except in two locations: At rkm 115 and between rkm 133 and rkm 144 where numerous seeps and springs were observed. Similarly, most of the Pend Oreille River was considered unconfined (>4) with the section from rkm 97 to rkm 122 having confinement values greater than 10. Additionally, there were no confined sections (<2) and only 9 km along the longitudinal profile were moderately unconfined, with most of these instances located near Box Canyon Dam and from rkm 83 to rkm 85.

The average number of tributaries per kilometer was 1.85 (SD ± 1.40). A maximum of 5 tributaries per kilometer were found at rkms 72, 84, and 87, but the longest section without tributary contributions was located from rkm 117 to 122, followed by the section from rkm 98 to 101. Most tributaries drained areas near or less than 50 km^2. From those tributaries that drained considerably larger areas, the largest tributary was Calispel Creek, followed by C6 and C11 (Figure 7).

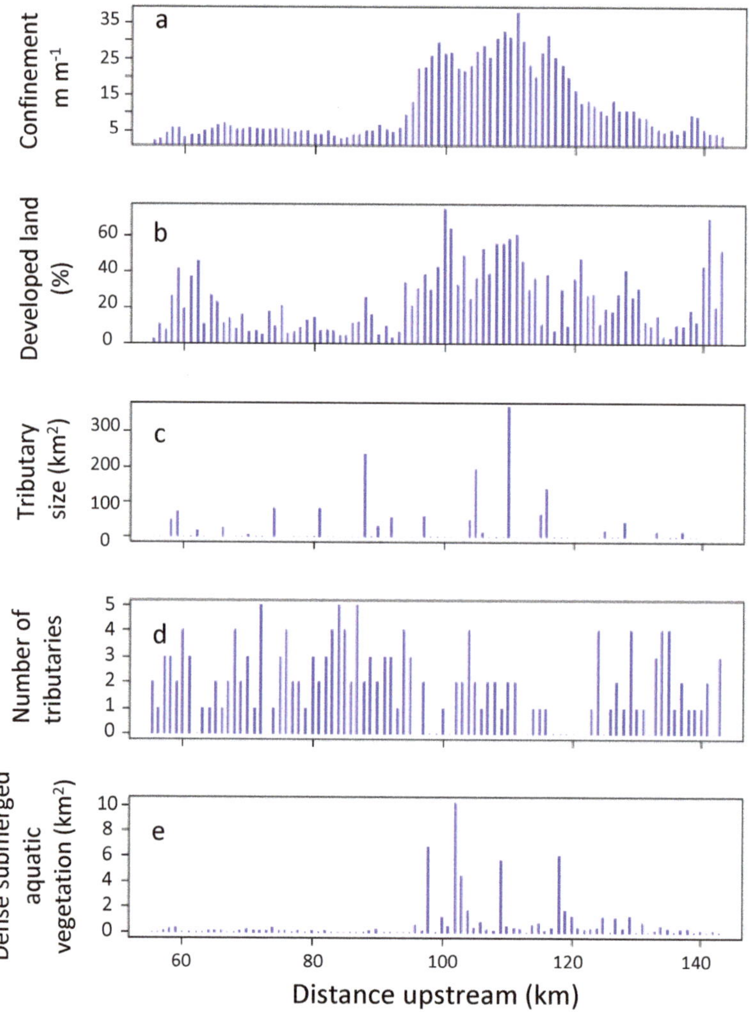

Figure 7. Longitudinal profiles of landscape metrics in the Pend Oreille River: Confinement (**a**), developed land (**b**), contributing tributary size (**c**), number of tributaries (**d**), and dense submerged aquatic vegetation (**e**). Bin width is 1 km.

3.3.2. Three-Dimensional Hydraulic Mapping and Channel Morphology

The z-scores for water velocity showed four distinct sections of the river along the longitudinal profile. The section immediately upstream of Box Canyon to rkm 78 and the section from rkm 95 to near rkm 125 were slower than the mean water velocity, and the sections from rkm 78 to 94 and the section nearest the Albeni Falls dam were faster than the mean velocity (Figure 8). The z-scores of the Froude number also showed like patterns to those observed for water velocity. As expected, the z-scores for wetted perimeter and channel width were similar and highly correlated (0.99, $p < 0.00001$). The z-scores of the cross-sectional area mirrored those of water velocity (Pearson correlation coefficient $= -0.90$, $p < 0.0001$). The z-scores of average depth and maximum depth indicated that the deepest locations in the river were near the Albeni Falls dam (rkm 131 and rkm 142), whereas most of the shallowest locations in the river were found between rkm 71 and rkm 87 with a few locations near Albeni Falls dam (Figure 8). Most of the river had slow, tranquil flow (e.g., 0.05–0.1 Fr), but some portions of the river in the upper reaches were faster and more turbulent (e.g., 0.4 Fr, depth < 1 m, and velocity 1 m s^{-1}).

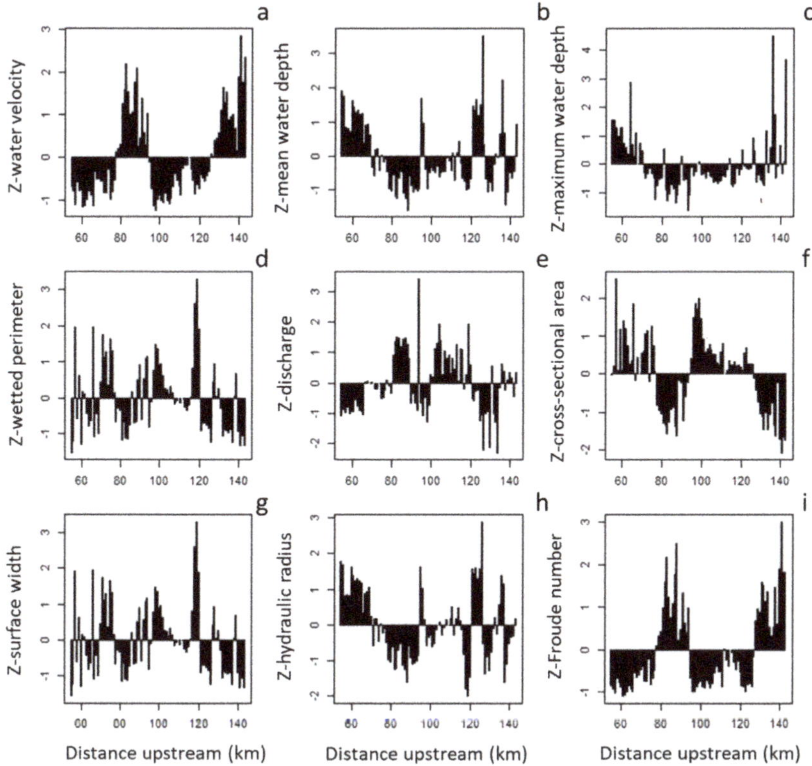

Figure 8. Z-scores for channel morphology metrics were either measured directly or estimated during the 3-D hydraulic mapping in September 2017 and March 2018: (**a**) Water velocity, (**b**) mean water depth, (**c**) maximum water depth or estimated, (**d**) wetted perimeter, (**e**) discharge, (**f**) cross-sectional area, (**g**) surface width, (**h**) hydraulic radius, and (**i**) Froude number. Bin width is 1 km.

3.3.3. Submerged Aquatic Vegetation

The most abundant dense patches of submerged aquatic vegetation were found from rkm 98 to 104. Aquatic vegetation density was low near both dams (rkm 55 to 71 and rkm 139 to 143) (Figure 7).

3.3.4. Associations between Hydrologic and Geomorphic Variables and TIR Imagery

The best reach-scale model within the set of models predicting median water temperature included confinement and the z-scores of cross-sectional area and maximum depth as explanatory variables (AIC$_c$ = 60.2, w$_i$ = 0.48, Table 5). Based on this model prediction, the z-scores of cross-sectional area, maximum depth, and confinement explained approximately 59% of the variability in the median water temperature at the reach scale (Table 6). The best reach-scale model within the set of models predicting the probability of occurrence of cool water included confinement and the z-scores of cross-sectional area as explanatory variables (AIC$_c$ = 65.0.1, w$_i$ = 0.20, Table 4). This model explained 51% of the variability (Table 6). Cross-sectional area had a positive relationship with median water temperature and a negative relationship with probability of occurrence of cool-water areas.

Table 5. Candidate models with reach and subcatchment scale explanatory variables predicting the median radiant water temperature and the probability of occurrence of cool water in the Pend Oreille River. For each model, the number of parameters (K), -2 log-likelihood ($-2\mathrm{logL}$), Akaike's information criterion adjusted for small sample size ($\mathrm{AIC_c}$), the difference in $\mathrm{AIC_c}$ between the given model and the best-performing model ($\Delta\mathrm{AIC_c}$), and Akaike weight (w_i) are shown. An asterisk (*) indicates a model selected based on parsimony in addition to $\mathrm{AIC_c}$ scores and weights.

Model	K	$-2\mathrm{LogL}$	$\mathrm{AIC_c}$	$\Delta\mathrm{AIC_c}$	w_i
Reach scale					
Median water temperature					
Confinement, Sum all dense vegetation, z-Area, z-Maximum depth	14	−12.98	60.0	0.00	0.52
Confinement, z-Area, z- Maximum depth	11	−16.09	60.2	0.16	0.48 *
Probability of occurrence of cool-water areas					
Confinement, z-Area, z-Maximum depth	6	−25.57	64.2	0.00	0.30
Confinement, Sum all dense vegetation, z-Area, z-Maximum depth	6	−25.57	64.2	0.00	0.30
Confinement, z-Area	4	−27.26	65.0	0.84	0.20 *
Confinement, Sum all dense vegetation, z-Area	4	−27.26	65.0	0.84	0.20
Subcatchment scale					
Median water temperature					
Distance from dam, Percent developed land	10	−5.25	34.3	0.00	0.28
Distance from dam, Percent developed land, Tributary area	10	−5.21	34.3	0.01	0.27
Distance from dam	9	−6.22	34.7	0.38	0.23 *
Distance from dam, Tributary area	10	−5.95	34.7	0.40	0.23
Probability of occurrence of cool water					
Distance from dam, Percent developed land	8	−19.87	59.6	0.00	0.50 *
Distance from dam, Percent developed land, Tributary area	8	−19.87	59.6	0.00	0.50

Table 6. Best generalized additive models for reach and subcatchment scales for each response variable, median water temperature, and probability of occurrence. Effective degrees of freedom (edf), reference degrees of freedom (Ref.df), F-values for the Gaussian distribution models, and chi-square values for the binomial distribution models, *p*-values, adjusted R-squared, and percent deviance explained by models.

Model	edf	Ref.df	F or Chi-Square	*p*-Value	R-sq. (adj)	Deviance (%)
Reach scale						
Median water temperature						
s (Confinement)	2.29	9	1.09	0.007	0.59	64
s (z-Area)	6.72	9	9.95	<0.001		
s (z-Maximum depth)	0.96	9	0.42	0.048		
Probability of occurrence of cool water						
s (Confinement)	2.92	9	8.74	0.02	0.51	50
s (z_Area)	0.99	9	16.63	<0.001		
Subcatchment scale						
Median water temperature						
s (Distance from dam)	7.78	9	21.07	<0.001	0.68	71
Probability of occurrence of cool water						
s (Distance from dam)	5.65	7	21.73	<0.01	0.63	64
s (Percent land developed)	0.77	9	2.49	0.07		

At the subcatchment scale, the best models included distance from the dam for the set of models predicting median radiant temperature (AIC$_c$ = 34.7, w$_i$ = 0.23, Table 5), and distance from the dam and percent land developed for the set of models predicting probability of cool water occurrence (AIC$_c$ = 59.6; w$_i$ = 0.50, Table 5). These models revealed that distance from Albeni Falls Dam explained approximately 68% of the variability of median water temperature and 63% of the probability of occurrence of cool-water areas. All the models explained at least half of the variability of the response variables, but the subcatchment models performed slightly better (Table 6).

4. Discussion

We combined the high spatial resolution of airborne thermal infrared (TIR) imagery with the high temporal resolution of instream measurements to describe broad-scale patterns of water temperature in relation to localized drivers of heterogeneity. Our findings demonstrated that (1) lateral contributions from tributaries along the longitudinal profile dominated thermal heterogeneity in the Pend Oreille River, (2) spatial and temporal thermal variability at confluences was approximately an order of magnitude greater than that of the main stem, (3) most of the potential cold-water refuges were found at confluences regardless of the refuge definition used, and (4) the probability of occurrence of cool areas and the median radiant water temperature were closely linked to geomorphic metrics at the reach scale and to the distance from the Albeni Falls Dam at the subcatchment scale. These results highlight the importance of using multiple approaches to describe thermal heterogeneity in large impounded rivers. TIR data alone provided an incomplete picture of the thermal variability found at tributary confluences and in areas of the main stem with vertical stratification. Instream thermographs at the bottom of most confluences recorded considerably cooler temperatures than surface loggers and revealed discontinuities in the longitudinal profile that were not observed in the TIR data. These results have important implications for the identification and restoration of cold-water refuges.

This study also expands understanding of thermal heterogeneity in large impounded rivers by complementing earlier work, mostly in Europe and North America (e.g., [8,32,33]). Overall, we observed a slight downstream increase in water temperature along the longitudinal profile with multiple peaks and troughs. A downstream increase in water temperature is a pattern that has been reported in many rivers and, until recently, was our conceptual model for rivers in general [26]. Like Arscott et al. [30], we found that lateral heterogeneity dominated thermal heterogeneity. Lateral heterogeneity in our study was an order of magnitude greater than longitudinal heterogeneity, but unlike Arscott et al. [30] where this variability was related to floodplain habitat, lateral variability in our study is associated with tributary confluences, as most floodplain habitat has been modified in the Pend Oreille River. Other studies [67–69] have investigated lateral heterogeneity at the segment and reach scales but longitudinal heterogeneity was not considered.

Because the rivers surveyed with TIR remote sensing are generally well mixed vertically except at isolated large features such as pools [16,26], previous studies have focused mostly on thermal variability in longitudinal and lateral dimensions [26,67,68]. Persistent thermal stratification in main river channels seldom occurs because streamflow is usually turbulent enough to cause complete mixing [70]. Like previous studies on riverine thermal stratification, we found that stratification in the main stem was only persistent at one site, P4, where the river was shallow, slow moving, unconfined, and had dense patches of aquatic vegetation. However, it is important to consider temporal variability of vertical thermal gradients because of their potential use by fish as ambient cold-water refuges. We found that 10 main stem sites had differences between surface and bottom water greater than 1 °C for at least one hour per day about 50% of the time. This difference appears small, but it may provide salmonids thermal refuge during migration while fish travel to and from cold-water refuges that better meet their physiological needs.

Our results corroborate those of Wawrzyniak et al. [71] who found that TIR-derived temperatures overestimated in situ measurements at most confluences where thermal stratification was regularly observed. In addition, four confluences were stratified persistently, near 100% of the time (C1, C3,

C4, and C5), two confluences were hardly ever stratified (C2, C13), and the rest of the confluences were stratified between 13 and 87% of the time. In a tributary of Three Gorges Reservoir in China, Jin et al. [72] found that the main drivers of stratification were reservoir operations (i.e., water level and water fluctuations) and air temperature which were not included in our study but are likely important factors in our study. We also speculate that temperature of the incoming tributary, volume of water, and water velocity at the confluences are important drivers influencing thermal stratification because the four confluences where we observed persistent stratification (95–100%) had low water velocity and temperature differences of at least 4 °C at the bottoms of the incoming tributary and the main stem.

These findings illustrate that the identification of a potential cold-water refuge depends on the definition used (i.e., the criteria) and the temporal scale observed. For instance, we did not observe main stem water temperatures cooler than 18 °C during our study; therefore, no main stem sites were classified as potential physiological cold-water refuges. However, when we assessed the main stem sites using our least restrictive 1 °C ambient criterion, all sites were considered cold-water refuges at depth during some portion of the day, albeit over different periods of time. Consequently, cold-water refuge definitions based solely on physiological thresholds or that exclude fine temporal and spatial scales overlook the importance of thermal heterogeneity for providing thermal refuge when temperatures exceed physiological thresholds even if it is for few hours a day [55]. Because most studies have focused on the spatial variability of cold-water refuges [1,16,45,59], cold-water refuge persistence has not been well studied. Dugdale et al. [18] documented in the Miramichi River that 61% of thermal refuges (≥0.5 °C colder than the ambient river channel temperature) were only identified once during multiple TIR surveys. In our study, we found that the persistence of confluence cold-water refuges was associated with the difference in 7DADM for the main stem and the confluence; larger differences in temperature created refuges that were persistent over longer periods, regardless of how refuges were defined.

We designed our study to describe water temperature patterns in the Pend Oreille River when conditions were warmest to understand how thermal heterogeneity during these stressful conditions influence cold-water refuge distribution for salmonids. Because water temperature was already above 20 °C when we started sampling in 2017, and water temperature in the main stem Pend Oreille River reached 18 °C during the first week of July [73], bull trout and, to a lesser extent, cutthroat trout were under thermal stress for most of the summer. Thus, using physiological thresholds alone for the main stem Pend Oreille was not very informative. However, comparing the spacing of cold-water refuges together with their persistence can help evaluate whether these refuges are adequate for salmonids migrating through the Pend Oreille River. Spacing along the longitudinal profile between persistent cold-water refuges (i.e., those that were present 100% of the time) that are formed by confluences ranged between 2 and 41 km (median = 7 km), but when we considered physiological cold-water refuges that were present at least 50% of the time, the spacing was reduced to less than 18 km, with an increase in the median separation distance (11 km). Additionally, areas that are 1–2 °C cooler than the median main stem bottom water temperature (main stem ambient cold-water refuges) but are still above the physiological thresholds can provide connectivity and can be used as stepping stones during migration [59]. Spatially continuous bottom water temperatures would be needed to determine the spacing of ambient cold-water refuges in the main stem along the longitudinal profile. Furthermore, we did not include every confluence in our analysis, so our numbers for cold-water refuge spacing are most likely overestimated, especially if localized groundwater inputs occur at depth or are too small to be detected at the surface using remotely sensed methods (e.g., TIR). Nevertheless, our spacing estimates for the confluences were within the range of variation of previous studies. Fullerton et al. [59], in a review of rivers in the Pacific Northwest, reported two to three times the median spacing of this study, but the spacing had high variability and ranged from 6 to 49 km, whereas Dzara et al. [28] in the Walker River in Nevada found that median spacing was 0.9 km and ranged from 0.3 to 37 km.

The high resolution of the geomorphic metrics measured along most of the Pend Oreille River provides a unique opportunity to examine how these metrics affect the occurrence of cool areas and median water temperature at the 1-km scale. The strong associations with channel confinement corroborated findings of previous studies [17,22,30,74]. Cross-sectional area, another variable with strong explanatory power for both reach-scale models, was also inversely correlated with water column velocity (Pearson correlation = −0.90). These robust associations support our hypotheses that channel confinement, water velocity, and cross-sectional areas explain different mechanisms that directly affect water temperature patterns. More specifically, at confined sections where the river is narrow and deeper, two alternative mechanisms may be at play depending on water velocity. At high velocities, water may be cooler as a result of mixing of the water column and topographic shading, whereas at low velocities, water may be warmer at the surface due to thermal stratification. In contrast, at unconfined sections, where channel width and cross-sectional area are largest, warmer water temperatures are the result of greater surface area being exposed to solar radiation.

At the subcatchment scale, our models suggest that the distance from Albeni Falls Dam was the most important factor determining the occurrence of cool water areas and explaining median water temperature. This result was unexpected for our study because of the weak longitudinal downstream patterns observed in the TIR data. However, our statistical results corresponded to those found in other studies when location in the network was considered [74,75]. The mechanism for this downstream increase was not entirely clear from our results. A regional study of the impact of dams on water temperature in Eastern Canada found that dams with impounded runoff index (storage capacity divided by median annual runoff) less than 10% had little impact on the thermal regime of medium-size rivers [76]. The Pend Oreille River is at least an order of magnitude larger than the rivers described by Maheu et al. [27], and has an impounded runoff index of less than 1%, which is typical of run-of-the-river systems. However, during most of the 2001 summer when the TIR data were collected, assuming no major withdrawals, the reservoir was storing water. The amount of time that the daily reservoir outflow was less than daily reservoir inflow was 82% in July, 70% in August, and 13% in September. The largest daily difference between outflow and inflow was 18% on August 7. Despite the Pend Oreille River having run-of-the-river characteristics on an annual basis, the reservoir appears to operate as a storage reservoir in the summer and is likely to experience a warming effect. Our statistical results did not point to the size or number of tributaries as explanatory variables. These results, together with TIR imagery from tributary plumes suggest that lateral contributions were operating at the reach scale, except at the lowest trough (rkm 80 to 90), where C6 and C5 were large enough to potentially have an impact on downstream water temperature patterns.

We recognize that our TIR survey was conducted in 2001 and our instream thermographs were deployed in 2017. Although, this mismatch could influence our interpretations, Fullerton et al. [26] found that spatial patterns of water temperature from repeat TIR surveys were consistent among years and, thus, comparisons with broad spatial patterns data collected in 2017 should be consistent over time given that the river morphology has not changed. Similarly, 3-D mapping was conducted at considerably different discharges: Average discharge from the second survey was almost three times the discharge from the first survey (268 m^3 s^{-1} versus 720 m^3 s^{-1}). We used z-scores to normalize these values to account for the temporal variability. The z-scores directly compare the patterns of variability rather than the magnitude of the variables and can be used as explanatory variables by removing any bias potential from sampling during different flow conditions. We also recognized that some confluences were highly spatially variable due to either their low stream bed stability (substrates were mostly sand or silt) or their dense aquatic vegetation, so placement directly below the thermal plume and foreseeing the path of the tributary was not possible. For these types of streams (e.g., C14, C8, C7, and C17), distributed temperature sensing (DTS) cables could have captured the plume path more consistently than instream loggers as the water elevation decreased. However, budget constraints and potential logistical and maintenance issues prevented our use of DTS. Incorporating this variability can be informative to management for potential cold-water refuges at these confluences with fluctuating conditions. Finally, our study was limited to one large, impounded river with specific geomorphic and hydrologic characteristics and flow operations. However, inferences from this study may still be relevant to other run-of-the-river reservoirs with similar physical characteristics.

Many questions remain about the thermal heterogeneity of large impounded rivers, and the implications for cold-water refuges. Because thermal heterogeneity is likely dampened in the main stem river, studies that consider flow management scenarios affecting the size, spacing, and persistence of cold-water refuges may be critical to develop management strategies for conserving, augmenting, and restoring areas suitable for salmonids. Our study also raises questions about the sufficiency of the potential cold-water refuges described here. Understanding the adequacy of cold-water refuges for bull trout and cutthroat trout migration is important for the survival of these species in the Pend Oreille River system. Suitability of cold-water refuges is not limited to water temperature; therefore, other factors such as dissolved oxygen and food availability should also be considered when determining the potential use of a cold-water refuge. For instance, we observed bull trout at C3 and C17 where dense abundant vegetation and night hypoxic conditions were present. Dense SAV may negatively impact these areas by decreasing dissolved oxygen at night near the bottom where it is cooler and may prevent fish from using these areas if SAV becomes so thick that it creates a barrier. Understanding the interactions between submerged aquatic vegetation, water temperature, and dissolved oxygen may provide management strategies at confluences where these conditions occur. Lastly, our findings suggest that the distance from Albeni Falls Dam was an important variable explaining occurrence of cool areas and median water temperature. Hence, understanding statistical associations of hydroclimatic variables related to conditions upstream of the dam, such as upstream water temperature, discharge, ratio of reservoir outflow to inflow, and air temperature to downstream water temperature could be used to examine whether Albeni Falls Dam affects downstream water temperature patterns.

Given that climate change will increase temperatures globally, thermal heterogeneity will become increasingly important for fish to behaviorally thermoregulate during times of thermal stress even as the ability of fish to locate cold-water areas is also diminished because of habitat homogenization [77]. Because unidentified refuges are at risk of being lost through habitat modification and degradation or unsuitable management [78], locating cold-water refuges is the first step to manage, restore, augment, and create thermal heterogeneity to conserve salmonid habitat [79].

5. Conclusions

Our study described thermal heterogeneity across space and time and identified potential cold-water refuges for salmonids in the Pend Oreille River, a large impounded river with an epilimnetic-release located in inland northwestern USA. Our findings demonstrate that the complexity of thermal patterns in impounded rivers require a combination of approaches, including assessment and modeling to understand the implications of thermal heterogeneity for cold-water fishes. These findings also highlight the importance of considering thermal heterogeneity when restoring and maintaining aquatic landscapes that can support native salmonids.

The results were consistent with our hypotheses that water temperature in the main stem of the Pend Oreille River increased as water moved downstream, but tributaries created discontinuities in the longitudinal profile. These lateral discontinuities (1) constituted most of the thermal heterogeneity in the river, (2) provided most of the potential cold-water refuges for salmonids, and (3) were more persistent in areas where thermal differences between the main stem and confluences were largest. In addition, we found that vertical thermal stratification was generally present at slow-moving confluences where vertical temperature gradients isolated cool water near the river bottom. Modeling results support mechanistic understanding of the interaction between confinement, cross-sectional area, water velocity, and depth to explain median water temperature and the probability of occurrence of cool-water areas. These results can be used to develop management and restoration strategies to enhance thermal heterogeneity.

Supplementary Materials: The following supplementary materials are available online at http://www.mdpi.com/ 2072-4292/12/9/1386/s1. Figure S1: Patterns of water temperature at selected confluences (C7, C8, and C14 (9

Remote Sens. **2020**, *12*, 1386

August–13 September 2017). Scripts S2: R code to generate graphics in Figure 4 for interactive visualization in three-dimensional space. Data S3: Data to generate graphics in Figure 4.

Additional Online Resources: Water temperature data used in this study are available at https://doi.org/10.5066/P9U5HJH9.

Author Contributions: Conceptualization, F.H.M., C.E.T., and J.R.M.; Data curation, F.H.M. and C.E.T.; Formal analysis, F.H.M. and C.E.T.; Funding acquisition, J.R.M.; Investigation, F.H.M., E.K.B., and M.S.L.; Methodology, F.H.M., C.E.T., and M.S.L.; Project administration, E.K.B., J.R.M., and C.E.T.; Resources, F.H.M., E.K.B., and M.S.L.; Software, F.H.M. and M.S.L.; Supervision, C.E.T.; Validation, F.H.M., C.E.T., and M.S.L.; Visualization, F.H.M., C.E.T., and M.S.L.; Writing—original draft, F.H.M.; Writing—review & editing, F.H.M., J.R.M., E.K.B., J.M.C., A.H.F., C.E.T., and J.L.E. All authors have read and agreed to the published version of the manuscript.

Funding: This research was funded by the Kalispel Tribe of Indians through a grant from the Bonneville Power Administration (Project Number 2011-018-00) and the U.S. Geological Survey.

Acknowledgments: Angela Klock (University of Washington) helped organized water temperature data, Jim LeMieux (Kalispel Tribe of Indians Natural Resources Department, KNRD) helped with land use analysis, and Raymond Ostlie (KNRD) helped with logistics. This paper improved as a result of critical reviews by Tyler King, Martin Fitzpatrick (USGS), and two anonymous reviewers. Any use of trade, product, or firm names is for descriptive purposes only and does not imply endorsement by the U.S. Government.

Conflicts of Interest: The authors declare no conflict of interest. The funders had no role in the design of the study; in the collection, analyses, or interpretation of data; in the writing of the manuscript, or in the decision to publish the results.

Disclaimer: The views expressed in this article are those of the author(s) and do not necessarily represent the views or policies of the U.S. Environmental Protection Agency.

References

1. Ebersole, J.L.; Liss, W.J.; Frissell, C.A. Thermal heterogeneity, stream channel morphology, and salmonid abundance in northeastern Oregon streams. *Can. J. Fish. Aquat. Sci.* **2003**, *60*, 1266–1280. [CrossRef]
2. Steel, E.A.; Beechie, T.J.; Torgersen, C.E.; Fullerton, A.H. Envisioning, quantifying, and managing thermal regimes on river networks. *Bioscience* **2017**, *67*, 506–522. [CrossRef]
3. Ward, J.V.; Stanford, J.A. The serial discontinuity of lotic ecosystems. In *Dynamics of Lotic Ecosystems*; Fontaine, T.D., Bartell, S.M., Eds.; Ann Arbor Science Publishers: Ann Arbor, MI, USA, 1983; pp. 29–42.
4. Stanford, J.A.; Ward, J.V.; Liss, W.J.; Frissell, C.A.; Williams, R.N.; Lichatowich, J.A.; Coutant, C.C. A general protocol for restoration of regulated rivers. *Regul. Rivers Res. Manag.* **1996**, *12*, 391–413. [CrossRef]
5. Olden, J.D.; Naiman, R.J. Incorporating thermal regimes into environmental flows assessments: Modifying dam operations to restore freshwater ecosystem integrity. *Freshw. Biol.* **2010**, *55*, 86–107. [CrossRef]
6. Caissie, D. The thermal regime of rivers: A review. *Freshw. Biol.* **2006**, *51*, 1389–1406. [CrossRef]
7. Webb, B.W.; Hannah, D.M.; Moore, R.D.; Brown, L.E.; Nobilis, F. Recent advances in stream and river temperature research. *Hydrol. Process.* **2008**, *22*, 902–918. [CrossRef]
8. Sabo, J.L.; Bestgen, K.; Graf, W.; Sinha, T.; Wohl, E.E. Dams in the cadillac desert: Downstream effects in a geomorphic context. *Ann. N. Y. Acad. Sci.* **2012**, *1249*, 227–246. [CrossRef]
9. Prats, J.; Armengol, J.; Marcé, R.; Sánchez-Juny, M.; Dolz, J. Dams and reservoirs in the lower Ebro river and its effects on the river thermal cycle. In *The Ebro River Basin*; Barceló, D., Petrovic, M., Eds.; Springer: Berlin/Heidelberg, Germany, 2011; pp. 77–95.
10. Graf, W.L. Downstream hydrologic and geomorphic effects of large dams on American rivers. *Geomorphology* **2006**, *79*, 336–360. [CrossRef]
11. Kondolf, G.M. Hungry water: Effects of dams and gravel mining on river channels. *Environ. Manag.* **1997**, *21*, 533–551. [CrossRef]
12. Sabo, J.L.; Caron, M.; Doucett, R.; Dibble, K.L.; Ruhi, A.; Marks, J.C.; Hungate, B.A.; Kennedy, T.A.; Siqueira, T. Pulsed flows, tributary inputs and food-web structure in a highly regulated river. *J. Appl. Ecol.* **2018**, *55*, 1884–1895. [CrossRef]
13. Brandt, S.A. Classification of geomorphological effects downstream of dams. *Catena* **2000**, *40*, 375–401. [CrossRef]
14. Poole, G.C.; Berman, C.H. An ecological perspective on in-stream temperature: Natural heat dynamics and mechanisms of human-caused thermal degradation. *Environ. Manag.* **2001**, *27*, 787–802. [CrossRef]
15. Carpenter, S.R.; Lodge, D.M. Effects of submersed macrophytes on ecosystem processes. *Aquat. Bot.* **1986**, *26*, 341–370. [CrossRef]

16. Torgersen, C.E.; Price, D.M.; Li, H.W.; McIntosh, B.A. Multiscale thermal refugia and stream habitat associations of chinook salmon in northeastern Oregon. *Ecol. Appl.* **1999**, *9*, 301–319. [CrossRef]

17. Torgersen, C.E.; Ebersole, J.L.; Keenan, D.M. *Primer for Identifying Cold-Water Refuges to Protect and Restore Thermal Diversity in Riverine Landscapes*; EPA 910-C-12-001; U.S. Environmental Protection Agency: Seattle, WA, USA, 2012; p. 91.

18. Dugdale, S.J.; Bergeron, N.E.; St-Hilaire, A. Temporal variability of thermal refuges and water temperature patterns in an Atlantic salmon river. *Remote Sens. Environ.* **2013**, *136*, 358–373. [CrossRef]

19. Rice, S.P.; Greenwood, M.T.; Joyce, C.B. Tributaries, sediment sources, and the longitudinal organisation of macroinvertebrate fauna along river systems. *Can. J. Fish. Aquat. Sci.* **2001**, *58*, 824–840. [CrossRef]

20. Vinson, M.R. Long-term dynamics of an invertebrate assemblage downstream from a large dam. *Ecol. Appl.* **2001**, *11*, 711–730. [CrossRef]

21. Benda, L.; Andras, K.; Miller, D.; Bigelow, P. Confluence effects in rivers: Interactions of basin scale, network geometry, and disturbance regimes. *Water Resour. Res.* **2004**, *40*, W05402. [CrossRef]

22. Dugdale, S.J.; Bergeron, N.E.; St-Hilaire, A. Spatial distribution of thermal refuges analysed in relation to riverscape hydromorphology using airborne thermal infrared imagery. *Remote Sens. Environ.* **2015**, *160*, 43–55. [CrossRef]

23. Knighton, D. *Fluvial Forms and Processes: A New Perspective*; Oxford University Press, Inc.: New York, NY, USA, 1998; p. 383.

24. Torgersen, C.E.; Gresswell, R.E.; Bateman, D.S.; Burnett, K.M. Spatial identification of tributary impacts in river networks. In *River Confluences, Tributaries and the Fluvial Network*; Rice, S.P., Roy, A.G., Rhoads, B.L., Eds.; John Wiley & Sons Inc.: Hoboken, NJ, USA, 2008; pp. 159–181.

25. Cristea, N.C.; Burges, S.J. Use of thermal infrared imagery to complement monitoring and modeling of spatial stream temperatures. *J. Hydrol. Eng.* **2009**, *14*, 1080–1090. [CrossRef]

26. Fullerton, A.H.; Torgersen, C.E.; Lawler, J.J.; Faux, R.N.; Steel, E.A.; Beechie, T.J.; Ebersole, J.L.; Leibowitz, S.G. Rethinking the longitudinal stream temperature paradigm: Region-wide comparison of thermal infrared imagery reveals unexpected complexity of river temperatures. *Hydrol. Process.* **2015**, *29*, 4719–4737. [CrossRef]

27. Maheu, A.; St-Hilaire, A.; Caissie, D.; El-Jabi, N. Understanding the thermal regime of rivers influenced by small and medium size dams in Eastern Canada. *River Res. Appl.* **2016**, *32*, 2032–2044. [CrossRef]

28. Dzara, J.R.; Neilson, B.T.; Null, S.E. Quantifying thermal refugia connectivity by combining temperature modeling, distributed temperature sensing, and thermal infrared imaging. *Hydrol. Earth Syst. Sci.* **2019**, *23*, 2965–2982. [CrossRef]

29. Vatland, S.J.; Gresswell, R.E.; Poole, G.C. Quantifying stream thermal regimes at multiple scales: Combining thermal infrared imagery and stationary stream temperature data in a novel modeling framework. *Water Resour. Res.* **2015**, *51*, 31–46. [CrossRef]

30. Arscott, D.B.; Tockner, K.; Ward, J.V. Thermal heterogeneity along a braided floodplain river (Tagliamento River, northeastern Italy). *Can. J. Fish. Aquat. Sci.* **2001**, *58*, 2359–2373. [CrossRef]

31. Imholt, C.; Soulsby, C.; Malcolm, I.A.; Hrachowitz, M.; Gibbins, C.N.; Langan, S.; Tetzlaff, D. Influence of scale on thermal characteristics in a large montane river basin. *River Res. Appl.* **2013**, *29*, 403–419. [CrossRef]

32. Steel, E.A.; Lange, I.A. Using wavelet analysis to detect changes in water temperature regimes at multiple scales: Effects of multi-purpose dams in the Willamette River basin. *River Res. Appl.* **2007**, *23*, 351–359. [CrossRef]

33. Wawrzyniak, V.; Piégay, H.; Poirel, A. Longitudinal and temporal thermal patterns of the French Rhône River using Landsat ETM+ thermal infrared images. *Aquat. Sci.* **2011**, *74*, 405–414. [CrossRef]

34. Ling, F.; Foody, G.; Du, H.; Ban, X.; Li, X.; Zhang, Y.; Du, Y. Monitoring thermal pollution in rivers downstream of dams with Landsat ETM+ thermal infrared images. *Remote Sens.* **2017**, *9*, 1175. [CrossRef]

35. Keefer, M.L.; Clabough, T.S.; Jepson, M.A.; Johnson, E.L.; Peery, C.A.; Caudill, C.C. Thermal exposure of adult Chinook salmon and steelhead: Diverse behavioral strategies in a large and warming river system. *PLoS ONE* **2018**, *13*, e0204274. [CrossRef]

36. Keefer, M.L.; Clabough, T.S.; Jepson, M.A.; Bowerman, T.; Caudill, C.C. Temperature and depth profiles of Chinook salmon and the energetic costs of their long-distance homing migrations. *J. Therm. Biol.* **2019**, *79*, 155–165. [CrossRef]

37. Snyder, M.N.; Schumaker, N.H.; Ebersole, J.L.; Dunham, J.B.; Comeleo, R.L.; Keefer, M.L.; Leinenbach, P.; Brookes, A.; Cope, B.; Wu, J.; et al. Individual based modeling of fish migration in a 2-D river system: Model description and case study. *Landsc. Ecol.* **2019**, *34*, 737–754. [CrossRef]

38. Adams, S.B.; Bjornn, T.C. Bull trout distributions related to temperature regimes in four central Idaho streams. In *Friends of the Bull Trout Conference*; MacKay, W.C., Brewin, M.K., Monita, M., Eds.; Bull Trout Task Force: Calgary, AB, Canada, 1997; pp. 371–380.

39. Selong, J.H.; McMahon, T.E.; Zale, A.V.; Barrows, F.T. Effect of temperature on growth and survival of bull Trout, with application of an improved method for determining thermal tolerance in fishes. *Trans. Am. Fish. Soc.* **2001**, *130*, 1026–1037. [CrossRef]

40. Bear, E.A.; Mcmahon, T.E.; Zale, A.V. Comparative thermal requirements of westslope cutthroat trout and rainbow 11-out: Implications for species interactions and development of thermal protection standards. *Trans. Am. Fish. Soc.* **2007**, *136*, 1113–1121. [CrossRef]

41. Mandeville, C.P.; Rahel, F.J.; Patterson, L.S.; Walters, A.W. Integrating fish assemblage data, modeled stream temperatures, and thermal tolerance metrics to develop thermal guilds for water temperature regulation: Wyoming case study. *Trans. Am. Fish. Soc.* **2019**, *148*, 739–754. [CrossRef]

42. Howell, P.J.; Dunham, J.B.; Sankovich, P.M. Relationships between water temperatures and upstream migration, cold water refuge use, and spawning of adult bull trout from the Lostine River, Oregon, USA. *Ecol. Freshw. Fish* **2010**, *19*, 96–106. [CrossRef]

43. Beechie, T.; Imaki, H. Predicting natural channel patterns based on landscape and geomorphic controls in the Columbia River basin, USA. *Water Resour. Res.* **2014**, *50*, 39–57. [CrossRef]

44. Watershed Sciences Inc. *Aerial Surveys in the Pend Oreille River Basin: Thermal Infrared and Color Videography*; Watershed Sciences Inc.: Corvallis, OR, USA, 2002; p. 39.

45. Torgersen, C.E.; Faux, R.N.; McIntosh, B.A.; Poage, N.J.; Norton, D.J. Airborne thermal remote sensing for water temperature assessment in rivers and streams. *Remote Sens. Environ.* **2001**, *76*, 386–398. [CrossRef]

46. Mustafa, M.; Barnhart, B.; Babbar-Sebens, M.; Ficklin, D. Modeling landscape change effects on stream temperature using the soil and water assessment tool. *Water* **2018**, *10*, 1143. [CrossRef]

47. Ebersole, J.L.; Wigington, P.J.; Leibowitz, S.G.; Comeleo, R.L.; Sickle, J.V. Predicting the occurrence of cold-water patches at intermittent and ephemeral tributary confluences with warm rivers. *Freshw. Sci.* **2015**, *34*, 111–124. [CrossRef]

48. Comeleo, R.L.; Wigington, P.J.; Leibowitz, S.G. *Creation of a Digital Aquifer Permeability Map for the Pacific Northwest*; EPA/600/R-14/431; U.S. Environmental Protection Agency: Corvallis, OR, USA, 2014; pp. 1–18.

49. Bisson, P.A.; Montgomery, D.R.; Buffington, J.M. Valley segments, stream reaches, and channel units. In *Methods in Stream Ecology*; Hauer, F.R., Lamberti, G.A., Eds.; Academic Press: Cambridge, MA, USA, 2017; Volume 1, pp. 21–47.

50. Benda, L.; Miller, D.; Andras, K.; Bigelow, P.; Reeves, G.; Michael, D. NetMap: A new tool in support of watershed science and resource management. *For. Sci.* **2007**, *53*, 206–219.

51. Lorang, M.S.; Tonolla, D. Combining active and passive hydroacoustic techniques during flood events for rapid spatial mapping of bedload transport patterns in gravel-bed rivers. *Fundam. Appl. Limnol. Arch. Hydrobiol.* **2014**, *184*, 231–246. [CrossRef]

52. Marotz, B.L.; Lorang, M.S. Pallid sturgeon larvae: The drift dispersion hypothesis. *J. Appl. Ichthyol.* **2018**, *34*, 373–381. [CrossRef]

53. Welty, E.Z.; Torgersen, C.E.; Brenkman, S.J.; Duda, J.J.; Armstrong, J.B. Multiscale analysis of river networks using the R package linbin. *N. Am. J. Fish. Manag.* **2015**, *35*, 802–809. [CrossRef]

54. U.S. Environmental Protection Agency. *EPA Region 10 Guidance for Pacific Northwest State and Tribal Temperature Water Quality Standards*; EPA 910-B-03-002; Office of Water: Seattle, WA, USA, 2003; p. 49.

55. Greer, G.; Carlson, S.; Thompson, S. Evaluating definitions of salmonid thermal refugia using in situ measurements in the Eel River, Northern California. *Ecohydrology* **2019**, *12*, e2101. [CrossRef]

56. Benjamin, J.R.; Heltzel, J.M.; Dunham, J.B.; Heck, M.; Banish, N. Thermal regimes, nonnative trout, and their influences on native bull trout in the Upper Klamath River Basin, Oregon. *Trans. Am. Fish. Soc.* **2016**, *145*, 1318–1330. [CrossRef]

57. Howell, P.J. Changes in native bull trout and non-native brook trout distributions in the upper Powder River basin after 20 years, relationships to water temperature and implications of climate change. *Ecol. Freshw. Fish* **2018**, *27*, 710–719. [CrossRef]

58. Barrows, M.G.; Anglin, D.R.; Sankovich, P.M.; Hudson, J.M.; Koch, R.C.; Skalicky, J.J.; Wills, D.A.; Silver, B.P. *Use of the Mainstem Columbia and Lower Snake Rivers by Migratory Bull Trout*; U.S. Fish and Wildlife Service, Columbia River Fisheries Program Office: Vancouver, WA, USA, 2016; p. 276.

59. Fullerton, A.H.; Torgersen, C.E.; Lawler, J.J.; Steel, E.A.; Ebersole, J.L.; Lee, S.Y. Longitudinal thermal heterogeneity in rivers and refugia for coldwater species: Effects of scale and climate change. *Aquat. Sci.* **2018**, *80*, 1–15. [CrossRef]

60. Guisan, A.; Edwards, T.C.; Hastie, T. Generalized linear and generalized additive models in studies of species distributions: Setting the scene. *Ecol. Model.* **2002**, *157*, 89–100. [CrossRef]

61. R Core Team. *R: A Language and Environment for Statistical Computing*; R Foundation for Statistical Computing: Vienna, Austria, 2016.

62. Wood, S.N. Fast stable direct fitting and smoothness selection for generalized additive models. *J. R. Stat. Soc. B* **2008**, *70*, 495–518. [CrossRef]

63. Wood, S.N. *Generalized Additive Models*; Chapman and Hall/CRC: New York, NY, USA, 2017; p. 496.

64. Amodio, S.; Aria, M.; D'Ambrosio, A. On concurvity in nonlinear and nonparametric regression models. *Statistica* **2014**, *74*, 85–98.

65. Barton, K. Mu-MIn: Multi-Model Inference. R Package Version 0.12.2/r18. Available online: http://R-Forge.R-project.org/projects/mumin/ (accessed on 3 April 2019).

66. Burnham, K.P.; Anderson, D.R. *Model Selection and Multimodel Inference. A Practical Information-Theoretic Approach*; Springer: New York, NY, USA, 2002; p. 488.

67. Tonolla, D.; Acuña, V.; Uehlinger, U.; Frank, T.; Tockner, K. Thermal heterogeneity in river floodplains. *Ecosystems* **2010**, *13*, 727–740. [CrossRef]

68. Wawrzyniak, V.; Piégay, H.; Allemand, P.; Vaudor, L.; Grandjean, P. Prediction of water temperature heterogeneity of braided rivers using very high resolution thermal infrared (TIR) images. *Int. J. Remote Sens.* **2013**, *34*, 4812–4831. [CrossRef]

69. Paillex, A.; Castella, E.; zu Ermgassen, P.; Gallardo, B.; Aldridge, D.C. Large river floodplain as a natural laboratory: Non-native macroinvertebrates benefit from elevated temperatures. *Ecosphere* **2017**, *8*, e01972. [CrossRef]

70. Turner, L.; Erskine, W.D. Variability in the development, persistence and breakdown of thermal, oxygen and salt stratification on regulated rivers of southeastern Australia. *River Res. Appl.* **2005**, *21*, 151–168. [CrossRef]

71. Wawrzyniak, V.; Piégay, H.; Allemand, P.; Vaudor, L.; Goma, R.; Grandjean, P. Effects of geomorphology and groundwater level on the spatio-temporal variability of riverine cold water patches assessed using thermal infrared (TIR) remote sensing. *Remote Sens. Environ.* **2016**, *175*, 337–348. [CrossRef]

72. Jin, J.; Wells, S.A.; Liu, D.; Yang, G.; Zhu, S.; Ma, J.; Yang, Z. Effects of water level fluctuation on thermal stratification in a typical tributary bay of Three Gorges Reservoir, China. *PeerJ* **2019**, *7*, e6925. [CrossRef]

73. Mejia, F.H.; Torgersen, C.E.; Berntsen, E.K. *Water Temperature Data from the Pend Oreille River, Washington and Idaho, 2016–2018*; U.S. Geological Survey: Reston, VA, USA, 2019. [CrossRef]

74. O'Sullivan, A.M.; Devito, K.J.; Curry, R.A. The influence of landscape characteristics on the spatial variability of river temperatures. *Catena* **2019**, *177*, 70–83. [CrossRef]

75. Monk, W.A.; Wilbur, N.M.; Curry, R.A.; Gagnon, R.; Faux, R.N. Linking landscape variables to cold water refugia in rivers. *J. Environ. Manag.* **2013**, *118*, 170–176. [CrossRef]

76. Maheu, A.; St-Hilaire, A.; Caissie, D.; El-Jabi, N.; Bourque, G.; Boisclair, D. A regional analysis of the impact of dams on water temperature in medium-size rivers in eastern Canada. *Can. J. Fish. Aquat. Sci.* **2016**, *73*, 1885–1897. [CrossRef]

77. Armstrong, J.B.; Schindler, D.E. Going with the flow: Spatial distributions of juvenile coho salmon track an annually shifting mosaic of water temperature. *Ecosystems* **2013**, *16*, 1429–1441. [CrossRef]

78. Reside, A.E.; Briscoe, N.J.; Dickman, C.R.; Greenville, A.C.; Hradsky, B.A.; Kark, S.; Kearney, M.R.; Kutt, A.S.; Nimmo, D.G.; Pavey, C.R.; et al. Persistence through tough times: Fixed and shifting refuges in threatened species conservation. *Biodivers. Conserv.* **2019**, *28*, 1303–1330. [CrossRef]

79. Kurylyk, B.L.; MacQuarrie, K.T.B.; Linnansaari, T.; Cunjak, R.A.; Curry, R.A. Preserving, augmenting, and creating cold-water thermal refugia in rivers: Concepts derived from research on the Miramichi River, New Brunswick (Canada). *Ecohydrology* **2015**, *8*, 1095–1108. [CrossRef]

 *remote sensing* 

Article

Empirical Assessment Tool for Bathymetry, Flow Velocity and Salinity in Estuaries Based on Tidal Amplitude and Remotely-Sensed Imagery

Jasper R. F. W. Leuven *, Steye L. Verhoeve, Wout M. van Dijk, Sanja Selaković and Maarten G. Kleinhans

Faculty of Geosciences, Utrecht University, Princetonlaan 8A, 3584 CB Utrecht, The Netherlands; s.l.verhoeve@students.uu.nl (S.L.V.); W.M.vanDijk@uu.nl (W.M.v.D.); S.Selakovic@uu.nl (S.S.); m.g.kleinhans@uu.nl (M.G.K.)
* Correspondence: j.r.f.w.leuven@uu.nl

Received: 8 November 2018; Accepted: 23 November 2018; Published: 30 November 2018

Abstract: Hydromorphological data for many estuaries worldwide is scarce and usually limited to offshore tidal amplitude and remotely-sensed imagery. In many projects, information about morphology and intertidal area is needed to assess the effects of human interventions and rising sea-level on the natural depth distribution and on changing habitats. Habitat area depends on the spatial pattern of intertidal area, inundation time, peak flow velocities and salinity. While numerical models can reproduce these spatial patterns fairly well, their data need and computational costs are high and for each case a new model must be developed. Here, we present a Python tool that includes a comprehensive set of relations that predicts the hydrodynamics, bed elevation and the patterns of channels and bars in mere seconds. Predictions are based on a combination of empirical relations derived from natural estuaries, including a novel predictor for cross-sectional depth distributions, which is dependent on the along-channel width profile. Flow velocity, an important habitat characteristic, is calculated with a new correlation between depth below high water level and peak tidal flow velocity, which was based on spatial numerical modelling. Salinity is calculated from estuarine geometry and flow conditions. The tool only requires an along-channel width profile and tidal amplitude, making it useful for quick assessments, for example of potential habitat in ecology, when only remotely-sensed imagery is available.

Keywords: estuary; morphology; rapid assessment; bathymetry; flow velocity; salinity; tool; remotely-sensed imagery

1. Introduction

Estuaries are characterised by fresh water inflow at the landward boundary and an open connection to the sea. Within these boundaries, tidal flows form dynamic patterns of channels and bars (Figure 1a). Many of these systems are managed to balance the needs of flood safety, access to harbours and ecological quality. The main channels in large systems such as the Western Scheldt, Elbe and Yangtze are dredged for access to harbours [1], while intertidal area on bars and estuarine shorelines forms valuable ecological habitat [2]. Depth, inundation time, flow velocity and salinity are the prime abiotic factors that affect living organisms in estuarine environments [2–4]. However, the general lack of information on these abiotic factors limits the prediction and sustainable management of habitat area in most estuaries [5]. Additionally, the effects of human interventions and rising sea-level on the hydrodynamic conditions, equilibrium morphology and, therefore, also on ecology are largely unknown. Unfortunately, data to study these effects are generally only available for a few economically exploited estuaries that are already under great pressure by human influence. Data are lacking for most other systems around the world. Here, we explore to what degree it is possible to globally estimate estuarine characteristics, such as bed elevation, inundation duration, flow velocity, and salinity based on limited but widely publicly available data: remotely-sensed imagery and tidal range at the estuary mouth. To aid application and further investigation, we make the toolbox available.

Currently, several alternative methods are available to study the equilibrium morphology and hydrodynamic conditions of estuaries: in situ measurements, numerical modelling [6–8], idealised form models [9], stability analyses [10–12] and stability relations [13–16]. Only numerical models, for example Delft3D, can predict the spatial patterns of bed elevation, flow velocity and salinity with considerable accuracy at the spatial detail that is required for the purpose of habitat estimation. However, morphological models are computationally expensive and require calibration and detailed measurement of boundary and initial conditions. In situ measurements are an alternative, but they are expensive and equally time-consuming. For example, Rijkswaterstaat typically covers 2 km^2/day in the Western Scheldt. While linear stability analysis may predict typical estuarine bar patterns, it is not straightforwardly applicable to fully developed bars and estuaries with irregular planforms.

For converging ideal estuaries, predictive relations are already available between width-averaged bed elevation and hydrodynamic conditions along the system [9,17,18] because of the imposed constraint that the tidal damping by friction is balanced by the gain in energy by convergence. While these concepts are useful when applied to end-members, they leave the more common estuaries with irregular planforms unexplored. We recently found that the concept of an ideal estuary can be used to predict bed elevation and bar patterns in estuaries with irregular planform shapes as a function of the deviation from the ideal shape [19,20]. However, we still lack empirical or theoretical relations that describe flow velocity in these non-ideal systems, while this is an essential input variable for salinity predictions [21–23] and habitat suitability [24,25]. Based on observations in the Western Scheldt, we hypothesise that a relation exists between the depth and tidal flow velocity amplitude (Figure 1a,b). Here, we derive a predictor for flow velocity as a function of bed elevation to complement the available relations for bed elevations, bar patterns and salinity in estuaries with irregular planforms.

Here, existing empirical relations and theoretical concepts were selected and combined in a fast-to-apply tool that predicts equilibrium morphology and hydrodynamic conditions. To assess the applicability, we test it against a range of estuaries and other end-member systems, characterised by varying degrees of tidal and river influence [9,26] as well as varying degrees of adaptation to boundary conditions [9,27] and varying degrees of bar presence (Figure 2). These systems are a river-dominated delta branch [26], a tide-dominated delta branch (an ideal estuary cf. Savenije [18]), an alluvial estuary filled with tidal bars [19,26], a relatively wide estuarine valley partially filled with bars [26,28], and a tidal basin lacking fluvial input [29–31] (Figure 2). In the discussion, we show applications for management of estuaries, palaeogeographic reconstructions and quantifying habitat area based on species-specific preferences.

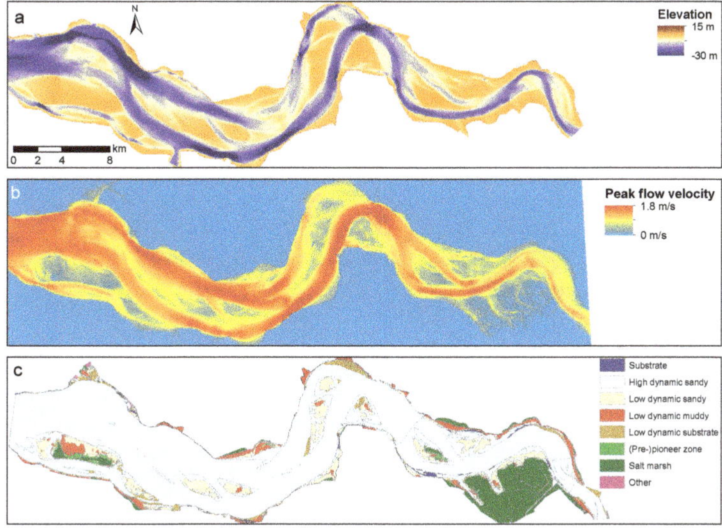

Figure 1. (**a**) bathymetry of the Western Scheldt in 2012 from the mouth (left) to the Dutch-Belgian border (right); (**b**) peak tidal flow velocities from a calibrated hydrodynamic model of Rijkswaterstaat (SCALWEST, [32]) for the bathymetry of the Western Scheldt in 2009 and (**c**) ecotope map of the Western Scheldt (2012). Bathymetry, hydrodynamic output and ecotope map were obtained from Rijkswaterstaat.

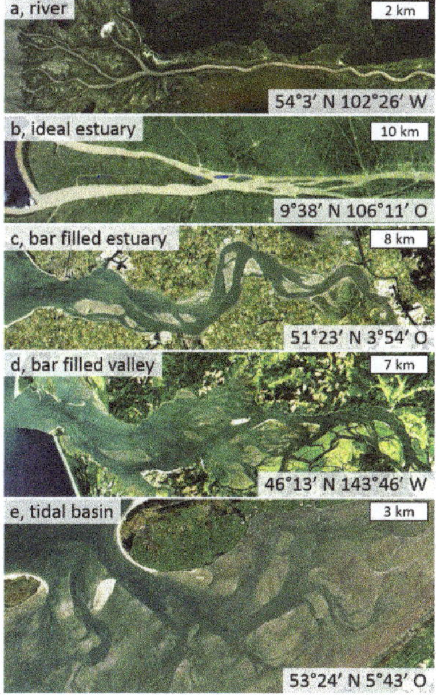

Figure 2. Examples of the systems used in this study. (**a**) a delta branch of a distributary of the Saskatchewan River (Canada); (**b**) the Tran De branch of the Mekong Delta (Vietnam); (**c**) the Western Scheldt estuary (The Netherlands); (**d**) the Columbia River estuary (USA); (**e**) the Ameland inlet and tidal basin (The Netherlands).

2. Methods

Below, first, the general approach and minimum data required to run the tool are described. Then, the general structure and functions used in the tool are presented. Finally, it is described how tool output was compared with measured data and numerical modelling for two estuaries filled with bars and three end-member systems (Figure 2), including the Western Scheldt and Columbia River. A general flowchart of the tool is given in Figure 3 and a list of symbols used is given in Appendix A Table A1. The full Python code of the tool and instructions on how to use it are available in the supplementary material and accessible on Github. The tool output comprises data exported in spreadsheet format and maps of bed elevation, inundation duration, peak flow velocity and salinity throughout the estuary.

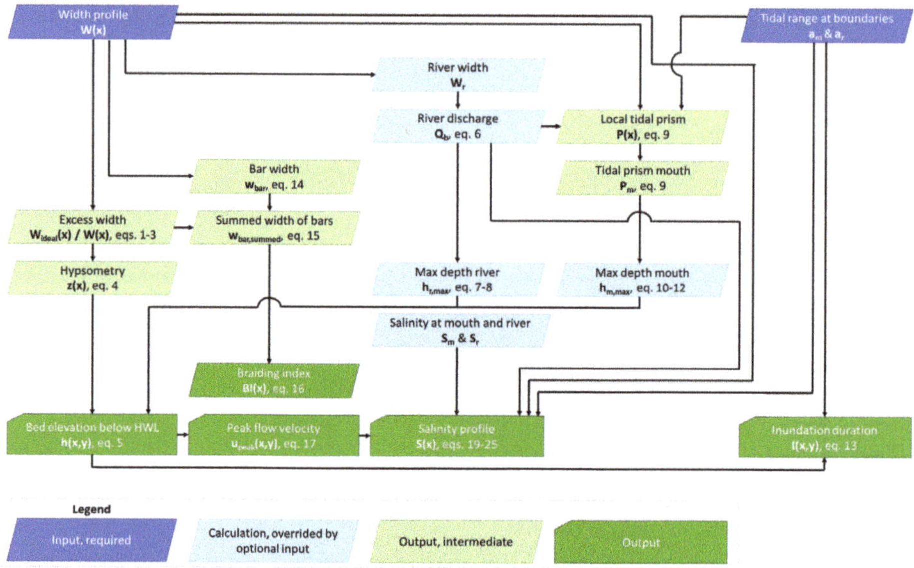

Figure 3. Flowchart of the calculation steps in the tool.

General assumptions underlying the entire method are as follows. There are no significant geological constraints on the depth along the estuary. At the landward and seaward boundary, we assume hydraulic geometry. Moreover, there is no significant effect of the offshore wave climate on the morphology of the mouth and bed elevations in the estuary. The along-channel tidal range is simplified as an along-channel profile that is constant or linearly increasing or decreasing. For calculations of inundation duration, a sinusoidal M2 tide is assumed.

2.1. Tool Approach

In general, the level of predictable detail depends on the available data. Data that are generally available for estuaries from all over the world are remotely-sensed imagery and a typical tidal range at the mouth, which means that our predictive relations should be based on and derivable from this information only. From remotely-sensed imagery, the along-channel width profile can be obtained (Section 2.2). It was previously found that the along-channel width profile correlates with the cross-sectional hypsometry [20]. The deviation from a perfectly converging planform translates in the occurrence of intertidal bars and therefore also in the shape of the hypsometric curve [33–35]. However, to scale dimensionless hypsometry to vertical bed elevation distributions, an along-channel depth profile is required. Estuary depth generally decreases linearly to near-linearly from the estuary mouth to the upstream tidal limit citepSavenije2015, enabling channel-depth estimation at any location in the estuary when the depth at the mouth of the estuary and upstream river are known. For the landward boundary, hydraulic geometry relations for rivers predict channel depth from width or bankfull discharge [36,37], while the geometry of the estuary mouth is related to local tidal prism [13,15,17,38,39]. The typical tidal range at the mouth in combination with the along-channel width profile gives an estimate of the local tidal prism. This means that bed elevations and inundation duration are predictable from only the along-channel width profile and tidal amplitude profile, which is the minimum input data required to run the tool.

In order to be able to use additional data when available, we allow some variables to override estimations by the above relations (Figure 3). Specifically, measured maximum mouth depth and maximum depth at the upstream boundary make their estimation based on tidal prism and hydraulic geometry redundant. A specified depth and width at the upstream boundary implicitly impose a user defined hydraulic geometry. Similarly, measured fresh water discharge may override the estimation of discharge based on channel width, which replaces the high uncertainty of the empirical relation with the lower measurement uncertainty. Finally, salinity at the mouth and landward boundary may be specified instead of the default values of 35 ppt at the mouth and 0 ppt at the landward boundary.

2.2. Bed Elevation Prediction

From remotely-sensed imagery, the first step is to obtain the along-channel width profile. Multiple options are available ranging from manually measuring the width at equally spaced transects to software tools that calculate river widths from remotely-sensed imagery (e.g., RivWidth [40,41]). Here, the following approach was used (following Leuven et al. [19]): first, we digitised the estuary planform in GIS software. Second, the polygon was imported in GIS software and the channel centreline was automatically extracted (e.g., Polygon To Centerline, ArcGIS function). Last, transects were drawn equally spaced on the centreline (e.g., Generate Transects Along Lines, ArcGIS function) and cropped with the planform polygon. The length of successive transects resulted in the along-channel width profile.

2.2.1. Estuary Shape Predicts Cross-Sectional Hypsometry

Estuaries typically narrow from the tidal inlet to the upstream river because the tidal wave generally dampens out in landward direction such that the combined discharge, or flow energy, of the river and the tides reduces. The estuary convergence is often characterised by a convergence length of width [18,42,43]. In ideal estuaries, the energy loss due to friction is balanced by channel convergence such that the tidal energy per unit of width remains constant up to the estuary head [17,44,45]. However, natural estuaries adapted in varying degrees to their initial and boundary conditions [9,27], resulting in planform shapes that deviate from the converging shape expected according to the theory for the ideal estuary state. The deviation in width from a converging shape was found to be a fairly accurate predictor of bar pattern and cross-sectional bed elevation distributions [19,20].

The convergence length (L_W) is the length over which the channel width reduces by a factor 2.72 (e) and is calculated as:

$$L_W = \frac{-s}{\ln\left(\frac{W_r}{W_m}\right)} \tag{1}$$

where W_m is the width of the estuary mouth, W_r is width of the river at the most landward boundary of the estuary and s is the along-channel distance measured between the seaward and landward boundary. We followed the same guidelines on the selection of the landward and seaward boundary as in Leuven et al. [20], given on p. 766.

The ideal width ($W_{ideal}(x)$) and excess width ($W_{excess}(x)$) are assumed to be related to this convergence length and the local width ($W(x)$) of the estuary:

$$W_{ideal}(x) = W_m e^{\frac{-x}{L_W}} \tag{2}$$

$$W_{excess}(x) = W(x) - W_{ideal}(x) \tag{3}$$

Here, x is the streamwise coordinate measured from the mouth along a centreline and y is the coordinate perpendicular to the centreline.

Along-channel variations in excess width translate into bed elevation with cross-sectional hypsometric curves. Hypsometric curves describe bed elevations with a cumulative profile. Multiple empirical relations have been proposed for the hypsometric shape of (partially) submerged bodies [34,35,46,47] and terrestrial landscapes [33]. The original formulation of Strahler [33] (Equation (5)) appeared capable to describe hypsometries that occur in estuaries [20]. Two parameters in the Strahler [33] equation can be tuned to modify the hypsometric shape: r sets the slope of curvature at the inflection point and z sets the concavity of the function, with lower values representing a more convex cross-sectional bed profile and higher values representing a more concave profile.

Reasonably accurate predictions are obtained in estuaries with r set to a constant value of 0.5, while the along-channel variation of best fitting z strongly depends on the along-channel variation in width [20]. Moreover, at locations where the estuary is much wider than expected from an ideal shape, bars are typically more abundant, the hypsometry is more convex and z-values are higher. The along-channel variation in concavity (z) is given by:

$$z(x) = 1.4 \left(\frac{W_{ideal}(x)}{W(x)} \right)^{1.2} \tag{4}$$

The resulting predictions are accurate within a factor 2 of the measured value for intertidal and subtidal area in multiple estuaries of various size and character. This means that it is possible to predict the hypsometric shape per cross-section based on the estuary shape and, in this study, its predictive capacity for end-member cases is assessed as well.

The along-channel variation in concavity of the hypsometric shape ($z(x)$) can subsequently be used in the general hypsometric curve, formulated by Strahler [33] as:

$$h_z(x,y) = \left(\frac{r}{r-1} \right)^{z(x)} \left[\frac{1}{(1-r)y+r} - 1 \right]^{z(x)} \tag{5}$$

in which $h_z(x,y)$ is the bed elevation above which proportion y of the width profile occurs. Alternatively said, $h_z(x,y)$ is the proportion of total section height and y the proportion of section width. r is set to a constant value of 0.5 and $z(x)$ is given by Equation (4).

The bed elevation ($h_z(x,y)$) and the width fraction (y) in Equation (5) are dimensionless, where $h_z(x,y)$ is normalised between the local high water level (HWL) and the maximum estuary depth for that cross-section. Values for y are dimensionalised with the local estuary width. Thus, to transform the predicted hypsometry to dimensional depth distributions, an along-channel profile of the maximum depth is required, which is estimated from the channel depth at the mouth and the depth at the landward boundary as explained in the next subsection.

2.2.2. Tidal Prism and River Discharge Predict Channel Geometry at the Mouth and Landward River

To calculate dimensional hypsometric profiles, an along-channel depth profile is required. Since the along-channel depth profile often shows a linear or almost linear profile [19,42], we fitted a linear along-channel profile on the maximum depth at the mouth of the estuary and the maximum depth at the landward river. When these values are known, the user can specify them, which overrides the estimation. When unspecified, the maximum and average depth at the mouth and at the upstream river are estimated with hydraulic geometry, using respectively the tidal prism and river discharge.

Hydraulic geometry is an empirical construct and should be used with care. However, it is often the best estimate in the absence of observed data. The resulting predictions are dependent on the specific region or river type for which they were developed. Here, we implemented, and used, the equation Hey and Thorne [37] as an example. Nevertheless, users of the tool can specify a measured channel depth and river discharge as input when available because it overrides the calculation with Equations (6) and (7) and therefore implicitly impose a hydraulic geometry. Alternatively, the user can calculate these values with a hydraulic geometry equation suitable for the type of river considered or implement their own equation in the source code.

When unknown, river discharge is estimated from the channel width at the landward boundary with hydraulic geometry. Here, we used Hey and Thorne [37], rewritten as:

$$Q_b = \left(\frac{W_r}{3.67} \right)^{1/0.45} \tag{6}$$

in which Q_b is the bankfull discharge and W_r is the width at the upstream river, measured at a location with almost zero tidal influence.

Subsequently, the width-averaged depth at the landward boundary is estimated from bankfull discharge as:

$$\overline{h_r} = 0.33 Q_b^{0.35} \tag{7}$$

The maximum depth at the landward boundary is a function of the average depth and the geometric shape of the channel. Therefore, the following equation was adopted:

$$h_{max,r} = s_r \overline{h_r} \tag{8}$$

in which s_r is the shape factor of the cross-section at the landward river, where 1 is a perfect rectangular and 2 a V-shaped cross-section. If unknown, a value of 1.85 is assumed for the landward boundary, based on geometry of five estuary mouths (Appendix A Figure A1).

The average depth at the mouth ($\overline{h_m}$) is calculated as the cross-sectional area at the mouth ($A_{csa,m}$, Equation (10)) divided by the local width (W_m), in which the cross-sectional area is estimated with the tidal prism. The local tidal prism is given as:

$$P(x) = \sum_{x}^{x+E} (2W(x)a(x)) + \frac{Q_b}{4} t \tag{9}$$

in which $P(x)$ is the local tidal prism resulting from the tidal amplitude ($a(x)$) and surface area upstream of location x. The contribution of river discharge (Q_r) to the local tidal prism is approximated as the bankfull discharge (Q_b) divided by 4. E is the tidal excursion length, which is the distance a water particle travels over half a tidal cycle. Since E is dependent on flow velocities, which are unknown at this point in the assessment tool, E has to be estimated at this point. The tidal excursion length is estimated by multiplying a typical tidal flow velocity with t, which is the duration of half a tidal period. According to Savenije [18], the order of magnitude of the tidal excursion length can be estimated even in ungauged estuaries because the peak velocity amplitude in alluvial estuaries from all over the world is remarkably similar, being approximately 1 ms^{-1}. Here, we use this typical flow velocity to obtain an estimate of the tidal excursion length, which will be evaluated in the results. In Leuven et al. [19], the upper limit for summation of the tidal prism was the upstream estuary boundary, but given that the input of the tool may comprise estuaries much longer than the tidal excursion length, the summation is here limited to the tidal excursion length.

Subsequently, the cross-sectional area at the mouth is related to the tidal prism. Earlier approaches derived these relations for tidal inlets or specific tidal systems [13,15,38,39,48–50]. Recently, we derived an empirical relation for multiple cross-sections within 35 estuaries (supplementary material of [19]). This equation is used to estimate the cross-sectional area at the mouth:

$$A_{csa}(x) = 0.13 \times 10^{-3} P(x) \tag{10}$$

in which $A_{csa}(x)$ is the along-channel cross-sectional area.

Now, the average depth at the mouth is calculated as:

$$\overline{h_m} = \frac{A_{csa,m}}{W_m} \tag{11}$$

and converted to a maximum depth at the mouth with

$$h_{max,m} = s_m \overline{h_m} \tag{12}$$

in which s_m is the shape factor of the cross-section at mouth. If unknown, a value of 1.65 is suggested (Appendix A Figure A1).

2.3. Inundation Duration Prediction

Dimensional depth distributions directly translate in inundation durations assuming a simple harmonic tide. Dimensional depth profiles are obtained from scaling the cross-sectional hypsometric profiles with the along-channel profile of maximum depth. Bed elevations below the low water level (below -1 times the amplitude) are always submerged and are therefore assigned a relative inundation duration of 1. For the intertidal zone, the inundation duration was calculated as:

$$I(x,y) = 0.5 - 0.5 \sin\left(\frac{0.5\pi}{a(x)} h_z(x,y)\right) \tag{13}$$

2.4. Bar Pattern Prediction: Bar Width and Braiding Index

The along-channel variation in bar pattern correlates with along-channel variations in width. Moreover, the summed width of bars is equal to the excess width and thus Braiding Index ($BI(x)$) can subsequently be calculated by dividing the excess width ($W_{excess}(x)$) by the predicted bar width ($w_{bar}(x)$) [19]:

$$w_{bar}(x) = 0.39 W(x)^{0.92} \tag{14}$$

$$w_{barssummed}(x) = W_{excess}(x) \tag{15}$$

$$BI(x) = \frac{w_{barssummed}(x)}{w_{bar}(x)} \tag{16}$$

2.5. Flow Velocity Prediction

A relation for typical flow conditions as a function of estuarine geometry was still lacking. Therefore, we explored this relation in a calibrated hydrodynamic model of Rijkswaterstaat (SCALWEST, [32]) for the bathymetry of the Western Scheldt in 2009. We hypothesised that a relation would exist between bed elevation and typical flow velocity (Figure 1b). Peak tidal flow velocity (u_{peak}) indeed showed a good correlation ($R^2 = 0.73$) with depth below high water level (Figure 4):

$$u_{peak}(x,y) = 0.10 + 0.91_{10} \log(h(x,y)) \tag{17}$$

Typical average flow velocity over half a tidal cycle was estimated depending on the deviation from the maximum along-channel tidal amplitude as follows (Appendix A Figure A5):

$$u_{avg}(x,y) = u_{peak}(x,y) \left[1 - \frac{a(x)}{\max(a(x))} \frac{2}{\pi}\right] \tag{18}$$

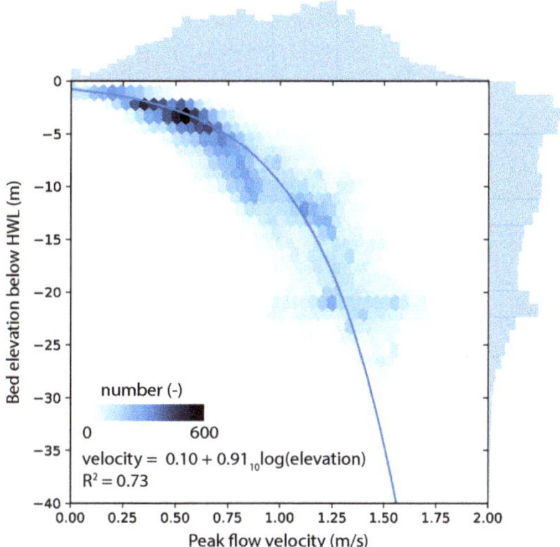

Figure 4. Relation between peak tidal flow velocity and depth below high water level from the Western Scheldt model (Figure 1a,b). Histograms indicate occurrence within bins and colours in the plot become more saturated with increasing point density.

2.6. Salinity Prediction

Salinity along the estuary depends on the freshwater discharge at the upstream boundary and the degree of mixing by the tides. Three along-channel salinity predictors were implemented in the assessment tool [21–23] and the quality of these predictors for the cases studied here will be described in the results.

Savenije [21] proposed a one-dimensional model that uses predictive equations for mixing and dispersion, such that it can be applied to systems for which limited information is available. The resulting model has been tested on a set of 15 estuaries. Gisen et al. [23] revised the predictive equation of Savenije [21], calibrated the model on 20 estuaries and validated it with another 10 estuaries. Brockway et al. [22] proposes a solution for advection-diffusion equations in converging estuaries and implements an empirical relation for longitudinal mixing based on data from the Incomati Estuary (Mozambique).

The output maps in the tool are based on the averaged prediction of Savenije [21] and Gisen et al. [23]. Nevertheless, the output for Brockway et al. [22] is also provided. The along-channel salinity profile of Brockway et al. [22] is typically exponentially decreasing to a salinity of 0, while the other two are generally more s-shaped with a strong gradient in the middle of the curve. Taking the average of all three predictors would thus not make sense because their characteristic shape would smoothen out. First, the latter is described because it has a simpler solution and less constitutive variables.

The salinity profile of Brockway et al. [22] ($S_b(x)$) is given as:

$$S_b(x) = S_m \exp\left\{ \frac{-Q_r}{L_{A_{csa}}^{-1} K A_{csa,m}} \left[e^{(L_{A_{csa}}^{-1} x)} - 1 \right] \right\} \tag{19}$$

in which S_m is the salinity at the mouth of the estuary (35 ppt), $A_{csa,m}$ is the cross-sectional area at the mouth, $L_{A_{csa}}$ is the convergence length of the cross-sectional area and K is an empirical equation for the longitudinal mixing coefficient based on average river discharge (Q_r) and tidal range ($a(x)$), being $K = 0.28 Q_r + 13\overline{a(x)}$.

The salinity profile of Savenije [21] ($S_s(x)$) is given as:

$$S_s(x) = (S_m - S_r)\left(\frac{D(x)}{D_m}\right)^{1/K} + S_r \tag{20}$$

in which S_r is the salinity at the fresh water river (0 ppt), K is the longitudinal mixing (Van der Burgh) coefficient (typically 0.07), $D(x)$ is the dispersion coefficient and D_m is the dispersion at the mouth of the estuary (typically 7000 m^{2}s^{-1}). The ratio between the along-channel dispersion and dispersion at the mouth is given in Savenije [21] as:

$$\frac{D(x)}{D_m} = 1 - \left(\frac{KLA_{csa}Q_r}{D_m A_{csa,m}}\right)\left(e^{\frac{x}{LA_{csa}}} - 1\right) \tag{21}$$

with

$$D_m = 220\sqrt{40}\frac{\overline{h_m}}{L_{A_{csa}}}\sqrt{N_r}\max(u_{peak}(mouth, y))E \tag{22}$$

in which

$$N_r = \frac{Q_r t/P(m)}{\rho_w \max(u_{peak}(mouth, y))^2/\Delta\rho g\overline{h_m}} \tag{23}$$

where ρ_w is the density of water, $\Delta\rho$ is the density difference between salt and fresh water and g is the gravitational acceleration (9.81 ms^{-2}).

The salinity profile of Gisen et al. [23] ($S_g(x)$) is equal to the profile of Savenije [21] for the part landward of the inflection point, which is the part used in this study, except for the equation for K. The longitudinal mixing coefficient (K) in Savenije [21] (K_s) is given as:

$$K_s = 0.16 \times 10^{-6}\frac{\overline{h_m}^{-0.69}g^{1.12}t^{2.24}}{(2a(m))^{0.59}L_W^{1.1}W_m^{0.13}} \tag{24}$$

while, in Gisen et al. [23], it (K_g) is given as:

$$K_g = 151.35 \times 10^{-6}\frac{W_r^{0.3}(2a(m))^{0.13}t^{0.97}}{W_m^{0.30}C_m^{0.18}[\max(u_{peak}(mouth, y))]^{0.71}L_W^{0.11}\overline{h_m}^{-0.15}r_s^{0.84}} \tag{25}$$

in which $a(m)$ is the tidal amplitude at the mouth, C_m is the Chezy roughness, estimated in the assessment tool as 42 m$^{0.5}$s^{-1}, which is adjustable in the tool code, $\max(u_{peak}(mouth, y))$ is the peak tidal flow velocity in the cross-section at the mouth, and r_s is the storage ratio, calculated as the width of intertidal area at the mouth divided by the width of the subtidal area at the mouth. The minimum value of K is 0 and the maximum value is 1.

2.7. Tool Validation

For purposes of illustration and validation, the tool was applied to the Columbia River estuary (USA), which is an example of a relatively wide valley that is partially filled with bars [26], and the Western Scheldt (NL), which is an alluvial estuary with large deviations from a converging along-channel width profile and filled with tidal bars [19,26] (Figure 2c,d). For validation, channel depth at the landward boundary was calculated in the tool with measured river discharge. Predicted velocities are validated on the Columbia River only because the regression was based on data for the Western Scheldt. Bed elevation and salinity predictions are validated on both systems. The predicted flow velocities from the tool were compared with results from numerical modelling for the Columbia River estuary [7,28] and from a calibrated hydrodynamic model of Rijkswaterstaat for the Western Scheldt (SCALWEST, [32]). Results from numerical modelling were used because field measurements of flow velocities do not provide full coverage in time and space. Salinity data were obtained from

Jay and Smith [51] for the Columbia River and from de Brauwere et al. [52] and Vroom et al. [53] for the Western Scheldt. Habitat preferences for potential habitat maps were obtained from the Marine Life Information Network (MarLIN), which is a database on the biology of species and the ecology of habitats.

2.8. Tool Application to End-Member Cases

We tested the tool's performance against end-member systems. The dominant factors controlling the tool output are (i) the along-channel width profile, (ii) river discharge at the landward boundary and (iii) tidal range at the seaward boundary. Given these three factors, we designed the three most extreme end-member cases (Figure 2a,b,e), which were tested in addition to two estuarine cases with variable width profile (the Western Scheldt and Columbia River) (Figure 2c,d). The first end-member is an estuary with a perfectly converging shape from the mouth to the upstream river, i.e., negligible width variation. This case was based on the dimensions of the Tran De branch of the Mekong Delta (Vietnam). The second end-member is a river with a constant channel width, i.e., no tidal range at the seaward boundary and negligible width variation. This case was based on the dimensions of a distributary of the Saskatchewan River, near Cumberland Lake (Saskatchewan, Canada). The third end-member is a tidal basin, i.e., no river discharge and landward increasing width instead of decreasing. This case was based on dimensions of the Ameland inlet and represented by a narrow mouth, an increasing channel width in landward direction and an infinitely small river at the landward boundary. It should be noted that, in the last case, along-channel width is measured perpendicular to the tidal channels, which means that the transects for an idealised tidal basin are semicircles. The predictions for the Tran De branch were compared with measured data [54–56] and the predictions for the tidal basin were compared with numerical model results [31,57].

3. Results

3.1. Tool Output

The tool output comprises maps of predicted bed elevation with respect to the high water level (HWL), relative inundation duration, flow velocity and salinity (Figures 5 and 6). In general, the bed deepens in seaward direction. At locations where the estuary is relatively wide, average bed elevations are higher and at locations where the estuary width is close to the ideal width, bed elevations are lower (Equations (4) and (5)). Zones where the excess width is large are the locations where stretches of intertidal area are wider and where inundation durations are less than 100%. Peak flow velocities increase in seaward direction with maxima around 1.8 ms^{-1} at the mouth. Peak flow velocities in the intertidal area are generally below 0.8 ms^{-1}. The salinity gradually decreases in landward directions and approaches river salinity at the landward boundary.

Before, we had to estimate a typical peak flow velocity because it was required for the estimation of the tidal excursion length (Equation (9)). At that point, a value of 1 ms^{-1} was assumed [18]. This assumption only partly influences the output flow velocity, which depends on bed elevation, which in turn is controlled by the estuary planform shape, the tidal range and the tidal excursion length. Nevertheless, output maps justify the assumption of 1 ms^{-1} for the purpose of calculating the tidal excursion length. The average peak tidal flow velocity is 1 ms^{-1} for the Western Scheldt and 1.2 ms^{-1} for the Columbia River when averaged over the surface area of a typical tidal excursion length (Figures 5 and 6). Median values of peak tidal flow velocity per transect are also remarkably close to the typical value proposed by Savenije [18] along the entire estuary (Figure 7a,c).

Additional output (Appendix A Figures A3 and A4) consists of predictions of bar width and braiding index per cross-section, maximum peak flow velocity and maximum average flow velocity per transect. Furthermore, four zones are calculated based on elevation with respect to the tidal range: subtidal deep ($<-2a$), subtidal low, intertidal low (more than 50% of time submerged) and intertidal high (less than 50% of time submerged). For each of these zones, the width of the zone, average depth and average and peak tidal flow velocities are calculated as output.

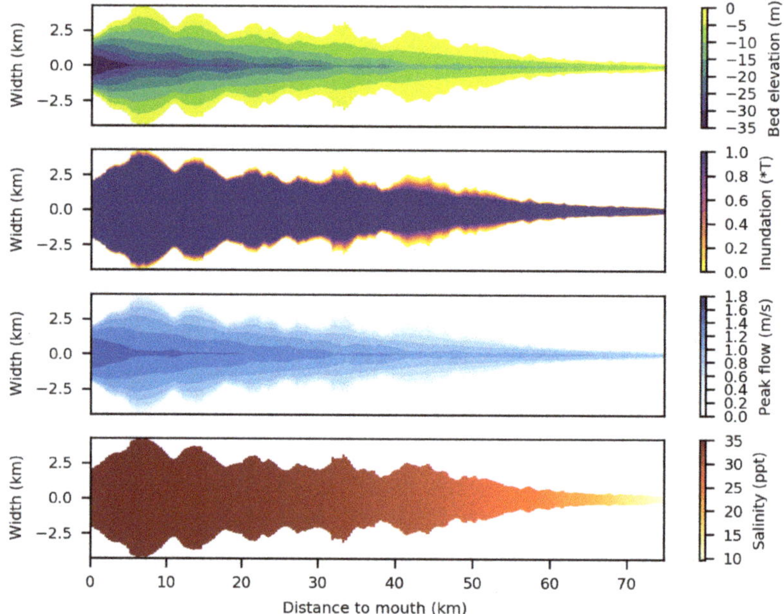

Figure 5. Resulting predictions of bed elevation with respect to the high water level, inundation duration, peak flow velocity and salinity for the Western Scheldt.

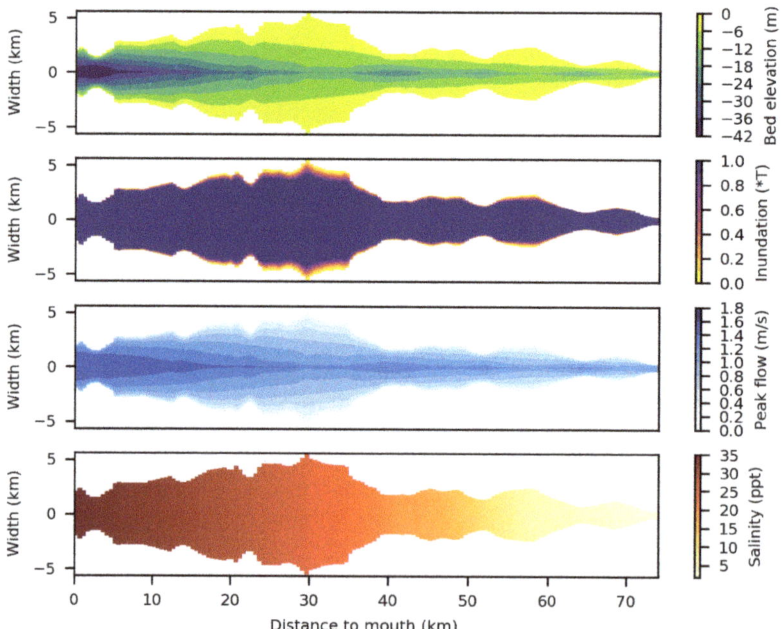

Figure 6. Resulting predictions of bed elevation with respect to the high water level, inundation duration, peak flow velocity and salinity for the Columbia River.

3.2. Tool Validation

3.2.1. Bed Elevations

To validate the predicted bed elevations, median and minimum values per transect were compared with measurements from bathymetry. Median and minimum bed elevations are generally within a factor 2 of the measured values from bathymetry (Figure 7a,b). Median bed elevations deviate the most in the landward part for the Western Scheldt (Figure 7a). In this section, a larger portion of the width is dredged, which causes the median depth to fall within the deeper dredged channel. The deviation between predicted and measured values lacks a trend with a position along the estuary (as indicated by the colours in Figure 7a,b). The largest outliers for the Columbia River are found at the mouth where the predicted median bed elevation is much lower than measured from bathymetry (Figure 7b). This is caused by the small embayments enclosed by groynes, located on the sides of the mouth area. In this zone, a larger proportion of high bed elevations causes a higher median bed elevation.

3.2.2. Flow Velocity

The range of predicted flow velocities is generally in the same range as the modelled values for both the peak and average flow velocities (Figure 7c,d). However, the along-channel variation in minimum and maximum flow velocities is larger for the models than for the tool output (Figure 8). We explain this by a simplification of some morphological elements in the tool. For example, the presence and location of scours and sills, which cause variation in the along-channel maximum depth, is not predicted by the tool. However, they do occur in natural systems and therefore also cause variations in along-channel flow velocity pattern. In the tool, the profile of maximum depth is linear along-channel, which results in lower variations in the maximum flow velocity. Nevertheless, the velocity predictions do capture the along-channel variability caused by the large-scale morphological variation in bed elevation, as illustrated with the median values per transect (Figure 8).

The tool does not reproduce the highest observed flow velocities in the Columbia River, which are well above 2 ms^{-1} in the first 15 km from the mouth. To obtain velocities above 2 ms^{-1} in the tool, bed elevation of at least 120 m below high water level are required (Equation (17)), which generally do not occur in natural systems. This means that the tool underpredicts flow velocities in cases where the cross-sectional area is relatively small compared to the local tidal prism, which might also suggest disequilibrium. Moreover, the regression between flow velocity and depth, simplifies the increasing scatter for larger depths and flow velocities, which is observed for the Western Scheldt (Figure 4).

Predicted velocity against modelled velocity shows a trend for the Columbia River with modelled velocities below 1 ms^{-1} being overpredicted by the tool and velocities above 1 ms^{-1} being underpredicted (Figure 7d). The degree of misprediction does not correlate with position along the estuary. The only explanation for the deviation is a different depth dependency of peak flow velocity caused by a different ratio between tidal prism and cross-sectional area.

Therefore, an alternative approach for calculating flow velocities was based on the local tidal prism and the cross-sectional channel geometry, i.e., continuity. For each cross-section, a typical velocity is calculated at the average depth of the cross-section (i.e., $P(x)/(T\overline{h(x)})$). Nevertheless, the quality of these predictions appeared to be less (Figure 9) for two reasons: (1) at the upstream boundary, velocities are mainly sensitive to river discharge. Here, local tidal prism approaches river discharge because it is obtained from integrating over the distance of a tidal excursion length upstream (Equation (9)) plus river discharge. In the case of the Columbia River, this leads to overprediction of velocities because river discharge is large, while, for the Western Scheldt, it would lead to underpredictions. (2) Flow velocities at the intertidal area and deepest parts still have to be extrapolated from the average prediction creating uncertainty in the most important parts for respectively ecology and shipping.

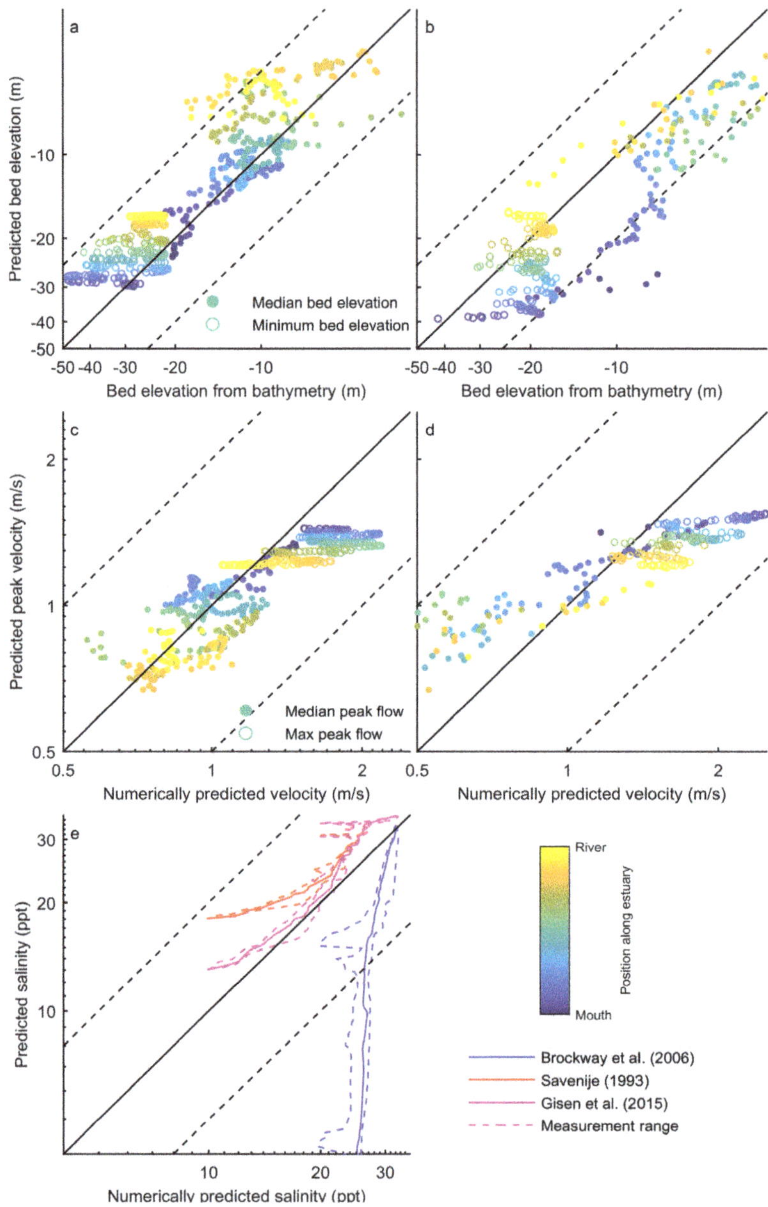

Figure 7. (**a**,**b**) predicted bed elevations against measured bed elevations taken from bathymetry for the Western Scheldt (**a**) and Columbia River (**b**); (**c**,**d**) predicted flow velocity from the tool against modelled flow velocity for numerical models of the Western Scheldt (**c**) and Columbia River (**d**). Filled circles indicate median values per cross-section, open circles indicate maximum values per cross-section. Dashed lines indicate a spread of a factor 2 around the $x = y$ line. Colours indicate the position along the estuary, with blue being the mouth and yellow being the most landward location; (**e**) predicted salinity for the three predictors against measured salinity in the Western Scheldt. Solid lines indicate mean measured salinity per transect against predicted salinity and dashed lines indicate maximum and minimum measured values.

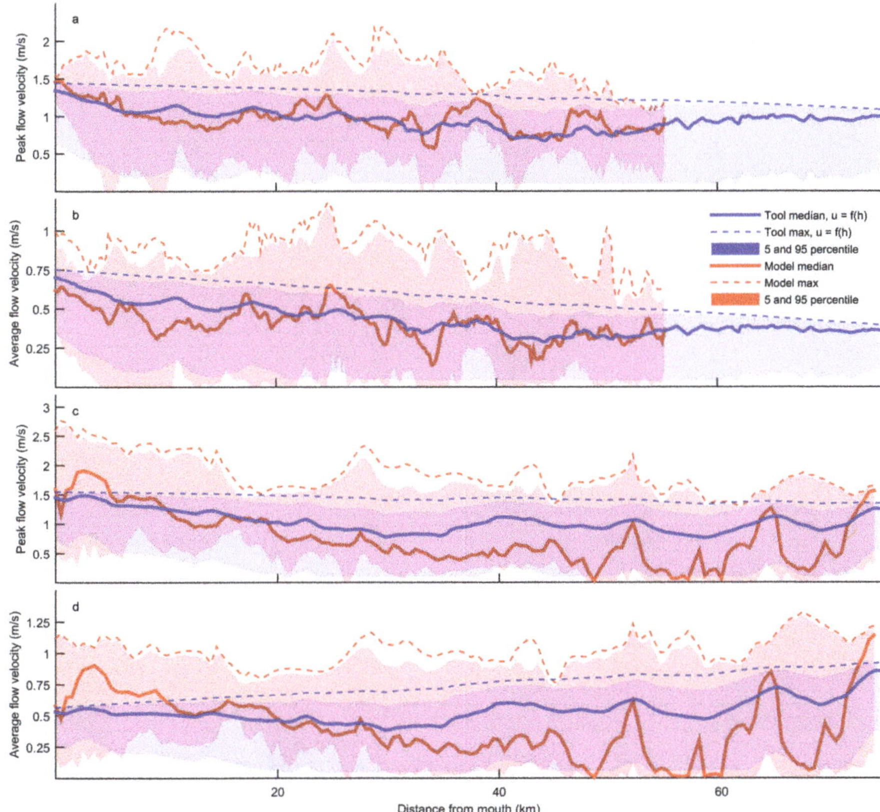

Figure 8. Comparison between modelled [red] and predicted [blue] flow velocities along the Western Scheldt (**a**,**b**) and the Columbia River (**c**,**d**). Shaded area shows the range of peak velocities (**a**,**c**) and range of average velocities over half a tidal cycle (**b**,**d**) bounded by the 5 and 95 percentile values. The solid line indicates the median and dashed line maximum value per transect.

Following this alternative approach, we anticipated that there would be a relation between the width variation and the misprediction based on depth only (Equation (17)). It was expected that a local increase in width would lead to flow expansion and therefore typically lower flow velocities than for a cross-section with similar tidal prism but lower width. However, the lack of trend between velocity deviation and width measures (Appendix A Figure A2) indicates that velocity predictions will not improve from including width in the regression.

The alternative approach shows that a prediction of flow velocities depending on tidal prism and depth does not reproduce modelled flow velocities better (Figure 9a,b,e,f). A predictive relation between flow velocity and depth is thus of value. This urges the need to compile a dataset with flow velocities and water depths of multiple systems. From that data, a relation that is better applicable to more systems could be derived, but this data is not yet available. Nevertheless, the present relation predicted flow velocity within a factor 2 of the modelled values even in the most extreme cases, which means it is of value for an estimate when limited data is available.

Predicted flow velocity and local depth result in a local tidal prism that is in the same range as the modelled values (Figure 10). Local tidal prism obtained from integrating a tidal range over the full length of the estuary is typically 1.5–2 times larger at the mouth than the tidal prism obtained from integrating over a typical tidal excursion length. This explains why using the latter is required to obtain accurate predictions of the depth at the estuary mouth and additionally is less sensitive to the length of the estuary.

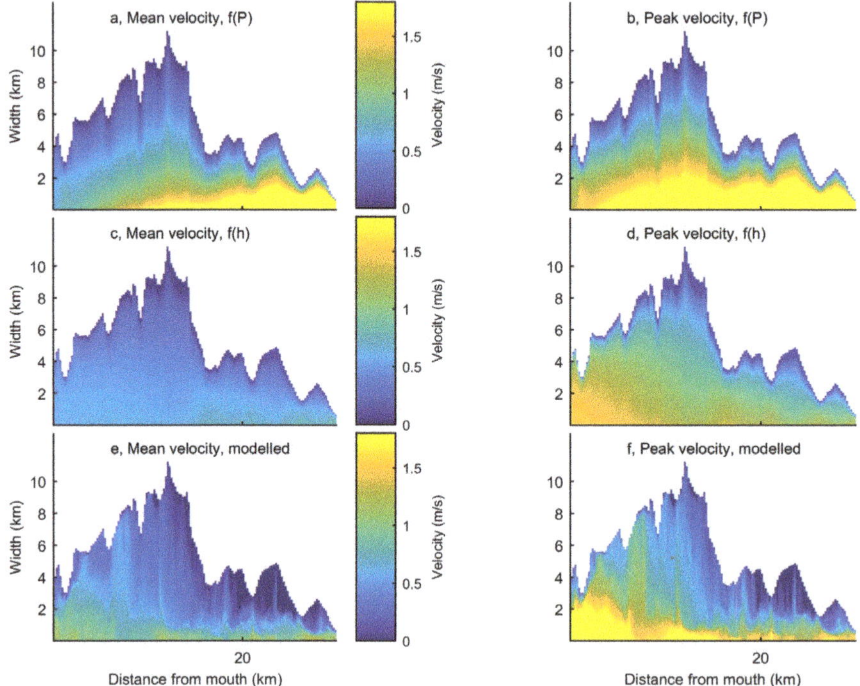

Figure 9. Mean velocity over a tidal cycle (left) and peak tidal flow velocity (right) as a function of local tidal prism (**a**,**b**), bed elevation (**c**,**d**) and numerically modelled values (**e**,**f**) for the Columbia River.

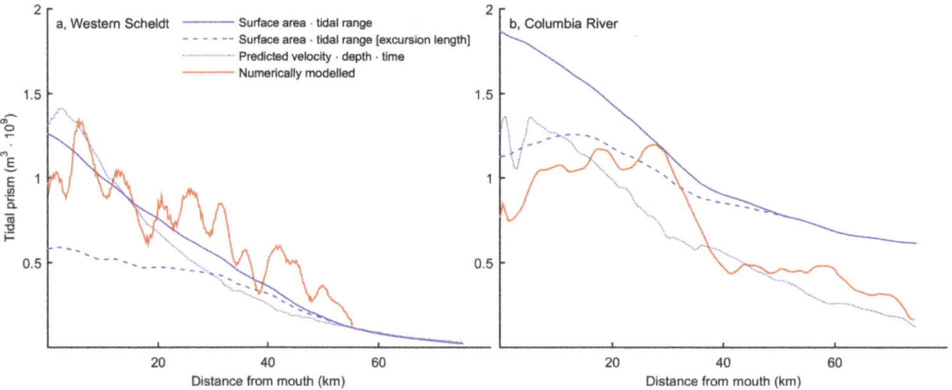

Figure 10. Resulting predictions of local tidal prism for the Western Scheldt (**a**) and Columbia River (**b**). Red lines are model results, blue lines are tool output.

3.2.3. Salinity

Predicted salinities are within the range of measured values [51] and values obtained with a numerical model with shallow-water and tracer-transport equations [52]. For the Western Scheldt, local salinity can vary between 4–8 ppt over a tidal cycle, with averages of 28 ppt at 10 km, 22 ppt at 30 km and 16 ppt at 45 km from the mouth [52]. In the Columbia River estuary, temporal salinity variation is even larger [51], with local variations of 20–25 ppt in the most seaward 15 km. Predicted salinities fall within the measured range, but salinities are overpredicted in the most landward reach, where measured values approach 0 ppt from approximately 45 km from the mouth [51].

The salinity predictor of Gisen and Savenije [50] is most accurate for the salinities observed in the Western Scheldt (Figure 7e), which was a system used for validation in Gisen and Savenije [50] and not for calibration. This is the only salinity predictor for which predicted salinity is within a factor 1.5 of the measured value for the entire along-channel section. Predictors of Savenije [21] and Brockway et al. [22] show large deviations at the landward side, where the predicted salinity gradient is respectively too weak and too strong compared to measured values (Figure 7e).

3.3. Applicability to End-Member Systems: A River, an Ideal Estuary, and a Tidal Basin

In the case of a river and ideal estuary, the local channel width will always be equal to the ideal channel width as obtained from a maximum fitting converging shape. This means that the shape of the cross-sectional hypsometry is concave and constant along-channel (Equations (4) and (5), Figure 11). The concave hypsometry results in a relatively constant along-channel bed elevation profile (Figure 11). Only the along-channel variation in maximum depth and increase in channel width in seaward direction cause a slight along-channel variation. The implication of an along-channel constant concave hypsometry is that intertidal area is very small and inundation duration is 100% for the largest part of the estuary. Typical flow velocities are also along-channel constant at approximately 1–1.2 ms^{-1}. A strong salinity gradient is predicted at the mouth of the estuary, with salinities dropping to $\approx$5 ppt at 30 km from the mouth.

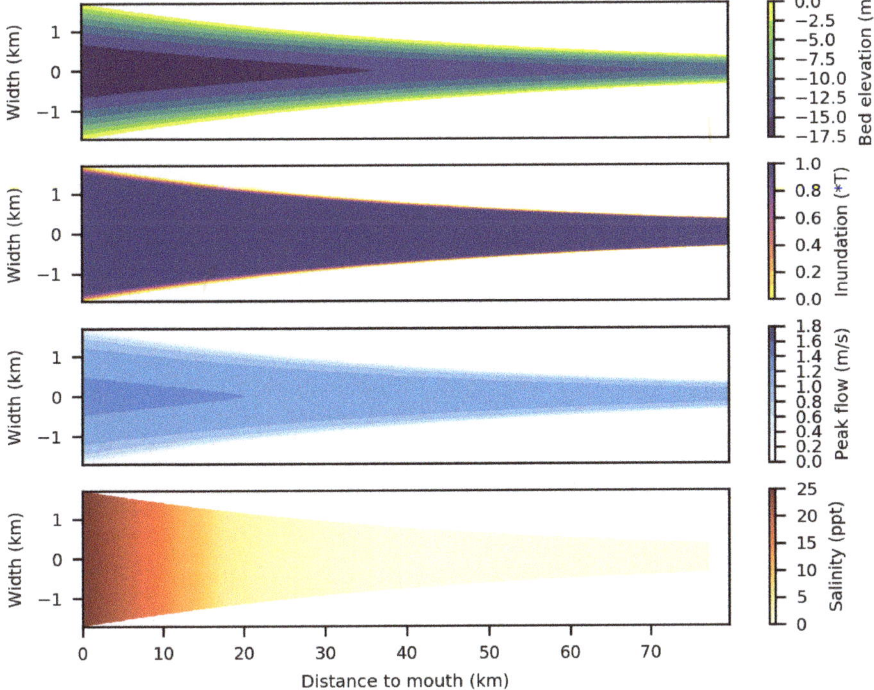

Figure 11. Resulting predictions of bed elevation with respect to the high water level, inundation duration, peak flow velocity and salinity for an ideal estuary based on dimensions of the Tran De branch of the Mekong river.

Ideal estuaries are said to be characterised by an along-channel constant velocity amplitude and depth [18,54] and these features are reproduced by the tool predictions. Moreover, the predictions correspond to observations in the Tran De branch of the Mekong: observed peak flow velocity is approximately 1 ms^{-1} along the entire branch [55,56] and salinity intrusion reaches up to 45–55 km from the mouth [54]. In addition, the along-channel predicted salinities fall within the range of observed salinities [54]. The only feature not completely reproduced is the along-channel constant depth. When the maximum depth at the mouth is calculated from tidal prism, predicted depth at the mouth is a factor 1.5 larger, 5 m deeper, than observed.

The prediction for depth at the mouth leads to a misprediction at the most downstream boundary for the river (Appendix A Figure A6). In that case, maximum depth at the landward boundary is controlled by river discharge (to the power of $\approx$0.35), the width of the river without tidal influence (to the power of $\approx$0.77) or if both of these are unspecified, the channel width at the landward boundary. Depth at the mouth is calculated with the tidal prism, which in the case of a river is equal to the river discharge times the duration of half a tidal cycle. Therefore, depth at the mouth is also a function of river discharge or landward river width, but to the power of $\approx$1. This means that the predicted channel depth at the mouth will always be deeper than at the upstream boundary, especially in the end-member case of a straight river.

An ideal tidal basin is characterised by a strong landward increase in width, while the ideal width approaches 0 at the landward end. This means that the ratio of $W_{ideal}(x)$ to $W(x)$ strongly decreases in landward direction, resulting in convex hypsometry (Equation (4)) and therefore large amounts of intertidal area within the basin (Figure A7). In this end-member case, the depth at the mouth is set by the surface area of the tidal basin multiplied with the tidal range. In the case of a tidal basin based on the Ameland inlet, the tool predicts a maximum depth at the mouth of 20 m below the high water level. This seems appropriate when compared with modelled [31,57] and measured values [29] (Figure A8). Flow velocities at the mouth are in the order of 1 ms^{-1}, which is comparable with measured and modelled values for tidal basins in equilibrium [57,58].

4. Discussion

4.1. Sensitivity of Tool Output to Tidal Range, River Discharge, and along-Channel Width Profile

The sensitivity of the input tidal range, river discharge, and along-channel width profile was tested on the following parameters: (1) the ratio between intertidal and subtidal area; (2) the ratio between low and high dynamic environment; and (3) peak flow velocity at the estuary mouth (Figure 12). Low dynamic environment, which forms potential habitat, is defined as intertidal area where flow velocities are below 0.8 ms^{-1}. All other areas are classified as a high dynamic area.

While most sensitive dependent variables are increasing or decreasing monotonously with the independent variable, the ratio of low and high dynamic intertidal area as a function of tidal amplitude shows an optimum. An increase in tidal amplitude increases the ratio between intertidal and subtidal area (Figure 12a) because the increased tidal range increases intertidal area. However, increases in tidal amplitude also generate a proportional increase in tidal prism. A larger tidal prism at the mouth will increase the predicted cross-sectional area and because the width is fixed; this will increase predicted depth at the mouth. Because maximum depth along the estuary is dependent on the maximum depth at the mouth and the maximum depth at the landward river, this increases the maximum depth along the entire estuary. The increased depths subsequently increase the predicted flow velocities (Figure 12g). This means that two opposing mechanisms affect the ratio between low dynamic and high dynamic environment: increased tidal range increases intertidal area which forms a more low-dynamic environment. In contrast, the deepening by increase in tidal prism increases flow velocities and therefore reduces the low dynamic area. Therefore, the combined effect results in an optimum curve for the ratio between low dynamic (velocity < 0.8 ms^{-1}) intertidal area and high dynamic area as a function of tidal range (Figure 12d).

An increase in the river discharge results in less intertidal area, less low dynamic area and larger peak flow velocities (Figure 12b,e,h). After all, increased river discharge increases the tidal prism, which increases channel depths.

The ratio between ideal width and excess width affects the distribution of bed elevations per cross-section and the tidal prism. In estuaries that adapt to the ideal shape, the excess width becomes lower. This results in more concave cross-sectional hypsometry and therefore narrower stretches of intertidal flats (Figure 12c). The tidal prism decreases as a result of reduced estuary area, resulting in lower maximum depths and therefore also lower flow peak velocities (Figure 12i).

A comparison between predicted and measured or modelled values showed that bed elevations, flow velocity and salinity can be predicted within a factor two of the measured or modelled values (Figure 7). The largest deviations between predicted and modelled flow velocities occur when they peak above 2 ms^{-1} in natural systems. Largest mispredictions for salinity occur at the transition from brackish to fresh water, where salinity is also the most dynamic over the seasons. Bed elevation predictions are most uncertain when the tidal range is low for the depth at the mouth, which is also sensitive to offshore wave climate that we have not considered here. This implies that the tool is capable to provide a first-order estimation for estuaries with limited data, but that it cannot replace measurements or numerical modelling.

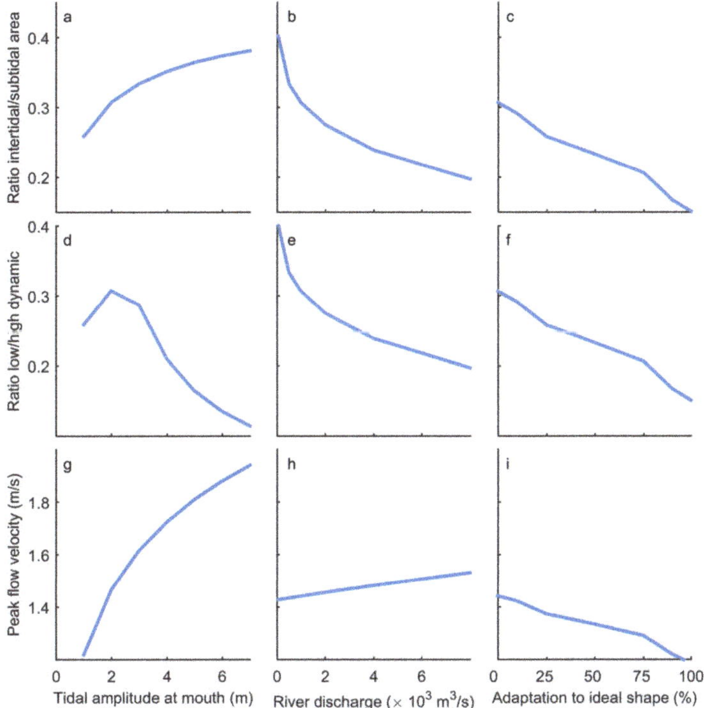

Figure 12. Sensitivity of tool output for the Western Scheldt to input tidal amplitude, river discharge and width profile. The ratio between intertidal and subtidal area increases with tidal amplitude (**a**) and decreases with river discharge and adaptation to ideal shape (**b,c**). The ratio between low dynamic (velocity < 0.8 ms^{-1}) intertidal area and high dynamic area has an optimum curve for tidal amplitude (**d**) and decreases with river discharge and adaptation to ideal shape (**e,f**). Peak flow velocity at the mouth of the estuary increases with tidal amplitude and river discharge (**g,h**) and decreases with adaptation to ideal shape (**i**).

4.2. Applications

4.2.1. Illustration of Application for Ecological Assessment

The major abiotic factors that determine habitat area of living organisms in estuaries are depth, flow velocity, salinity and inundation time [2–4,59]. Habitat suitability models (e.g., HABITAT, [24]) can translate these abiotic factors into a characteristic habitat suitability index, which for example has been applied to study the effect of climate change and dam construction on potential habitat in rivers [25]. Multiple case studies of individual systems or individual species are also available for estuaries [60–64], but application to more systems worldwide and application to scenarios of future climate change and human engineering are hampered by a lack of data.

The tool presented in this study can provide typical abiotic factors that are input for habitat suitability models within minutes. Therefore, the output can lead to a first-order estimation of the potential habitat (Figure 13) based on species specific preferences (Table 1). The potential habitat area based on tool output generally overpredicts the measured habitat. First, the tool always predicts a zone with intertidal area because the predicted hypsometry is stretched between the high water line and maximum channel depth. Second, other constraining biotic and abiotic factors were excluded, such as the competition between species and the water quality. Lastly, the settling and growth conditions may deviate from the conditions under which species occur in a later life stage. For example, seedlings of salt marsh species generally only settle when peak flow is below 0.25 ms^{-1} and, in their initial life stage, they may only sustain flows up to 0.5 ms^{-1} [65]. Nevertheless, the application of the tool for ecological assessment is of value because it is capable of providing abiotic factors of individual systems and scenarios within minutes, rather than days to months for numerical modelling and in situ measurements.

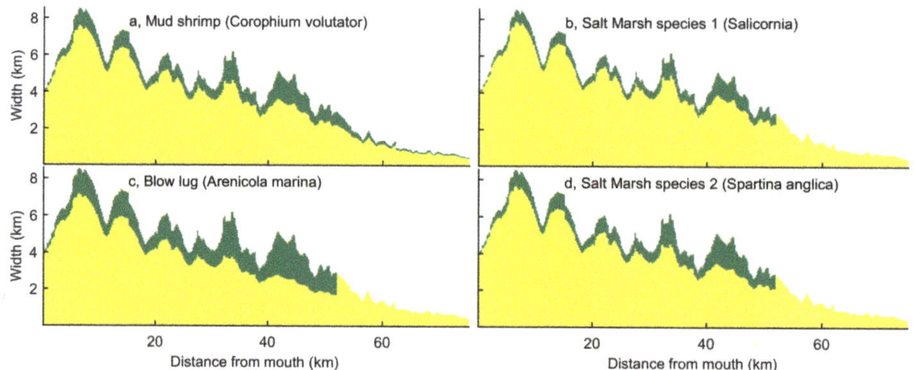

Figure 13. Potential habitat area (in green) as part of total width for the species in Table 1.

Table 1. Habitat preferences of species used in this study based on MarLIN database.

Name	Depth Range (-)	Flow Velocity (ms^{-1})	Salinity (ppt)
Mud shrimp (*Corophium volutator*)	Intertidal	<0.5	2–40
Blow lug (*Arenicola marina*)	Intertidal	<4.0	18–40
Salt marsh species 1 (*Salicornia*)	Intertidal above Mean Sea Level	<0.5	18–40
Salt marsh species 2 (*Spartina anglica*)	Intertidal above −0.5a	<1.0	18–40

4.2.2. Application for Management of Estuaries

The main channels in estuaries are often used as shipping fairways, while the intertidal area forms important ecological habitat [2]. Given a certain channel width and depth required for shipping, the empirical predictions of bed elevations that result from this study can estimate volumes that need to be dredged and typical locations where this is necessary. While, for many of the present-day shipping fairways, bathymetry is available, this application may be relevant for quick assessments by consultancy in developing countries with growing economies for the construction of fairways, harbours or studies on land loss in these areas [66,67].

Another implication from this study is that the wider an estuary is relative to the fitted converging shape, the wider the zone of intertidal area will be. Bed elevations and basin hypsometry form an indication of possible habitat composition [35] and low-dynamical intertidal areas are generally labelled as valuable areas [68] because they are suitable for settling and feeding. The output can indicate which areas satisfy these conditions (e.g., suitable flow conditions) and which areas can be transformed into habitat by disposal of dredged sediment. For example, in the Western Scheldt, dredged sediment has been disposed on the edges of tidal flats to maintain intertidal area [68], which aids in achieving *European Union Natura 2000* regulations [69]. This approach requires bathymetry, which may not be available at locations where new fairways or harbours are planned. The predictive tool can provide a first-order estimation in these cases.

4.2.3. Application in Palaeogeographic Reconstruction

Our empirical method is capable of making quantitative estimates of the bar dimensions, the number of bars and channels (Braiding Index), channel dimensions, flow conditions and salinities. The tool is applicable for palaeogeographical reconstructions [70,71] where the shape of tidal channel belts or estuaries has been identified, but wherein bar configuration is typically not recognisable and typical flow conditions are unknown. If the along-channel width profile is known and the tidal range can be estimated, the tool presented in this study can be applied. This application may be particularly relevant for high-stand estuaries that are filled with sediment. Besides the forward approach of palaeogeographical reconstructions based on estuary outline, the tool and equations also allow a reverse approach. A single measurement or preferably a few measurements of bar or channel dimensions [12,72–76] would be enough to reconstruct typical channel width and local tidal prism and therefore also typical flow conditions.

Additionally, the combination with ecological habitat analyses may improve the understanding of the evolution of estuaries. It is known that vegetation succession stabilises tidal sand bars [65,77–79], but it remains an open question to what extent this succession confines the width of estuaries on the longer term. Additionally, it is unknown what the equilibrium shape of an estuary will be under a given set of stabilising and destabilising eco-engineering species.

5. Conclusions

A comprehensive set of empirical relations allows estimating bathymetry, flow velocity, salinity and inundation durations for estuaries for which limited data are available. Morphological characteristics are based on the concept that the degree to which an estuary deviates from a maximum fitting converging shape is reflected in the locations where intertidal bars are found, which also determines the bed elevation distributions. This study provides a new correlation between depth below high water level and peak tidal flow velocity, based on numerical model outcomes, which complements the bed elevation predictions and allows for salinity predictions. The resulting predictions are within a factor two of the measured or modelled values, but generally better. Largest deviations between predicted and modelled or measured occur for flow velocities when they peak above $2\ \mathrm{ms}^{-1}$, for salinity at the transition from brackish to fresh water and for bed elevations when the tidal range is low for the depth at the mouth. The end-member cases of river-dominated delta branches and tide-dominated delta branches, i.e., rivers and ideal estuaries, are most sensitive to depth at both boundaries. In these cases, when possible, it should be measured in a single cross-section and specified as input to prevent misprediction.

We conclude that the tool is suitable for alluvial estuaries filled with tidal bars, for relatively wide valleys that partially filled with bars and for tidal basins. The results are usable to quantify valuable ecological habitat area, to estimate dredging volumes in management of estuaries or to make palaeographic reconstructions when limited data are available.

Supplementary Materials: The following is available at http://www.mdpi.com/2072-4292/10/12/1915/s1, A zip file with tool code and instructions to use the tool. The ZIP contains: (1) 'Instructions.pdf' with instructions on how to use the tool; (2) 'Input_variables.xls', which is file where input parameters need to be specified; (3) '.csv' files with examples of along-channel width profiles; (4) 'Model_v1_0.py', which is the tool code that needs to be run in Python; (5) 'Examples' folder with examples of the output for the Western Scheldt.

Author Contributions: Contributions were in following proportions to conception and design, tool programming, data collection and analysis, conclusions and manuscript preparation: J.R.F.W.L. (65%, 50%, 60%, 85%), S.L.V. (5%, 50%, 20%, 5%), W.M.v.D. (5%, 0%, 10%, 5%), S.S. (5%, 0%, 5%, 0%), and M.G.K. (20%, 0%, 5%, 5%).

Funding: J.R.F.W.L., W.M.v.D. and M.G.K. were supported by the Dutch Technology Foundation TTW (grant Vici 016.140.316/13710 to M.G.K., which is part of the Netherlands Organisation for Scientific Research (NWO), and is partly funded by the Ministry of Economic Affairs). S.L.V. was supported by Future Deltas, Utrecht University (seed grant to J.R.F.W.L.). S.S. was supported by an ERC Consolidator grant (647570) to M.G.K.

Acknowledgments: This work is part of the PhD research of J.R.F.W.L. and MSc research of S.L.V. We thank three anonymous reviewers for their comments on the manuscript. We thank Rijkswaterstaat for providing bathymetry of the Western Scheldt and Edwin Elias for providing the model of the Columbia River. Discussion with Sepehr Eslami Arab, Barend van Maanen and Tjeerd Bouma helped to improve the manuscript. The code of the tool is made available open access online with the manuscript and through GitHub (https://github.com/JasperLeuven/EstuarineMorphologyEstimator). Other data sources have been referenced in the text.

Conflicts of Interest: The authors declare no conflict of interest.

Appendix A. Supplementary Figures and Table

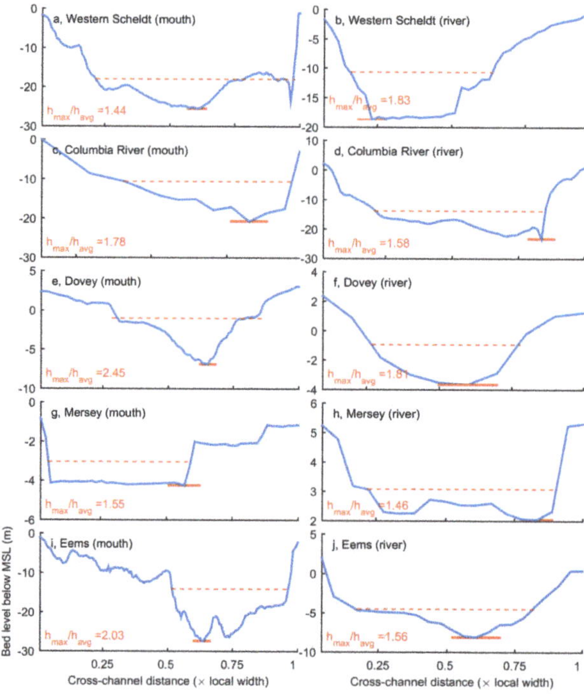

Figure A1. Cross-sectional profiles at the mouth (**left**) and landward river boundary (**right**) for five estuaries. The maximum depth (solid red line) and an approximation of the average water depth (dashed line) are indicated. The ratio between maximum depth and average depth, which is used in Equations (12) and (8), is given in the figures. Average ratio for the geometry of the estuary mouth is 1.85 and for the landward river boundary 1.65.

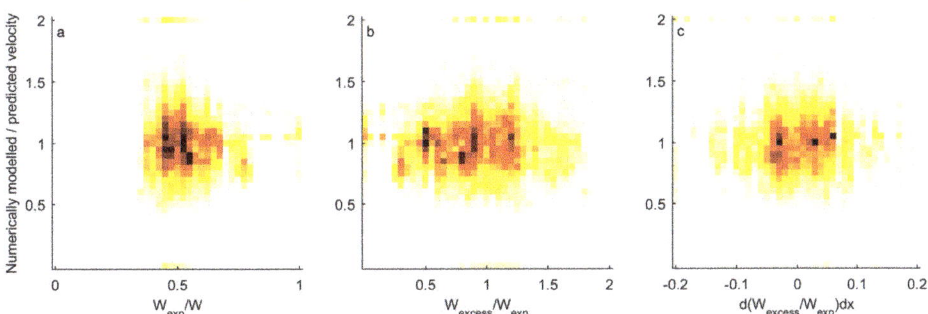

Figure A2. Measured velocity from the model divided by predicted velocity from the tool against three expressions of excess width. Lack of trend between velocity deviation and width measures indicates that velocity predictions will not improve from including width in the regression. (**a**) ideal width divided by local width, (**b**) excess width divided by ideal width and (**c**) along-channel derivative of excess width divided by ideal width. Colours in the plot become more yellow with increasing point density.

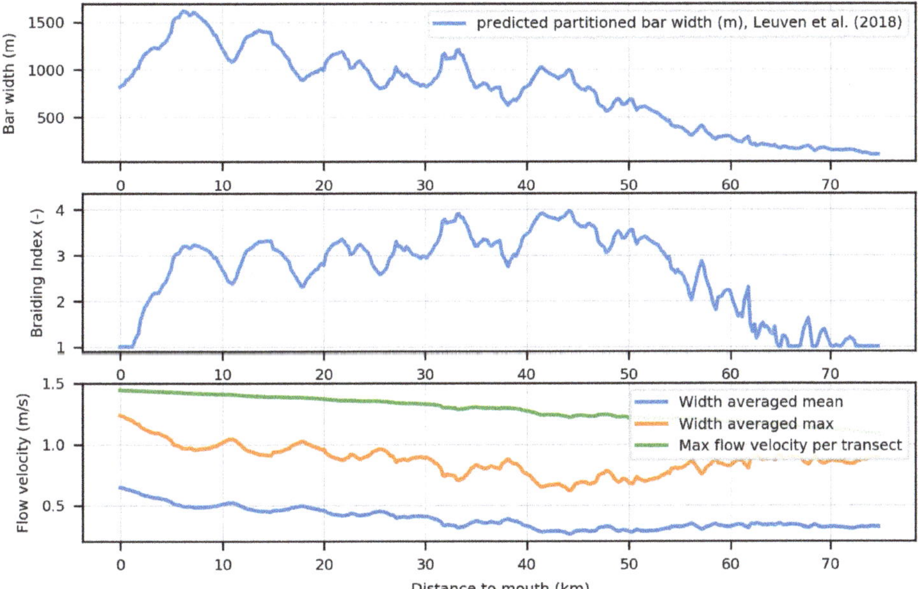

Figure A3. Tool output for the Western Scheldt. (**top**) predicted along-channel bar width, (**middle**) predicted along-channel braiding index and (**bottom**) the cross-sectional maximum and average of the predicted along-channel peak tidal flow velocity and the cross-sectional average of the mean flow velocity over half a tidal cycle.

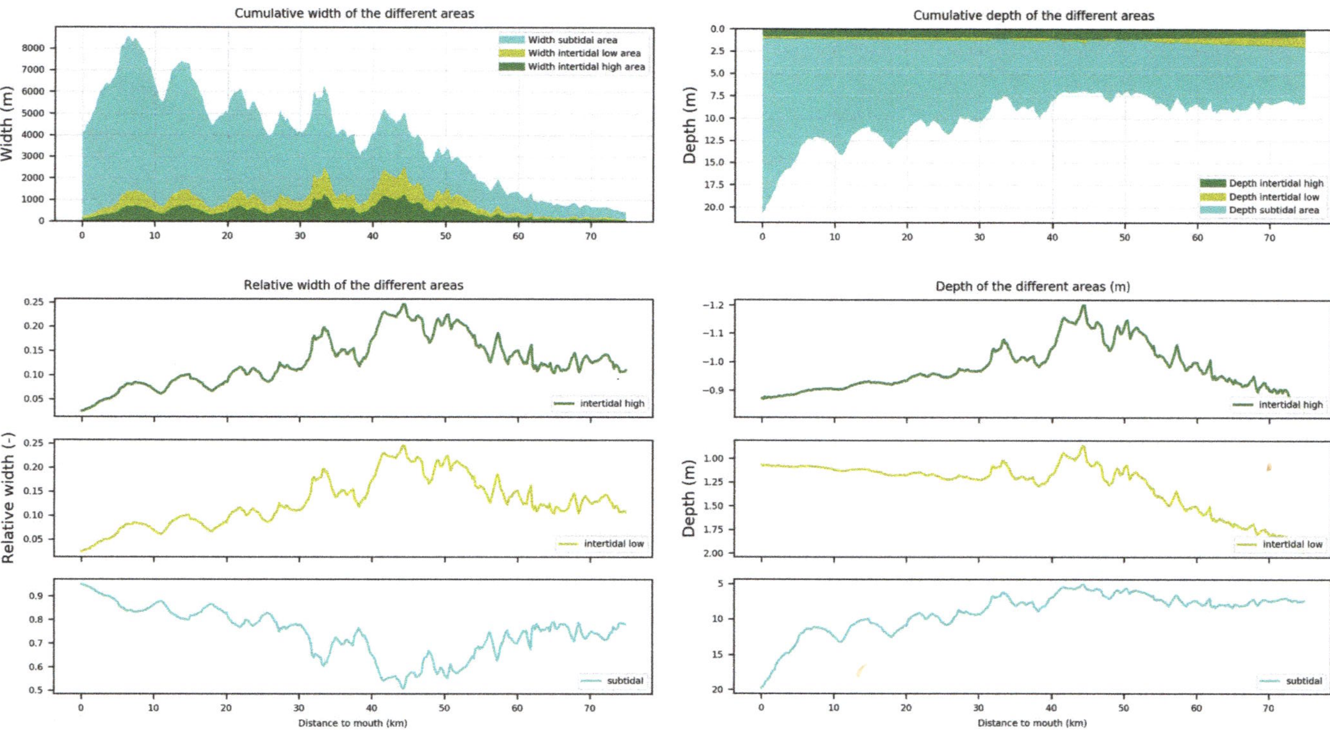

Figure A4. *Cont.*

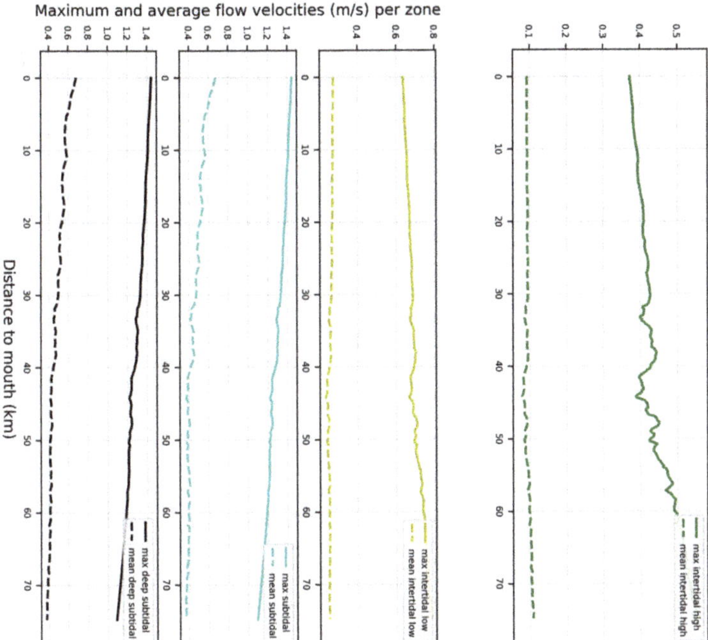

Figure A4. Tool output for the Western Scheldt. (**Top**) along-channel width of subtidal, intertidal low and intertial high zones; (**center**) average depth of subtidal, intertidal low and intertial high zones; (**bottom**) average and maximum flow velocity of subtidal, intertidal low and intertial high zones.

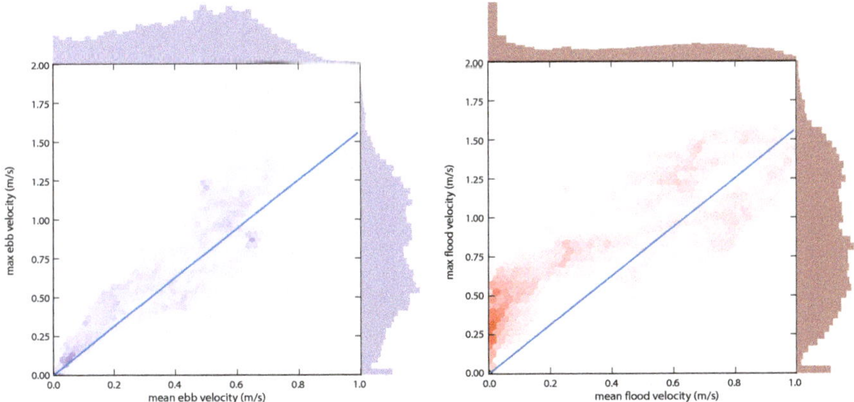

Figure A5. (**Left**) maximum ebb velocity as a function of mean ebb velocity; (**Right**) maximum ebb velocity as a function of mean ebb velocity. Histograms indicate occurrence within bins and colours in the plot become darker with increasing density. Drawn lines indicate the relation $x = 2/\pi y$, which is the expected average velocity when calculated as the average of half a sine function with an amplitude equal to the peak velocity.

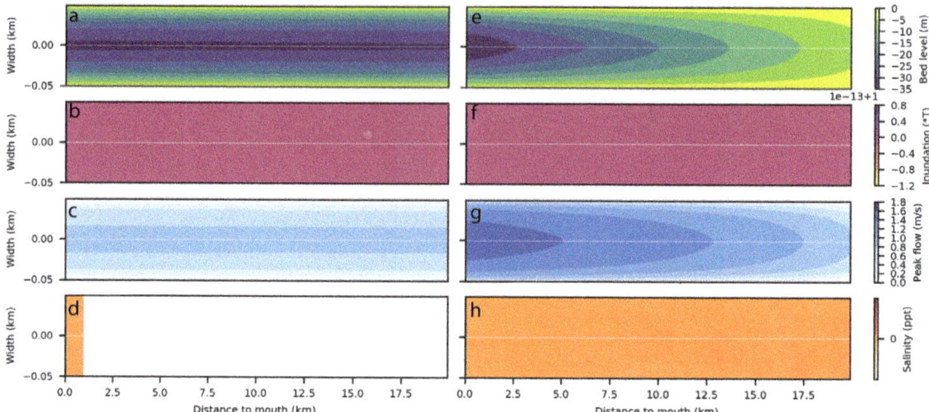

Figure A6. Resulting predictions of bed elevation with respect to the high water level, inundation duration, peak flow velocity and salinity for a river based on dimensions of a distributary of the Saskatchewan River, Cumberland Lake (Saskatchewan (Canada), coordinates: 54°04′N 102°22′W) (Figure 2a). In the left panels (**a–d**) the boundary condition for maximum depth at the mouth and the landward end are both based on hydraulic geometry for rivers (Equations (6)–(8)). In the right panels (**e–h**), the boundary condition for maximum depth at the mouth is based on the tidal prism relation (Equations (10) and (12)) and boundary condition at the landward end is based on hydraulic geometry for rivers (Equations (8)–(6)). See Bolla Pittaluga et al. [80] for measurements in this river.

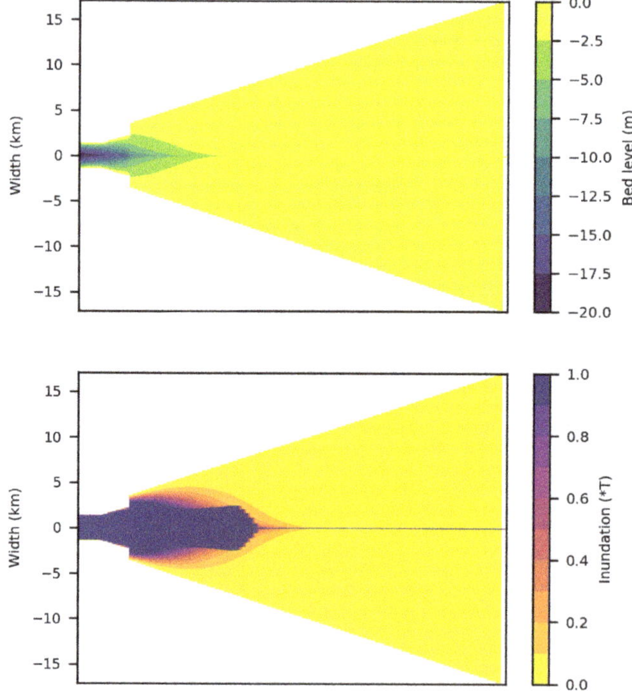

Figure A7. *Cont.*

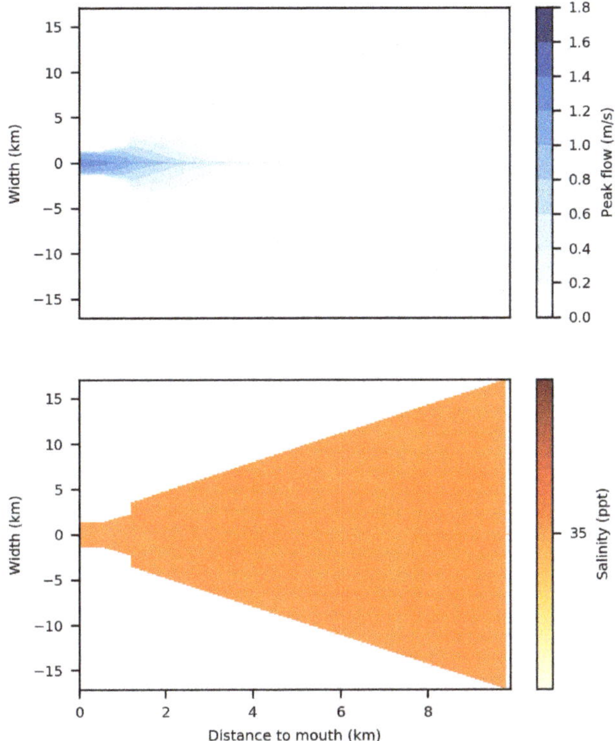

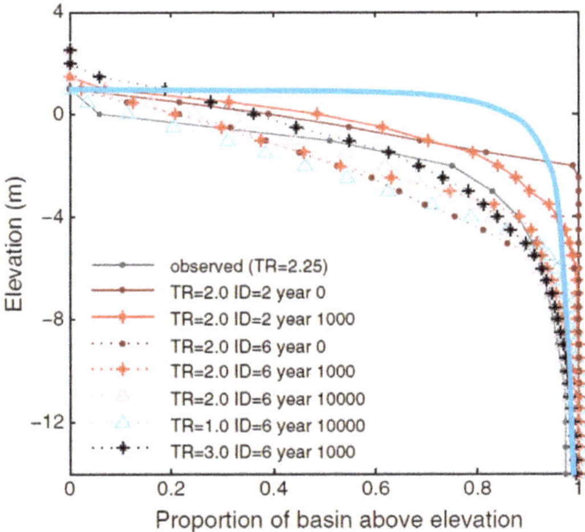

Figure A7. Resulting predictions of bed elevation with respect to the high water level, inundation duration, peak flow velocity and salinity for a tidal basin with dimensions based on the Ameland inlet.

Figure A8. The blue solid line indicates the resulting basin hypsometry from the prediction for a tidal basin with dimensions based on the Ameland inlet (Figure A7) plotted in Figure 7 of van Maanen et al. [31].

Table A1. List of symbols used.

Symbol	Units	Variable
$A_{csa,m}$	[m^2]	Cross-sectional area at the mouth
$A_{csa}(x)$	[m^2]	Cross-sectional area at coordinate x
$a(m)$	[m]	Tidal range at the mouth
$a(x)$	[m]	Tidal range at the mouth
$BI(x)$	[-]	Braiding Index at coordinate x
C_m	[m$^{0.5}$s^{-1}]	Chezy roughness at the estuary mouth
D_m	[m^{2}s^{-1}]	Dispersion coefficient at the mouth of the estuary
$D(x)$	[m^{2}s^{-1}]	Dispersion coefficient at coordinate x
$\Delta\rho$	[-]	Density difference between salt and fresh water
E	[m]	Tidal excursion length
g	[ms^{-2}]	Gravitational acceleration
$h_{max,r}$	[m]	Maximum channel depth at the landward boundary
$\overline{h_r}$	[m]	Average channel depth at the landward boundary
$h_{max,m}$	[m]	Maximum channel depth at the estuary mouth
$\overline{h_m}$	[m]	Average channel depth at the estuary mouth
$h_z(x,y)$	[-]	Dimensionless bed elevation at coordinate (x,y)
$h(x,y)$	[m]	Dimensional bed elevation at coordinate (x,y)
$I(x,y)$	[× t]	Inundation duration
K	[m^{2}s^{-1}]	Empirical equation for the longitudinal mixing coefficient
K_s	[-]	Longitudinal Van der Burgh mixing coefficient in Savenije [21]
K_g	[-]	Longitudinal Van der Burgh mixing coefficient in Gisen et al. [23]
L_W	[m]	Width convergence length
$L_{A_{csa}}$	[m]	Cross-sectional area convergence length
N_r	[-]	Estuarine Richardson number
$P(x)$	[m^3]	Local tidal prism at coordinate x
$P(m)$	[m^3]	Local tidal prism at the estuary mouth
Q_b	[m^{3}s^{-1}]	Bankfull river discharge
Q_r	[m^{3}s^{-1}]	Average river discharge
r	[-]	Coefficient in the Strahler [33] equation
r_s	[-]	Storage ratio (intertidal area/subtidal area)
ρ_w	[kgm^{-3}]	Water density
S_m	[ppt]	Salinity at the estuary mouth
S_r	[ppt]	Salinity at the landward boundary
$S_b(x)$	[ppt]	Salinity at coordinate x [22]
$S_g(x)$	[ppt]	Salinity at coordinate x [23]
$S_s(x)$	[ppt]	Salinity at coordinate x [21]
s	[s]	Distance between the mouth and landward boundary
s_m	[-]	Shape factor of the channel at the estuary mouth
s_r	[-]	Shape factor of the channel at the landward boundary
t	[s]	Duration of half a tidal cycle
$u_{avg}(x,y)$	[ms^{-1}]	Tidal average flow velocity at coordinate (x,y)
$u_{peak}(x,y)$	[ms^{-1}]	Peak tidal flow velocity at coordinate (x,y)
$u_{peak}(mouth,y)$	[ms^{-1}]	Peak tidal flow velocity at the estuary mouth
$W_{excess}(x)$	[m]	Excess width at coordinate x
$W_{ideal}(x)$	[m]	Ideal width at coordinate x
W_m	[m]	Width at the estuary mouth
W_r	[m]	Width at the landward boundary
$W(x)$	[m]	Local width at coordinate x
$w_{bar}(x)$	[m]	Predicted bar width at coordinate x
$w_{barssummed}(x)$	[m]	Summed width of bars at coordinate x
x	[m]	Streamwise coordinate measured from the mouth along the centreline
y	[m]	Coordinate perpendicular to the centreline
z	[-]	Coefficient in the Strahler [33] equation
$z(x)$	[-]	Value of z-coefficient at coordinate x

References

1. de Vriend, H.J.; Wang, Z.B.; Ysebaert, T.; Herman, P.M.; Ding, P. Eco-morphological problems in the Yangtze Estuary and the Western Scheldt. *Wetlands* **2011**, *31*, 1033–1042. doi:10.1007/s13157-011-0239-7. [CrossRef]
2. Bouma, H.; de Jong, D.; Twisk, F.; Wolfstein, K. *A Dutch Ecotope System for Coastal Waters (ZES. 1). To Map the Potential Occurence of Ecological Communities in Dutch Coastal And Transitional Waters*; Technical Report, Rijkswaterstaat, Report RIKZ/2005.024; Rijksinstituut voor Kust en Zee/RIKZ: Middelburg, The Netherlands, 2005.
3. de Jong, D. *Ecotopes in the Dutch Marine Tidal Waters: A Proposal for a Classification of Ecotopes and a Method to Map Them*; Technical Report, Rijkswaterstaat, Report RIKZ/99.017; Rijksinstituut voor Kust en Zee/RIKZ: Middelburg, The Netherlands, 1999.
4. Gurnell, A.M.; Bertoldi, W.; Corenblit, D. Changing river channels: The roles of hydrological processes, plants and pioneer fluvial landforms in humid temperate, mixed load, gravel bed rivers. *Earth-Sci. Rev.* **2012**, *111*, 129–141. doi:10.1016/j.earscirev.2011.11.005. [CrossRef]
5. Cozzoli, F.; Smolders, S.; Eelkema, M.; Ysebaert, T.; Escaravage, V.; Temmerman, S.; Meire, P.; Herman, P.M.; Bouma, T.J. A modeling approach to assess coastal management effects on benthic habitat quality: A case study on coastal defense and navigability. *Estuar. Coast. Shelf Sci.* **2017**, *184*, 67–82. doi:10.1016/j.ecss.2016.10.043. [CrossRef]
6. van der Wegen, M.; Roelvink, J. Reproduction of estuarine bathymetry by means of a process-based model: Western Scheldt case study, the Netherlands. *Geomorphology* **2012**, *179*, 152–167. doi:10.1016/j.geomorph.2012.08.007. [CrossRef]
7. Elias, E.P.; Gelfenbaum, G.; Van der Westhuysen, A.J. Validation of a coupled wave-flow model in a high-energy setting: The mouth of the Columbia River. *J. Geophys. Res. Oceans* **2012**, *117*, C09011. doi:10.1029/2012JC008105. [CrossRef]
8. Braat, L.; van Kessel, T.; Leuven, J.R.F.W.; Kleinhans, M.G. Effects of mud supply on large-scale estuary morphology and development over centuries to millennia. *Earth Surf. Dyn.* **2017**, *5*, 617–652. doi:10.5194/esurf-5-617-2017. [CrossRef]
9. Townend, I. The estimation of estuary dimensions using a simplified form model and the exogenous controls. *Earth Surf. Process. Landf.* **2012**, *37*, 1573–1583. doi:10.1002/esp.3256. [CrossRef]
10. Seminara, G.; Tubino, M. Sand bars in tidal channels. Part 1. Free bars. *J. Fluid Mech.* **2001**, *440*, 49–74. doi:10.1017/S0022112001004748. [CrossRef]
11. Schramkowski, G.; Schuttelaars, H.; De Swart, H. The effect of geometry and bottom friction on local bed forms in a tidal embayment. *Cont. Shelf Res.* **2002**, *22*, 1821–1833. doi:10.1016/S0278-4343(02)00040-7. [CrossRef]
12. Leuven, J.R.F.W.; Kleinhans, M.G.; Weissher, S.A.H.; van der Vegt, M. Tidal sand bar dimensions and shapes in estuaries. *Earth-Sci. Rev.* **2016**, *161*, 204–233. doi:10.1016/j.earscirev.2016.08.004. [CrossRef]
13. O'Brien, M.P. Equilibrium flow areas of inlets on sandy coasts. In *Joumal of the Waterways and Harbors Division: Proceedings of the American Society of Civil Engineers (ASCE)*; American Society of Civil Engineers: New York, NY, USA, 1969; pp. 43–52. doi:10.1061/9780872620087.039. [CrossRef]
14. Wang, Z.; Jeuken, C.; De Vriend, H. *Tidal Asymmetry and Residual Sediment Transport in Estuaries*; Delft Hydraulics Report Z2749; Deltares (WL): Delft, The Netherlands, 1999.
15. Lanzoni, S.; D'Alpaos, A. On funneling of tidal channels. *J. Geophys. Res. Earth Surf.* **2015**, *120*, 433–452. doi:10.1002/2014JF003203. [CrossRef]
16. Zhou, Z.; Coco, G.; Townend, I.; Gong, Z.; Wang, Z.; Zhang, C. On the stability relationships between tidal asymmetry and morphologies of tidal basins and estuaries. *Earth Surf. Process. Landf.* **2018**, *43*, 1943–1959. doi:10.1002/esp.4366. [CrossRef]
17. Langbein, W. The hydraulic geometry of a shallow estuary. *Hydrol. Sci. J.* **1963**, *8*, 84–94. doi:10.1080/02626666309493340. [CrossRef]
18. Savenije, H.H. *Salinity and Tides in Alluvial Estuaries*; Elsevier: Amsterdam, The Netherlands, 2006.
19. Leuven, J.R.F.W.; Haas, T.; Braat, L.; Kleinhans, M.G. Topographic forcing of tidal sand bar patterns for irregular estuary planforms. *Earth Surf. Process. Landf.* **2018**, *43*, 172–186. doi:10.1002/esp.4166. [CrossRef]

20. Leuven, J.R.F.W.; Selaković, S.; Kleinhans, M.G. Morphology of bar-built estuaries: empirical relation between planform shape and depth distribution. *Earth Surf. Dyn.* **2018**, *6*, 763–778. doi:10.5194/esurf-6-763-2018. [CrossRef]

21. Savenije, H.H. Predictive model for salt intrusion in estuaries. *J. Hydrol.* **1993**, *148*, 203–218. doi:10.1016/0022-1694(93)90260-G. [CrossRef]

22. Brockway, R.; Bowers, D.; Hoguane, A.; Dove, V.; Vassele, V. A note on salt intrusion in funnel-shaped estuaries: Application to the Incomati estuary, Mozambique. *Estuar. Coast. Shelf Sci.* **2006**, *66*, 1–5. doi:10.1016/j.ecss.2005.07.014. [CrossRef]

23. Gisen, J.; Savenije, H.; Nijzink, R. Revised predictive equations for salt intrusion modelling in estuaries. *Hydrol. Earth Syst. Sci.* **2015**, *19*, 2791–2803. doi:10.5194/hess-19-2791-2015. [CrossRef]

24. Haasnoot, M.; van de Wolfshaar, K. Combining a conceptual framework and a spatial analysis tool, HABITAT, to support the implementation of river basin management plans. *Int. J. River Basin Manag.* **2009**, *7*, 295–311. [CrossRef]

25. van Oorschot, M.; Kleinhans, M.; Buijse, T.; Geerling, G.; Middelkoop, H. Combined effects of climate change and dam construction on riverine ecosystems. *Ecol. Eng.* **2018**, *120*, 329–344. doi:10.1016/j.ecoleng.2018.05.037. [CrossRef]

26. Dalrymple, R.W.; Zaitlin, B.A.; Boyd, R. Estuarine facies models: Conceptual basis and stratigraphic implications: Perspective. *J. Sediment. Res.* **1992**, *62*. doi:10.1306/D4267A69-2B26-11D7-8648000102C1865D. [CrossRef]

27. de Haas, T.; Pierik, H.; van der Spek, A.; Cohen, K.; van Maanen, B.; Kleinhans, M. Holocene evolution of tidal systems in The Netherlands: Effects of rivers, coastal boundary conditions, eco-engineering species, inherited relief and human interference. *Earth-Sci. Rev.* **2017**, *177*, 139–163. doi:10.1016/j.earscirev.2017.10.006. [CrossRef]

28. Stevens, A.; Gelfenbaum, G.; MacMahan, J.; Reniers, A.; Elias, E.; Sherwood, C.; Carlson, E. *Oceanographic Measurements and Hydrodynamic Modeling of the Mouth of the Columbia River, Oregon and Washington, 2013*; Technical Report; U.S. Geological Survey: Reston, VA, USA, 2017. doi:10.5066/F7NG4NS1.

29. Marciano, R.; Wang, Z.B.; Hibma, A.; de Vriend, H.J.; Defina, A. Modeling of channel patterns in short tidal basins. *J. Geophys. Res. Earth Surf.* **2005**, *110*, F01001. doi:10.1029/2003JF000092. [CrossRef]

30. Wang, Z.B.; Hoekstra, P.; Burchard, H.; Ridderinkhof, H.; De Swart, H.E.; Stive, M.J.F. Morphodynamics of the Wadden Sea and its barrier island system. *Ocean Coast. Manag.* **2012**, *68*, 39–57. doi:10.1016/j.ocecoaman.2011.12.022. [CrossRef]

31. van Maanen, B.; Coco, G.; Bryan, K.R. Modelling the effects of tidal range and initial bathymetry on the morphological evolution of tidal embayments. *Geomorphology* **2013**, *191*, 23–34. doi:10.1016/j.geomorph.2013.02.023. [CrossRef]

32. Verbeek, H.; Van der Male, K.; Jansen, M. *Het SCALWEST-model (in Dutch)*; Technical Report, Technical Report RIKZ/OS/2000.814; Rijksinstituut voor Kust en Zee/ RIKZ: Middelburg, The Netherlands, 2000.

33. Strahler, A.N. Hypsometric (area-altitude) analysis of erosional topography. *Geol. Soc. Am. Bull.* **1952**, *63*, 1117–1142. doi:10.1130/0016-7606(1952)63[1117:HAAOET]2.0.CO;2. [CrossRef]

34. Boon, J.D.; Byrne, R.J. On basin hyposmetry and the morphodynamic response of coastal inlet systems. *Mar. Geol.* **1981**, *40*, 27–48. doi:10.1016/0025-3227(81)90041-4. [CrossRef]

35. Townend, I.H. Hypsometry of estuaries, creeks and breached sea wall sites. *Proc. Inst. Civ. Eng.-Marit. Eng.* **2008**, *161*, 23–32. doi:10.1680/maen.2008.161.1.23. [CrossRef]

36. Leopold, L.B.; Maddock, T., Jr. *The Hydraulic Geometry of Stream Channels and Some Physiographic Implications*; Technical Report, U.S. Geological Survey, Professional Paper 252; U.S. Geological Survey: Reston, VA, USA, 1953.

37. Hey, R.D.; Thorne, C.R. Stable channels with mobile gravel beds. *J. Hydraul. Eng.* **1986**, *112*, 671–689. doi:10.1061/(ASCE)0733-9429(1986)112:8(671). [CrossRef]

38. Eysink, W. Morphologic response of tidal basins to changes. *Coast. Eng. Proc.* **1990**, *1*, 1948–1961. doi:10.1061/9780872627765.149. [CrossRef]

39. Friedrichs, C.T. Stability shear stress and equilibrium cross-sectional geometry of sheltered tidal channels. *J. Coast. Res.* **1995**, *11*, 1062–1074.

40. Pavelsky, T.M.; Smith, L.C. RivWidth: A software tool for the calculation of river widths from remotely sensed imagery. *IEEE Geosci. Remote Sens. Lett.* **2008**, *5*, 70–73. doi:10.1109/LGRS.2007.908305. [CrossRef]

41. Donchyts, G.; Schellekens, J.; Winsemius, H.; Eisemann, E.; van de Giesen, N. A 30 m resolution surface water mask including estimation of positional and thematic differences using landsat 8, srtm and openstreetmap: A case study in the Murray-Darling Basin, Australia. *Remote Sens.* **2016**, *8*, 386. doi:10.3390/rs8050386. [CrossRef]

42. Savenije, H.H. Prediction in ungauged estuaries: An integrated theory. *Water Resour. Res.* **2015**, *51*, 2464–2476. doi:10.1002/2015WR016936. [CrossRef]

43. Davies, G.; Woodroffe, C.D. Tidal estuary width convergence: Theory and form in North Australian estuaries. *Earth Surf. Process. Landf.* **2010**, *35*, 737–749. doi:10.1002/esp.1864. [CrossRef]

44. Pillsbury, G. *Tidal Hydraulics*, rev. ed.; US Army Corps of Engineers: Washington, DC, USA, 1956.

45. Dronkers, J. Convergence of estuarine channels. *Cont. Shelf Res.* **2017**, *144*, 120–133. doi:10.1016/j.csr.2017.06.012. [CrossRef]

46. Wang, Z.; Jeuken, M.; Gerritsen, H.; De Vriend, H.; Kornman, B. Morphology and asymmetry of the vertical tide in the Westerschelde estuary. *Cont. Shelf Res.* **2002**, *22*, 2599–2609. [CrossRef]

47. Toffolon, M.; Crosato, A. Developing macroscale indicators for estuarine morphology: The case of the Scheldt Estuary. *J. Coast. Res.* **2007**, *23*, 195–212. [CrossRef]

48. Jarrett, J.T. *Tidal Prism-Inlet Area Relationships*; Technical Report, DTIC Document WES-GITI-3; DTIC: Fort Belvoir, VA, USA, 1976.

49. Shigemura, T. Tidal prism–throat area relationships of the bays of Japan. *Shore Beach* **1980**, *48*, 30–35.

50. Gisen, J.I.A.; Savenije, H.H. Estimating bankfull discharge and depth in ungauged estuaries. *Water Resour. Res.* **2015**, *51*, 2298–2316. doi:10.1002/2014WR016227. [CrossRef]

51. Jay, D.A.; Smith, J.D. Circulation, density distribution and neap-spring transitions in the Columbia River Estuary. *Prog. Oceanogr.* **1990**, *25*, 81–112. doi:10.1016/0079-6611(90)90004-L. [CrossRef]

52. de Brauwere, A.; De Brye, B.; Blaise, S.; Deleersnijder, E. Residence time, exposure time and connectivity in the Scheldt Estuary. *J. Mar. Syst.* **2011**, *84*, 85–95. doi:10.1016/j.jmarsys.2010.10.001. [CrossRef]

53. Vroom, J.; De Vet, L.; Van der Werf, J. *Validatie Waterbeweging Delft3D-NeVla model Westerscheldemonding*; Technical Report, Rapport 1210301-001-ZKS-0001; Deltares, The Nederland, 2015.

54. Nguyen, A.; Savenije, H. Salt intrusion in multi-channel estuaries: A case study in the Mekong Delta, Vietnam. *Hydrol. Earth Syst. Sci. Discuss.* **2006**, *10*, 743–754. doi:10.5194/hess-10-743-2006. [CrossRef]

55. Nowacki, D.J.; Ogston, A.S.; Nittrouer, C.A.; Fricke, A.T.; Van, P.D.T. Sediment dynamics in the lower Mekong River: Transition from tidal river to estuary. *J. Geophys. Res. Oceans* **2015**, *120*, 6363–6383. doi:10.1002/2015JC010754. [CrossRef]

56. Xing, F.; Meselhe, E.; Allison, M.; Weathers, H., III. Analysis and numerical modeling of the flow and sand dynamics in the lower Song Hau channel, Mekong Delta. *Cont. Shelf Res.* **2017**, *147*, 62–77. doi:10.1016/j.csr.2017.08.003. [CrossRef]

57. van Maanen, B.; Coco, G.; Bryan, K.R. A numerical model to simulate the formation and subsequent evolution of tidal channel networks. *Aust. J. Civ. Eng.* **2011**, *9*, 61–72. doi:10.1080/14488353.2011.11463969. [CrossRef]

58. Dissanayake, D.; Roelvink, J.; Van der Wegen, M. Modelled channel patterns in a schematized tidal inlet. *Coast. Eng.* **2009**, *56*, 1069–1083. doi:10.1016/j.coastaleng.2009.08.008. [CrossRef]

59. Dame, R. Estuaries. In *Encyclopedia of Ecology*; Jörgensen, S.E.; Fath, B.D., Eds.; Academic Press: Oxford, 2008; pp. 1407–1413. doi:10.1016/B978-008045405-4.00329-3.

60. Kemp, W.M.; Batleson, R.; Bergstrom, P.; Carter, V.; Gallegos, C.L.; Hunley, W.; Karrh, L.; Koch, E.W.; Landwehr, J.M.; Moore, K.A.; et al. Habitat requirements for submerged aquatic vegetation in Chesapeake Bay: Water quality, light regime, and physical-chemical factors. *Estuaries* **2004**, *27*, 363–377. doi:10.1007/BF02803529. [CrossRef]

61. Vinagre, C.; Fonseca, V.; Cabral, H.; Costa, M.J. Habitat suitability index models for the juvenile soles, Solea solea and Solea senegalensis, in the Tagus estuary: Defining variables for species management. *Fish. Res.* **2006**, *82*, 140–149. doi:10.1016/j.fishres.2006.07.011. [CrossRef]

62. Barnes, T.; Volety, A.; Chartier, K.; Mazzotti, F.; Pearlstine, L. A habitat suitability index model for the eastern oyster (*Crassostrea virginica*), a tool for restoration of the Caloosahatchee Estuary, Florida. *J. Shellfish Res.* **2007**, *26*, 949–959. doi:10.2983/0730-8000(2007)26[949:AHSIMF]2.0.CO;2. [CrossRef]

63. Degraer, S.; Verfaillie, E.; Willems, W.; Adriaens, E.; Vincx, M.; Van Lancker, V. Habitat suitability modelling as a mapping tool for macrobenthic communities: An example from the Belgian part of the North Sea. *Cont. Shelf Res.* **2008**, *28*, 369–379. doi:10.1016/j.csr.2007.09.001. [CrossRef]

64. Feyrer, F.; Newman, K.; Nobriga, M.; Sommer, T. Modeling the effects of future outflow on the abiotic habitat of an imperiled estuarine fish. *Estuar. Coasts* **2011**, *34*, 120–128. doi:10.1007/s12237-010-9343-9. [CrossRef]

65. Lokhorst, I.; Braat, L.; Leuven, J.R.F.W.; Baar, A.W.; van Oorschot, M.; Selaković, S.; Kleinhans, M.G. Morphological effects of vegetation on the fluvial-tidal transition in Holocene estuaries. *Earth Surf. Dyn. Discuss.* **2018**, *2018*, 1–28. doi:10.5194/esurf-2018-29. [CrossRef]

66. Lotze, H.K.; Lenihan, H.S.; Bourque, B.J.; Bradbury, R.H.; Cooke, R.G.; Kay, M.C.; Kidwell, S.M.; Kirby, M.X.; Peterson, C.H.; Jackson, J.B. Depletion, degradation, and recovery potential of estuaries and coastal seas. *Science* **2006**, *312*, 1806–1809. doi:10.1126/science.1128035. [CrossRef] [PubMed]

67. Yap, W.Y.; Lam, J.S.L. 80 million-twenty-foot-equivalent-unit container port? Sustainability issues in port and coastal development. *Ocean Coast. Manag.* **2013**, *71*, 13–25. doi:10.1016/j.ocecoaman.2012.10.011. [CrossRef]

68. Depreiter, D.; Sas, M.; Beirinckx, K.; Liek, G.J. *Flexible Disposal Strategy: Monitoring as a Key to Understanding And Steering Environmental Responses to Dredging and Disposal in the Scheldt Estuary*; Technical Report; International Marine & Dredging Consultants: Antwerp, Belgium, 2012.

69. Vikolainen, V.; Bressers, H.; Lulofs, K. A shift toward building with nature in the dredging and port development industries: Managerial implications for projects in or near natura 2000 areas. *Environ. Manag.* **2014**, *54*, 3–13. doi:10.1007/s00267-014-0285-z. [CrossRef] [PubMed]

70. Vos, P.; de Koning, J.; Van Eerden, R. Landscape history of the Oer-IJ tidal system, Noord-Holland (The Netherlands). *Neth. J. Geosci.* **2015**, *94*, 295–332. doi:10.1017/njg.2015.27. [CrossRef]

71. Pierik, H.; Cohen, K.; Stouthamer, E. A new GIS approach for reconstructing and mapping dynamic late Holocene coastal plain palaeogeography. *Geomorphology* **2016**. doi:10.1016/j.geomorph.2016.05.037. [CrossRef]

72. Wood, L.J. Predicting tidal sand reservoir architecture using data from modern and ancient depositional systems, integration of outcrop and modern analogs in reservoir modeling. *AAPG Mem.* **2004**, *80*, 45–66.

73. Yoshida, S.; Johnson, H.D.; Pye, K.; Dixon, R.J. Transgressive changes from tidal estuarine to marine embayment depositional systems: The Lower Cretaceous Woburn Sands of southern England and comparison with Holocene analogs. *AAPG Bull.* **2004**, *88*, 1433–1460. doi:10.1306/05140403075. [CrossRef]

74. Jackson, M.D.; Yoshida, S.; Muggeridge, A.H.; Johnson, H.D. Three-dimensional reservoir characterization and flow simulation of heterolithic tidal sandstones. *AAPG Bull.* **2005**, *89*, 507–528. doi:10.1306/11230404036. [CrossRef]

75. Steel, R.J.; Plink-Bjorklund, P.; Aschoff, J. Tidal deposits of the Campanian Western Interior Seaway, Wyoming, Utah and Colorado, USA. In *Principles of Tidal Sedimentology*; Springer: Berlin, Germany, 2012; pp. 437–471. doi:10.1007/978-94-007-0123-6_17.

76. Saha, S.; Burley, S.D.; Banerjee, S.; Ghosh, A.; Saraswati, P.K. The morphology and evolution of tidal sand bodies in the macrotidal Gulf of Khambhat, western India. *Mar. Pet. Geol.* **2016**, *77*, 714–730. doi:10.1016/j.marpetgeo.2016.03.028. [CrossRef]

77. Gray, A.J. *Saltmarshes: Morphodynamics, Conservation and Engineering Significance*; Allen, J., Pye, K., Eds.; Cambridge University Press: Cambridge, UK, 1992; pp. 63–79.

78. Thomson, A.; Huiskes, A.; Cox, R.; Wadsworth, R.; Boorman, L. Short-term vegetation succession and erosion identified by airborne remote sensing of Westerschelde salt marshes, The Netherlands. *Int. J. Remote Sens.* **2004**, *25*, 4151–4176. doi:10.1080/01431160310001647688. [CrossRef]

79. van der Wal, D.; Wielemaker-Van den Dool, A.; Herman, P.M. Spatial patterns, rates and mechanisms of saltmarsh cycles (Westerschelde, The Netherlands). *Estuar. Coast. Shelf Sci.* **2008**, *76*, 357–368. doi:10.1016/j.ecss.2007.07.017. [CrossRef]

80. Bolla Pittaluga, M.; Coco, G.; Kleinhans, M.G. A unified framework for stability of channel bifurcations in gravel and sand fluvial systems. *Geophys. Res. Lett.* **2015**, *42*, 7521–7536. doi:10.1002/2015GL065175. [CrossRef]

MDPI
St. Alban-Anlage 66
4052 Basel
Switzerland
Tel. +41 61 683 77 34
Fax +41 61 302 89 18
www.mdpi.com

Remote Sensing Editorial Office
E-mail: remotesensing@mdpi.com
www.mdpi.com/journal/remotesensing